Figure 1   The classic intercrop agroecosystem of corn, beans, and squash in the tropical lowlands of Tabasco, Mexico. This cropping system has deep roots in the local Maya culture of the region, as well as strong foundations in agroecological science. (*Credit:* Steve Gliessman.)

Figure 2   A diverse shade-coffee agroecosystem in the village of Cantagallo, in the Las Segovias region of northwestern Nicaragua. This multifunctional agroforestry system provides coffee as a key cash crop, but also generates food, firewood, alternative market products, soil and water protection, biodiversity conservation, and even carbon sequestration. (*Credit:* Steve Gliessman.)

Figure 3 Smallholder coffee farmer's house showing banana plants, fruit trees, and a homegarden, in the highlands of Canta Gallo, Estelí, Nicaragua. Research highlighted in the first chapter of this book showed that households reporting the presence of fruits trees experienced shorter periods of seasonal hunger. (*Credit:* Chris Bacon.)

Figure 4 Margarita Fernandez, a contributor of this volume (Chapter 10), presents the results of her participatory action research (PAR) dissertation to the leadership of the CESMACH smallholder coffee cooperative, in Chiapas, Mexico. (*Credit:* Ernesto Méndez.)

Figure 5 Vermont farmer Eric Noel discusses soil management of his organic pasture at Health Hero Farm, in South Hero VT. This was part of a training event of the Vermont Agricultural Resilience in a Changing Climate Initiative (VAR), which is highlighted in Chapter 1. (*Credit:* Ernesto Méndez.)

Figure 6 Smallholders in coffee growing regions, such as the ones highlighted in Chapters 1 and 13, do more than cultivate coffee bushes. In this picture, a farmer in a mountain village near the town of San Ramón, Nicaragua, seeds the *milpa* (corn and bean plot), with a variety that he claims is adapted to the cool and moist conditions of this environment. (*Credit:* Chris Bacon.)

Figure 7 A youth leader in San Ramon, Nicaragua, trained in participatory action research methodologies and collecting survey data regarding changes in food security among participants of the Youth Leadership and Food Sovereignty Project, highlighted in Chapter 13. (*Credit*: Community Agroecology Network-CAN.)

Figure 8 Participants in CAN's Annual International Agroecology Shortcourse visit coffee producers in Cantagallo, Nicaragua, as part of a training in participatory action research (PAR) methodologies regarding the development of sustainability indicators. (*Credit:* Community Agroecology Network.)

Agroecology

A Transdisciplinary, Participatory and Action-oriented Approach

Advances in Agroecology

Series Editor: Clive A. Edwards

Advisory Board

Agroecology

A Transdisciplinary, Participatory and Action-oriented Approach

Edited by
V. Ernesto Méndez
University of Vermont, Burlington, USA
Christopher M. Bacon
Santa Clara University, California, USA
Roseann Cohen
Community Agroecology Network,
Santa Cruz, California, USA
Stephen R. Gliessman
University of California, Santa Cruz, USA

CRC Press
Taylor & Francis Group
Boca Raton   London   New York

CRC Press is an imprint of the
Taylor & Francis Group, an **informa** business

CRC Press
Taylor & Francis Group
6000 Broken Sound Parkway NW, Suite 300
Boca Raton, FL 33487-2742

© 2016 by Taylor & Francis Group, LLC
CRC Press is an imprint of Taylor & Francis Group, an Informa business

No claim to original U.S. Government works

Printed on acid-free paper
Version Date: 20151009

International Standard Book Number-13: 978-1-4822-4176-1 (Hardback)

Visit the Taylor & Francis Web site at
http://www.taylorandfrancis.com

and the CRC Press Web site at
http://www.crcpress.com

Contents

Chapter 1
Introduction: Agroecology as a Transdisciplinary, Participatory, and Action-oriented Approach 1

V. Ernesto Méndez, Christopher M. Bacon, and Roseann Cohen

Chapter 2
Agroecology: Roots of Resistance to Industrialized Food Systems.................................23

Stephen R. Gliessman

Chapter 3
Transformative Agroecology: Foundations in Agricultural Practice, Agrarian Social
Thought, and Sociological Theory 37

Graham Woodgate and Eduardo Sevilla Guzmán

Chapter 4
Political Agroecology: An Essential Tool to Promote Agrarian Sustainability 55

Manuel González de Molina

Chapter 5
Learning Agroecology through Involvement and Reflection 73

**Charles Francis, Edvin Østergaard, Anna Marie Nicolaysen, Geir Lieblein,
Tor Arvid Breland, and Suzanne Morse**

Chapter 6
Complexity in Tradition and Science: Intersecting Theoretical Frameworks in
Agroecological Research 99

John Vandermeer and Ivette Perfecto

Chapter 7
Agroecology, Food Sovereignty, and the New Green Revolution 113

Eric Holt-Giménez and Miguel A. Altieri

Chapter 8
The Intercultural Origin of Agroecology: Contributions from Mexico 123

Francisco J. Rosado-May

Preface

The spark that ignited this book, and other related publications, started with conversations among Steve Gliessman, Ernesto Méndez, and Manuel González de Molina while teaching at the agroecology graduate program of the International University of Andalucía (UNIA), Spain, in 2011. The discussions centered on how an increasing number of publications on agroecology were starting to appear in the scientific and gray literatures, and our perception that not all agroecological perspectives were being adequately discussed. This led to the idea of preparing a special edited issue to inaugurate the launch of the renaming of the *Journal of Sustainable Agriculture* with the new name of *Agroecology and Sustainable Food Systems,* the first international English language journal with "agroecology" in the title. After further discussions with Chris Bacon and Roseann Cohen, they joined Ernesto as guest editors, and the special issue was published in 2013. Our second project was to produce an open access, Spanish translation of the special issue, which appeared as a number in the Spanish journal *Agroecología* in 2014. This edited book represents the third contribution of this conceptual project. Steve Gliessman joined us as an editor with a goal of bringing together fully revised contributions to the special issue as well as additional conceptual and empirical chapters. The motivations behind this collection of work are to (1) more explicitly and critically discuss the different perspectives that are present in the growing field of agroecology and (2) provide conceptual and empirical material of an agroecology that aspires to be transdisciplinary, participatory, and action-oriented. We hope that this volume will provide an inspiration for others who are working to innovate and transform our current agri-food system into one that is more sustainable for all people, ecologies, and landscapes.

V. Ernesto Méndez
Agroecology and Rural Livelihoods Group (ARLG), University of Vermont

Christopher M. Bacon
Department of Environmental Studies, Santa Clara University

Roseanne Cohen
Community Agroecology Network

Stephen R. Gliessman
Community Agroecology Network and University of California at Santa Cruz

Foreword

As a contribution to the science of agronomy, agroecology aims to reduce the use of external fossil-based inputs, to recycle waste, and to combine different elements of nature in the process of production, in order to maximize synergies between them. But agroecology is more than a range of agronomic techniques that present some of these characteristics. It is both a certain way of thinking of our relationship to nature, and it is growing as a social movement.

Agroecology invites us to embrace the complexity of nature; it sees such complexity not as a liability, but as an asset. The farmer, in this view, is a discoverer as he or she proceeds experimentally, by trial and error, observing what consequences follow from which combinations and learning from what works best—even though the ultimate "scientific" explanation may remain elusive. This is empowering; the farmer is put in the driver's seat, constructing the knowledge that works best in the local context in which he or she operates. In contrast, so-called "modern" agriculture, which is, in fact, the twentieth-century agriculture, did the exact opposite—it sought to simplify nature. What to do in the field was defined by whatever was prescribed by "science" developed in laboratories. The path from research to practice was unidirectional and it was seen as unproblematic. Since solutions were based on science, they were considered universally applicable. The experiential knowledge of the farmer was irrelevant at best; at worst, it was treated as "prejudice" and as an obstacle to the top-down implementation of sound scientific prescriptions from "experts." In this view from twentieth-century science, the complexity of nature is a problem; simplify it if you can, and never mind if this means robbing the farmer of the opportunity to develop his or her art and of transforming that art into the literacy of reading instructions on spray bottles and seed bags.

If agroecology stems from a renewed understanding of nature and our relationship with nature, it naturally follows that it is also a social movement. This movement encourages peer-to-peer exchanges of information between farmers. It prioritizes local solutions relying on local resources. And it transforms the relationship between the farmer and the "expert," be it from a department of agriculture or from an international agency. This is not done in order to reverse it and to replace one hierarchy with another, but to move toward the co-construction of knowledge. This is very clearly illustrated by participatory plant breeding examples.

It is only if we see agroecology as something other than a particular set of agronomic techniques that we can understand the opposition that it faces. Indeed, as a branch of agronomy that borrows from ecology to undertake the act of farming within the ecosystems in which that act takes place, agroecology is particularly well suited to meet the challenges of the day. In our still dominant industrial farming system, it takes about 10 calories of fossil energy to produce one calorie of food, a clearly unsustainable approach as we reach peak gas and peak oil. This system is a huge emitter of greenhouse gases at least 13.5% of total human-generated greenhouse gas emissions come from agriculture. This rises up to one-third once we factor into that calculation the deforestation to create pastures and expand cultivated areas as well as the various stages of food processing, packaging, transport, and retail. Small-scale farms are systematically put at a disadvantage, and this is because they are less well-equipped to mechanize and to achieve economies of scale. This is because they are less competitive in a world in which farmers are asked to become suppliers of raw commodities—of large volumes of uniform "stuff"—for the food processing industry. The impacts on rural development are considerable, as small family farms are disappearing in huge numbers. Moreover, as it has been shaped in the past, industrial food systems have encouraged the shift to highly processed foods, including ready-to-eat "convenience" foods and ultraprocessed "junk" foods. The consequences of such modern approaches are well-known. Worldwide, the prevalence of obesity doubled between 1980 and 2008. More than 1 billion adults are now overweight, and another 400 million are obese. Combined with more sedentary lifestyles and tobacco and alcohol consumption, inadequate diets are resulting in the rise of noncommunicable diseases. Type 2 diabetes, heart disease, or gastrointestinal cancers—all directly related to diets—are now growing fast in all regions, and not only in rich countries as was the case in the past.

Agroecology provides a number of answers. It favors a gradual transition away from the fossil energy-based farming of the earlier generation, and it seeks to preserve soil health and reduce soil erosion. In fact, it is mostly because of its environmental benefits that agroecology is now of interest to governments and international agencies. Although it can be practiced on a large scale, its insistence on intercropping techniques and on various combinations between plants, trees, and animals—in order to reestablish the agro-silvo-pastoral complementarities that "modern" agriculture has negated—make it especially suitable when practiced on relatively smaller farms. As such, increased support to agroecology shall contribute to rebalancing a competition between large, industrial-sized farms and smaller farms. This balance is currently significantly skewed in favor of the former agricultural model. Agroecology favors better nutrition, both because greater diversity on the farm results in greater diversity on the plates for the communities who produce their own food, and because of the proven benefits to health. Organic crops, recent studies show, have an up to 60% higher number of key antioxidants than conventionally grown ones, and of course show much lower levels of pesticide residues and toxic heavy metals, such as cadmium, than industrially grown crops. Most importantly, agroecology represents a shift away from the quasi-exclusive focus on growing large areas of cereals in monocultures, which over the past 30 years had in fact reduced the diversity of the plants on which our diets are based, and has favored an ever-increasing reliance on heavily processed foods that are richer in saturated fats and in added sugars and salt. The health benefits of an agroecological revolution would be significant.

Why is it, then, that despite all the benefits it may provide, agroecology remains marginalized? Four major lock-ins still form considerable obstacles to the agroecological revolution. First, technologies and infrastructures are biased in favor of achieving economies of scale through large monocultures that can be more easily mechanized. Second, dominant agribusiness actors—the large commodity buyers and food processing companies—are better positioned to supply markets with low-priced foodstuffs against which other actors who use the other more sustainable modes of production are unable to compete. Until industrial farming methods are obliged to fully internalize the social and environmental costs they impose on the collectivity, this will not change. Third, our lifestyles have evolved with the industrial way of producing food that we have been encouraging. People today have less time to cook, they have relegated food to a secondary place in their lives, and many families have lost even the most basic cooking skills. This culinary knowledge is required to reduce the dependency on heavily processed foods, including the convenience foods that we have become so accustomed to. Fourth and finally, political obstacles remain. Large agribusiness actors veto any significant change that would threaten their position in the system and that would question, in particular, the relegation of the farmer to the position of a captive buyer of inputs and a provider of raw materials to the food processing industry.

These obstacles are formidable. This is why food democracy—the ability for people to make real choices about how to produce food, what to produce, and how to eat—is a key to unlocking the system. The agroecological revolution is much needed. It will succeed, however, only if we overcome the political economic obstacles to change. I welcome this volume as an important contribution to this ambitious and urgent undertaking.

Olivier De Schutter
United Nations Special Rapporteur on the Right to Food (2008–2014)
Member of the UN Committee on Economic, Social and Cultural Rights.

Acknowledgments

As with most agroecological endeavors, this book is the result of a highly collaborative process. We are grateful to John Sulzycki and Jill Jurgensen from CRC Press/Taylor and Francis Group for their enthusiastic support of this book from beginning to end. We are also indebted to Rachel Schattman, doctoral candidate at the UVM Agroecology and Rural Livelihoods Group (ARLG), for her inquisitive editorial work and formatting of all of the chapters in this volume. We deeply appreciate the effort of all the contributing authors who have generously shared their work. V.E. Méndez would also like to thank his wife, Karen, and children, Adriel and Sofia, for their unconditional love, joy, and support through this and all projects, and acknowledge the members of the ARLG for providing motivation, good humor, and support as an agroecological "community of practice." Steve Gliessman thanks his partner, Robbie Jaffe, for her persistence in insisting that the social and ecological components of agroecology must be fully linked and integrated for effective food system transformation to occur. Rose Cohen thanks her daughter, Emma Sofia, for inspiring her to work toward a better future; her husband, Alan, for his companionship and support through project after project; and the Community Agroecology Network for sharing in the challenges and successes of putting agroecology into practice. Chris Bacon offers a profound thanks to daughter, Rosalía, for her creative and loving spark and wife, Maria Eugenia Flores Gómez, for her support, thoughtful ideas, and loving presence.

Editors

V. Ernesto Méndez is an associate professor of Agroecology and Environmental Studies at the University of Vermont's (UVM) Environmental Program and Department of Plant and Soil Science, Burlington, Vermont. At UVM, he leads the Agroecology and Rural Livelihoods Group (ARLG), a community of practice that studies and contributes to develop practical solutions to key issues in our current agrifood system. His empirical work is mostly with smallholder coffee farmers and cooperatives in Mesoamerica and a variety of growers in Vermont. His research uses agroecology as a transdisciplinary, participatory, and action-oriented approach, focusing on the interactions among agriculture, food, farmer livelihoods, and environment. Most of his work utilizes a participatory action research (PAR) approach to directly support agroecological practice and farmer livelihoods. A native of El Salvador, he has more than 20 years of experience doing research and development work with smallholder farmers in Mexico and Central America. He holds a BS in Crop Science from California Polytechnic State University at San Luis Obispo, California, an MS in Tropical Agroforestry from the Tropical Agriculture Research and Higher Education Center (CATIE), Turrialba, Costa Rica, and a PhD in agroecology and environmental studies from the University of California at Santa Cruz, California.

Christopher M. Bacon is an assistant professor with the Department of Environmental Studies at Santa Clara University, Santa Clara, California. After serving as a Peace Corps volunteer in Nicaragua, he completed a PhD in Environmental Studies at the University of California, Santa Cruz, California, and an S.V. Ciriacy-Wantrup Postdoctoral Fellowship affiliated with the Geography Department at University of California, Berkeley, California. His primary research involves smallholders, cooperatives, and food security in the context of market and climatic change in northern Nicaragua. It examines the political ecology of conventional and alternative food systems and their impacts on rural development and change. He often uses community-based participatory action research and agroecology to study questions, such as, how do changes in the governance of fair trade coffee commodity chains relate to rural livelihoods, seasonal hunger, and ecosystem services in Latin America's diversified farming systems? In addition to continued work in Central America, a second line of research focuses on environmental and food justice in California. Previous work has been published in *Global Environmental Change*, the *Journal of Peasant Studies*, *Ecology and Society*, and *World Development*.

Roseann Cohen holds a PhD from the Environmental Studies Department at the University of California, Santa Cruz, California, with a specialization in Latin American and Latino Studies. Her research focuses on the sociocultural significance of farmers' relationship to their crops and land, as well as the impacts of insecure land tenure, forced migration, and violence on farming communities. She has worked in Colombia and is now expanding her research to migrant farmworker communities in California engaged in urban community gardens. After completing a fellowship at the Agrarian Studies Program at Yale University, New Haven, Connecticut, she is currently the executive director for the Community Agroecology Network (CAN), Santa Cruz, California, a nonprofit committed to sustaining rural livelihoods and landscapes in the global south through the integration of collaborative research, agroecological capacity-building, and locally informed development strategies.

Stephen R. Gliessman holds graduate degrees in botany, biology, and plant ecology from the University of California, Santa Barbara, California. He has accumulated more than 40 years of teaching, research, and production experience in the field of agroecology. His international experiences in tropical and temperate agriculture, small-farm and large-farm systems, traditional and conventional farm management, hands-on and academic activities, nonprofit and business employment, and organic and synthetic chemical farming approaches have provided a unique combination of experiences and perspectives to his formation as an agroecologist. He has been a W.K. Kellogg Foundation Leadership Fellow and a Fulbright Fellow. He was the founding director of the Agroecology Program at the University of California, Santa Cruz (UCSC), California, one of the first formal agroecology programs in the world, and was the Alfred and Ruth Heller Professor of Agroecology in the Department of Environmental Studies at UCSC until his retirement in 2012. He is the cofounder of the nonprofit Community Agroecology Network (CAN), Santa Cruz, California, and currently serves as president of its board of directors. His textbook, *Agroecology: The Ecology of Sustainable Food Systems*, is in its third edition and has been translated into many languages. He is the editor of the international journal *Agroecology and Sustainable Food Systems* and dry farms organic wine grapes and olives with his family in northern Santa Barbara County, California.

Contributors

Antonio M. Alonso
Universidad Internacional de Andalucía
Baeza, Spain

Miguel A. Altieri
Department of Environmental Science, Policy
    and Management
University of California
Berkeley, California

Christopher M. Bacon
Department of Environmental Studies and
    Sciences
Santa Clara University
Santa Clara, California

Tor Arvid Breland
Department of Plant Science
Norwegian University of Life Sciences
Ås, Norway

Roseann Cohen
Community Agroecology Network (CAN)
Santa Cruz, California

Heather Darby
Northwest Crops and Soils Program
University of Vermont Extension
St. Albans, Vermont

Margarita Fernandez
Agroecology and Rural Livelihoods Group
    (ARLG), Department of Plant and Soil
    Science
University of Vermont
Burlington, Vermont

Charles Francis
Department of Agronomy and Horticulture
University of Nebraska
Lincoln, Nebraska

Stephen R. Gliessman
Community Agroecology Network (CAN)
Santa Cruz, California

Alba González-Jácome
Coordinator of Technical Assistance; Dirección
    General
SEPE/USET
Tlaxcala, México

Manuel González de Molina
Agroecosystem History Laboratory
Universidad Pablo de Olavide
Sevilla, Spain

Vern Grubinger
University of Vermont Extension and Northeast
    Sustainable Agriculture Research and
    Education (SARE)
Brattleboro, Vermont

Isabel A. Gutiérrez-Montes
Academic Program in Development Practice,
    Education Division, Tropical Agricultural
    Research and Higher Education Center
    (CATIE)
Turrialba, Costa Rica

Gloria I. Guzmán
Agroecosystem History Laboratory
Universidad Pablo de Olavide
Sevilla, Spain

Debra Heleba
Northwest Crops and Soils Program and
    Northeast Sustainable Agriculture Research
    and Education (SARE)
University of Vermont Extension
St. Albans, Vermont

Eric Holt-Giménez
Institute for Food and Development Policy
    (Food First)
Oakland, California

Roberta M. Jaffe
Community Agroecology Network (CAN)
Santa Cruz, California

Geir Lieblein
Department of Plant Science
Norwegian University of Life Sciences
Ås, Norway

Daniel López
Agroecosystem History Laboratory
Universidad Pablo de Olavide
Sevilla, Spain

Teresa Mares
Department of Anthropology
University of Vermont
Burlington, Vermont

V. Ernesto Méndez
Department of Plant and Soil Science
University of Vermont
Burlington, Vermont

Suzanne Morse
College of the Atlantic
Bar Harbor, Maine

Anna Marie Nicolaysen
Department of Plant Science
Norwegian University of Life Sciences
Ås, Norway

Edvin Østergaard
Department of Plant Science, Department of
    Mathematical Sciences and Technology
Norwegian University of Life Sciences
Ås, Norway

Ivette Perfecto
School of Natural Resources and Environment
University of Michigan
Ann Arbor, Michigan

Heather Putnam
Community Agroecology Network (CAN)
Santa Cruz, California

Felicia Ramírez Aguero
Academic Program in Development Practice,
    Education Division, Tropical Agricultural
    Research and Higher Education Center
    (CATIE)
Turrialba, Costa Rica

Lara Román
Universidad Internacional de Andalucía
Baeza, Spain

Francisco J. Rosado-May
Universidad Intercultural Maya de
    Quintana Roo
José María Morelos,
Quintana Roo, México

Rachel Schattman
Agroecology and Rural Livelihoods Group
    (ARLG), Department of Plant and Soil
    Science
University of Vermont
Burlington, Vermont

Eduardo Sevilla Guzmán
Instituto de Sociología y Estudios Campesinos
Universidad de Córdoba
Córdoba, Spain

John Vandermeer
Department of Ecology and Evolutionary
    Biology
University of Michigan
Ann Arbor, Michigan

Graham Woodgate
UCL Institute of the Americas
University College London
London, UK

CHAPTER 1

Introduction
Agroecology as a Transdisciplinary, Participatory, and Action-oriented Approach

V. Ernesto Méndez, Christopher M. Bacon, and Roseann Cohen

CONTENTS

1.1 INTRODUCTION

Agroecology has emerged as an approach that helps us to better understand the ecology of traditional farming systems and respond to the mounting problems resulting from an increasingly globalized and industrialized agri-food system (Altieri 1987). In its early stages, agroecology mainly focused on "applying ecological concepts and principles to the design of sustainable agricultural systems" (Altieri 1987; Gliessman 1990). This was followed by a more explicit integration of concepts and methods from the social sciences, which were necessary to better understand the complexity of agriculture that emerges from unique sociocultural contexts (Guzmán-Casado et al. 1999; Hecht 1995). In the last decade, the number of publications and initiatives that people describe as agroecological has increased exponentially (Wezel and Soldat 2009). The result is the emergence of several distinct standpoints, which, in this paper, we refer to as different agroecological perspectives or "agroecologies." As can be expected in any field of science or knowledge, we can observe some important differences between specific agroecologies. Hence, the objectives of this introductory chapter are to (1) discuss the implications of the increasing use and adoption of agroecology in unprecedented scientific, social, and political spaces; (2) examine the evolution of the field of agroecology into distinct perspectives or "agroecologies;" (3) illustrate the application of an agroecological perspective grounded in transdisciplinary, participatory, and action-oriented approaches, including two case studies; (4) discuss the issue of scalability in agroecology; and (5) introduce the reader to the objectives and contents of this edited volume.

1.2 AGROECOLOGICAL MAINSTREAMING

Agroecology has reached a high level of prominence in a diversity of academic, policy, and advocacy spaces worldwide (Guzmán-Casado et al. 1999; IAASTD 2009; Wezel and Soldat 2009). An important example of this was the recently held International Symposium on Agroecology for Food Security and Nutrition in September 2014 (http://www.fao.org/about/meetings/afns/en/), organized by the United Nations Food and Agriculture Organization (FAO). This was the first event on agroecology organized by the FAO in its history, and it was attended by several high-ranking officials and agriculture ministers from France, Brazil, Costa Rica, Senegal, Algeria, Japan, and the European Commission. In addition, through persistent, long-term efforts, agroecologists have been able to institutionalize the field in academic organizations, including the establishment of a growing number of agroecology programs and degrees at universities of both developed and developing countries (Francis et al. 2003; http://sustainableaged.org/projects/degree-programs/). Other applications of agroecology are more recent, but just as important. These include the adoption of the field by policy-oriented actors, as well as a wider use of agroecology within rural social movements and farmer or peasant organizations.

The appearance of agroecology in international food and agricultural policy debates is not new. However, until recently it was mostly used in the context of nongovernmental organizations focusing on sustainable agriculture and rural development topics and more specifically those oriented toward empowering small-scale farmers and resource-poor rural communities (e.g., Food First).

The turning point for the inclusion of agroecology at higher policy circles probably came with the publication of the International Assessment of Agricultural Knowledge, Science, and Technology for Development (IAASTD), and its recognition that the field represented a promising alternative approach to resolve the interrelated global problems of hunger, rural poverty, and unsustainable development (IAASTD 2009).* Subsequently, Oliver De Schutter, who was appointed as the United Nations Special Rapporteur on the Right to Food between 2008 and 2014, continually advocated for the use of an agroecological approach to confront global food insecurity and advance the right to food. De Schutter did this through policy-oriented presentations

---

* The IAASTD is a high-profile report commissioned by the World Bank, the United Nations, and the World Health organization, which sought to direct research and development policy solutions to the issues of global hunger, poverty, and sustainable agricultural development. It brought together hundreds of scientists and institutions from all regions of the world over a seven-year period. It is considered by many as the agricultural equivalent of the Intergovernmental Panel for Climate Change (IPCC) reports. On the other hand, other scientists have expressed serious doubts about the rigor of the report. Its findings remain somewhat controversial.

and lectures, publications geared for a broad audience, and an interactive website (see [De Schutter 2010; De Schutter and Vanloqueren 2011]; http://www.srfood.org/). The aforementioned FAO symposium represents another key event in the advancement of agroecology in international policy circles, as it could be influential on national and international agriculture and rural development actors. Most recently, an International Panel of Experts on Sustainable Food Systems (IPES-Food) has been established, with the goal of using agroecology and a political economy approach to transform food systems to sustainability (www.ipes-food.org).

1.3 AN EXAMINATION OF THE DIFFERENT "AGROECOLOGIES"

A comprehensive review by Wezel et al. (2009) interpreted agroecology as a field that has expressions as a science, a social movement, a practice, or a combination of all three. The authors concluded that there is "certain confusion in the use of the term 'agroecology'" (Wezel et al. 2009: 10), and that the different uses of the term are affected by a variety of factors related to to geographic, scientific, and contextual backgrounds. We disagree with the notion that there are no clear lines between existing agroecological perspectives. Rather, we argue that a persistent depiction of agroecology as unclear explicitly ignores important aspects of its evolution as a field of knowledge. Furthermore, presenting the agroecological approach as confusing justifies the application of narrow definitions. This interpretation is favored by those who view agroecology solely as a new form of natural science endeavor, devoid of its transdisciplinary nature and its links to farmer knowledge, social sciences, and rural social movements (see Figure 1.1).

Although there are a wide diversity of interpretations and applications of agroecological approaches, we have identified two predominant perspectives. The first one tends to apply agroecology as a framework to reinforce, expand, or develop scientific research, firmly grounded in the Western tradition and the natural sciences (Wezel et al. 2009; Wezel and Soldat 2009; e.g., see Martin and Sauerborn 2013). These agroecological approaches represent important endeavors for

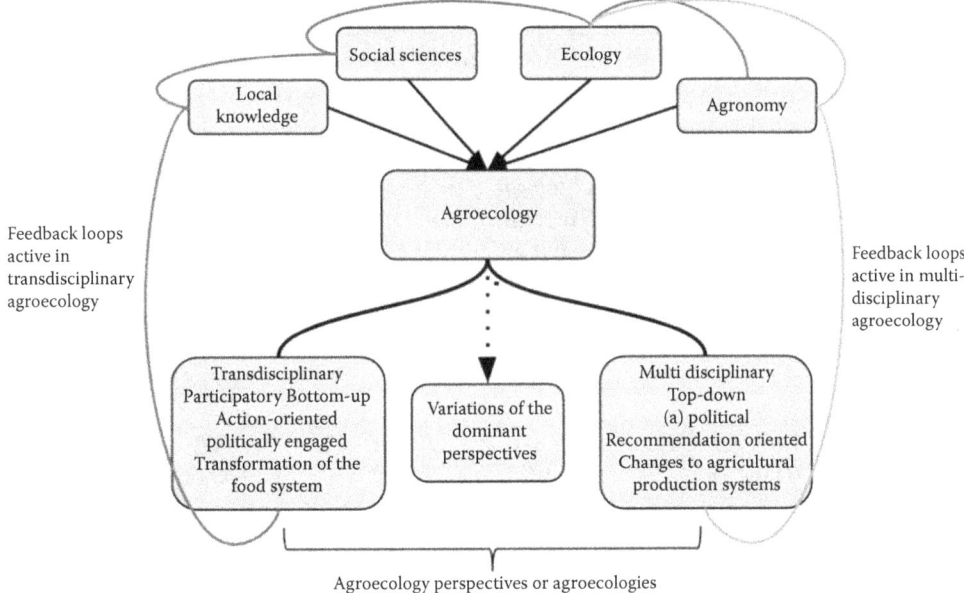

Figure 1.1 Schematic representation of the evolution of different types of agroecologies. (Adapted from Méndez et al. 2013.)

advancing findings on agronomic and ecological processes, and for improving the management of farms and landscapes. The information they generate can contribute to redirect agricultural production and management toward an ecologically based approach. However, although these standpoints may seek to impact broader agri-food systems, their approach remains largely grounded in natural science research with a primary focus on ecological analyses at different scales of *agricultural landscapes* (i.e., farm, region) not of agri-food system (i.e., from farm to table). These perspectives provide useful ecological information to inform agroecological applications, but, on their own, they cannot accomplish a comprehensive understanding of agriculture as a complex socioecological system, especially its social, cultural, and political dimensions. An agroecology that focuses only on the ecology of agricultural systems/landscapes runs the risk of silencing contributions of knowledge from within the social sciences and constructed outside of the Western scientific paradigm (i.e., local, traditional, cultural knowledge, and practice). The exclusion of the social science disciplines and local knowledge also diminishes the possibility of creating feedback loops that enrich these approaches through insights gained from this broader agroecological approach (see Figure 1.1).

In contrast to the perspective outlined in the previous paragraph, some agroecological scholars, often trained in natural science disciplines (e.g., entomology, ecology, and agronomy) and frequently cross-trained in critical social science approaches, have pursued a path that seeks to simultaneously deepen conceptual inquiry within specific subfields, while expanding and redefining a broader agroecological perspective; one that engages with the social sciences and broader agri-food system issues. This agroecological approach developed from firm roots in the sciences of ecology and agronomy into a framework that integrates transdisciplinary, participatory, and action-oriented approaches, as well as critically engages political–economic issues that affect agri-food systems (Gliessman 2006; Méndez 2010; Sevilla-Guzmán 2006b; Wezel et al. 2009).

The use of terms such as "transdisciplinary," "participatory," and "action oriented" can be interpreted and applied in a diversity of ways, some of which can be controversial (Francis et al. 2008; Kindon et al. 2007a). Others may consider that these terms are overly optimistic and unrealistic in terms of research and applications. However, we perceive that the evolution of this particular form of agroecology has explicitly embraced these characteristics through an in-depth, and frequently challenging, process of research, reflection, and action. We are not arguing that all scientific endeavors should be transdisciplinary, participatory, and action oriented. In fact, we believe that the best-case scenario would be to have basic, discipline-oriented science actively informing and interacting with this reflexive perspective that seeks to be more participatory (by collaborating with and including knowledge from multiple actors) and with an explicit bias toward generating knowledge that can contribute to direct actions for agri-food system transformation.

In the previous paragraphs, we argue that there are two predominant agroecological perspectives. However, it is important to recognize that in between these two broader approaches there exists a gradient of interpretations and applications that may lean more toward one or the other, or seek a relatively balanced position between the two (see Figure 1.1). For a recent example of an agroecological perspective located in between the two dominant ones, see a review by Tomich et al. (2011).

1.4 AGROECOLOGY AS A TRANSDISCIPLINARY, PARTICIPATORY, AND ACTION-ORIENTED APPROACH

In this section, we discuss an agroecological perspective with the following characteristics: (1) it originated from a predominantly ecological and agronomic interpretation of the field in the early 1970s; (2) it has evolved toward an approach grounded in transdisciplinary and participatory research through engagement with social scientists, agricultural communities, and nonscientific knowledge systems; (3) it incorporates a critique of the role of prevalent political–economic structures in the construction of the current agri-food system; and (4) as an action-oriented effort, it seeks to directly contribute to redirect current agri-food systems toward sustainability. This particular agroecological perspective has been advanced by some of the most influential academics in the field, including Stephen R. Gliessman (Gliessman 2015), Miguel Altieri (Altieri and Toledo 2011), John Vandermeer (Vandermeer 2009), Ivette Perfecto (Perfecto et al. 2009), and Eduardo Sevilla-Guzmán (Sevilla-Guzmán 2006b). In this section, we undertake an in-depth examination of the key characteristics of this perspective.

1.4.1 Agroecology and Transdisciplinarity

We consider transdisciplinary approaches as those that value and integrate different types of knowledge systems, which can include information from scientific or academic disciplines, as well as experiential, local, indigenous, or other forms of knowledge. In addition, transdisciplinary approaches often adopt a problem-based focus (Aeberhard and Rist 2009; Belsky 2002; Francis et al. 2008; Godemann 2008). An appreciation for farmer-generated knowledge challenges conventional approaches to agricultural research and related policymaking that privileges Western epistemologies of knowledge production (Cuéllar-Padilla and Calle-Collado 2011). Since the 1980s, some agroecologists have been valued and sought to better understand the experiential agroecological knowledge of farmers as a necessary component to develop a more sustainable agriculture. This was clearly illustrated in Stephen Gliessman's work in the Mexican tropics, in the 1970s and 1980s, which focused on understanding the ecological bases of traditional Mexican agriculture (Gliessman 1980, 1982, 1978; Gliessman et al. 1981), and which drew from the scholarship of Efraím Hernández-Xolocotzi. This empirical information, based on observation and practice, and which also integrates cultural aspects, was viewed as a source of knowledge to conceptualize and apply agroecology. More recently, the Universidad Intercultural Maya de Quintana Roo, Mexico, has institutionalized agroecological teaching and research through the concept of interculturality (http://www.uimqroo.edu.mx/). This approach is based on a platform for knowledge exchange and collaboration under conditions of mutual respect among cultures and knowledge systems (i.e., Maya and Western-based) that are crucial for applying both participatory and transdisciplinary approaches (see Chapter 8 in this volume). The incorporation of local and/or farmer-generated knowledge is an important component of this particular type of agroecological thought and practice.

1.4.2 Participatory and Principles-based Approaches in Agroecology

An increasing interest in participatory and action-oriented research is evident in a variety of fields, such as ecology (Whitmer et al. 2010) and several disciplines in the social sciences (Fals-Borda and Rahman 1991; Greenwood and Levin 1998; Stringer 1999), health (Minkler and Wallerstein 2008), natural resources (Castellanet and Jordan 2002; Fortmann 2008), geography (Kindon et al. 2007b), and agroecology (Guzmán-Casado et al. 1999; Snapp and Pound 2008; Uphoff 2002). Participatory action research (PAR) and related approaches seek to involve a diversity of actors as active participants in a cyclical, iterative process that integrates research, reflection, and action, and which seeks to include or amplify those voices that have been traditionally excluded from the research process (Figure 1.2; Bacon et al. 2005; Kindon et al. 2007b).

Agroecological approaches that have sought to integrate farmer knowledge into research and outreach fit well with the PAR approach. In the last decade, an increasing number of studies have combined agroecology with participatory approaches in different ways. For example, graduate students and professors at the University of California at Santa Cruz collaborated in a participatory project involving coffee farming communities of Mexico and Central America, which yielded a variety of outcomes. These ranged from direct actions in coffee cooperatives to research studies and academic publications. A key academic product of this work was an edited book on the coffee

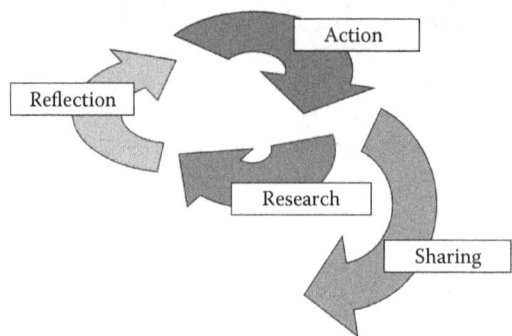

Figure 1.2 Participatory action research cycle. (Adapted from Bacon, C., et al., Participatory Action-Research and Support for Community Development and Conservation: Examples From Shade Coffee Landscapes of El Salvador and Nicaragua, Center for Agroecology and Sustainable Food Systems [CASFS], University of California, Santa Cruz, CA, 2005.)

price crisis (Bacon et al. 2008), while action-oriented projects and outreach were mostly channeled through the Community Agroecology Network (CAN, http://www.canunite.org/). A similar trajectory can be observed in Andalucía, Spain, where researchers, professors, and extension agents associated with the International University of Andalucia's graduate program in agroecology have worked with a diversity of family farmers in southern Spain (Cuéllar-Padilla and Calle-Collado 2011; Guzmán-Casado et al. 1999; Guzmán-Casado and Alonso-Mielgo 2007, 2008; Sevilla-Guzmán 2006a, b). Likewise, in Brazil, agroecologists have worked with the Landless Peasant Movement (MST) and La Via Campesina to support the incorporation of agroecology into these social movements (Altieri and Toledo 2011).

Participatory approaches in agroecology tend to adhere to a common set of principles associated with PAR. Not surprisingly, these principles share substantial overlap with an evolving set of agroecological principles that help define the field and unite different perspectives (Altieri 2000; Gliessman 2015). Table 1.1 summarizes selected and overlapping principles from both PAR and agroecology. A more complete list of the principles of agroecology and sustainability can be found at http://agroecology.org/Principles_List.html.

Table 1.1   Comparison of Selected Participatory Action Research and Agroecological Principles

Participatory Action Research Principles	Agroecology Principles
PAR prioritizes empowerment as community-based partners contribute to define the research agenda.	Agroecologists work with farmers, eaters, communities, agricultural ministries, food advocates, and others to support empowering people.
PAR processes are context dependent as they bring together trans/interdisciplinary teams responding to stakeholder aspirations.	Agroecology establishes farming and food systems that adjust to local environments.
PAR research processes inform action at multiple scales for positive social change.	Agroecology offers principles and analysis toward the creation of more sustainable agriculture and food systems.
PAR processes deepen as long-term relationships are formed and multiple iterations of the cycle occur.	Agroecology seeks to develop strategies to maximize long-term benefits.
PAR processes listen to a diversity of voices and knowledge systems to democratize the research and social change processes.	Agroecology incorporates farmer voices and knowledge into the research process and seeks to diversify biota, landscapes, markets, and institutions.

Sources: Modified from Bacon, C., et al., Participatory Action-Research and Support for Community Development and Conservation: Examples From Shade Coffee Landscapes of El Salvador and Nicaragua, Center for Agroecology and Sustainable Food Systems (CASFS), University of California, Santa Cruz, CA, 2005 and http://www.agroecology.org/Principles_List.html.

Like agroecology, PAR approaches in agriculture often involve researchers, farmers, and other organizations (e.g., governmental or nongovernmental organizations [NGOs]). The process values the collaborative definition, implementation, and interpretation of research that includes different forms of knowledge, as well as people's diverse aspirations in the design of research agendas and transitions toward collectively defined goals. Processes of empowerment are complex, uneven, and require attention to the formal and informal exercise of power, as well as critical reflections about the intersection of access to resources, privileges, and identities (Fox 2005; Minkler and Wallerstein 2008). The PAR projects tend to be highly negotiated processes, and it is unlikely that actors will continue to participate if they are not perceiving benefits or advancing their interests.

The final two principles listed in Table 1.1, for both agroecology and PAR, concern temporal- and diversity-related issues. Although researchers are aware of their own professional needs and pressing theoretical questions within their academic fields, these priorities do not often align with the needs of farmers and other social actors (Fox 2006). Instead of predetermining a project and then asking nonresearch partners to sign off, PAR collaboration should begin at the earliest stages of the research process. Ideally, partners work through a mutual, iterative dialogue to arrive at a project proposal that harmonizes stakeholder needs, capacities, and methods. However, it is also common for PAR processes to develop from different starting points. For example, a PAR process can emerge from a more conventional research project when the partners shift to a more inclusive dialogue and more even power relations (Bacon et al. 2005). The iterative nature of PAR results in shifts between the emphases placed on research, reflection, or action. This usually occurs through a negotiated dialogue, where each actor advocates for his/her interests, but is also willing to reach compromises for the benefit of all partners. Through this exchange, the researcher and other participants have a clear understanding of project expectations and potential challenges and benefits. The conversation must also be linked to action, thus creating a praxis—or an ongoing iterative process of reflection and action (Freire 2000). After an action is taken, the context may shift, and this is when the longer-term partnership often becomes more important, as both the researcher and the other partners have learned from the first cycle and may decide to continue in a different or similar direction in future iterations. The agroecological principle of maximizing long-term benefits suggests multiple considerations, such as efforts to:

- Maximize intergenerational benefits, not just annual profits.
- Maximize livelihoods and quality of life.
- Facilitate intergenerational transfers.
- Use long-term strategies, such as developing plans that can be adjusted and reevaluated through time.
- Incorporate long-term sustainability into overall agroecosystem design and management.
- Build soil fertility over the long term.

The principle of recognizing, learning from, and engaging social and ecological diversity is among the most important for linking PAR with an action-oriented agroecology. The PAR approach calls for greater attention to a wider diversity of voices, especially those that are frequently marginalized by mainstream society (e.g., farm workers, smallholder farmers, indigenous groups, and rural women). This suggests the need to create the time and space for deeper listening and identification of strategies that use human diversity as a source of innovation. The principle of diversity, as seen through an agroecological lens, is no less profound as it directs analytic attention to the domains of biota, landscapes, and social institutions. Examples of farm- and plot-level management of diversity include intercrops, crop rotations, polycultures, and the integration of animals, cultivars, and genetic diversity. At the landscape scale, one must consider issues such as buffer zones, forest fragments, rotational grazing, and contour and strip tillage. The important point is not simply the presence of a wide diversity of species or agricultural practices, but the way they interact to provide critical ecosystem services (e.g., pollination, pest control, and nutrient cycling) that support agricultural production and farmer livelihoods (Kremen et al. 2012). The social domains of diversity encourage agroecologists to consider multiple forms of farmer organization, government regulation, identities, and the many different types of markets and alternative networks that constitute agri-food systems (Goodman et al. 2011). The presence of alternative distribution systems and the diversity of social institutions and economic relations in agriculture, such as farmer's markets, community-supported agriculture, cooperatives, and production for both subsistence and sale, offer several important incentives that could be coupled with an enabling policy environment (Iles and Marsh 2012). Together, these related strategies could contribute to a transformation of current agri-food systems into ones that prioritize ecological and human health at all stages and integration among the multiple, interacting system components.

1.4.3 Toward Transformative Agroecology

A transformative agroecology incorporates a critique of the political economic structures that shape the current agri-food system (see Holt-Giménez and Altieri, this issue and González de Molina, this issue). It is explicitly committed to a more socially just and sustainable future by reshaping power relations from farm to table. This view requires that agroecologists move beyond the farm scale to consider the broader forces—such as market and government institutions—that undermine farmers' cultural practices, economic self-sufficiency, and the ecological resource base. In part, agroecology as a field of study emerged in response to concerns about the social and ecological impacts generated by the industrialization of agriculture and the implementation of Green Revolution technologies (Patel 2013). Narrow approaches that reduce agroecology to an ecologically sensitive agronomic science have disregarded the influence of critical social science research and theory, as well as broader social concerns, as part of the field's development. An "agroecology as natural science" perspective tends to privilege positivist science and Cartesian reductionism over other ways of knowing (e.g., holistic, indigenous, or local knowledge), and thus risks producing research that is not appropriate to local contexts and which ignores the larger power structures that impact farmer livelihood strategies.

The transformative agroecology that we propose has continued to develop a more holistic approach to the science and practice of agroecology in close dialogue with critiques of rural development put forth by academics, practitioners, and social movements. Political ecologists, in particular, have analyzed how power-laden relationships operating at the international, national, and regional level influence local agricultural practices, livelihoods, and landscapes. For instance, Blaikie and Brookefield's landmark study on soil degradation demonstrated how social marginalization, rather than maladaptation (i.e., in need of modernization), shaped farmers' land management practices (Blaikie and Brookfield 1987). This was a crucial shift in perspective that emphasized a multiscalar analysis to articulate local social and ecological phenomena to regional and global forces (Paulson et al. 2003). In short, political ecologists draw attention to the power relations that govern natural resources, often leaving farmers, due to their class, gender, or ethnic position, with a lack of access to productive assets (Peet and Watts 2004; Rocheleau et al. 1996). If farmers cannot access the resources they need, often dispersed within a surrounding territory and governed by overlapping power structures, they cannot continue to maintain or develop sustainable agroecosystems. A politically engaged agroecology considers the complex challenges, both social and ecological, that smallholders face in their efforts to transition toward sustainability (see González de Molina, in this issue).

The connection between agroecological practice, equitable distribution of resources, and self-determination has been made explicit by marginalized communities demanding justice through food sovereignty (Holt-Giménez and Altieri, in this issue). Ecological sustainability has become central to demands made in defense of rural livelihoods and culturally specific ways of life. These ways of living are increasingly at risk owing to the deepening of capitalist relations that turn people into labor and nature into resources (Carruthers 1996; Grueso et al. 2003). Agroecologists are aptly positioned to contribute to these struggles by participating in a creative process of knowledge production with farmers and other agri-food system actors. This requires a broader understanding of knowledge and learning as a community of practice that involves farmer scientists, university-trained researchers, and other members of civil society (Kloppenburg 1991; Thomas-Slayter et al. 1996). Agroecology, through its parallel development as a science and social movement, provides an apt space to construct relevant research and practice that addresses asymmetrical power relations.

1.4.4 Challenges

Developing and applying transdisciplinary, participatory, and action-oriented approaches for research and practice presents a series of challenges that we would like to discuss. First, the two inspirations for our work, transdisciplinary approaches to agroecology and PAR are emerging and dynamic frameworks of research and *praxis*. Most of the individuals and collaborative groups seeking to apply these approaches are only recently starting to do so, and hence face a set of new and emerging challenges. In addition, although interesting to many established academic and funding institutions, support for transdisciplinary research and PAR (and agroecology for that matter) is still severely limited. Partly, this is due to the nature of these approaches, which integrate a variety of activities that have traditionally operated and been funded separately (i.e., research, extension, and direct applications or actions). Hence, many institutions and funders are structured to treat them independently; even when they are interested in supporting the integration of these initiatives, many times they do not know how to do it. The same applies for research and nonresearch partners who are interested in participating in both transdisciplinary and PAR processes. Many of these actors and institutions have developed around disciplinary and professional norms and find it difficult to cross into other disciplines, knowledge systems, timelines, and other dynamics that are different from their own. These points are further illustrated in the case studies presented in the next section. We believe it is important to emphasize that doing the work we are proposing requires facing serious institutional, funding, and individual challenges. However, we are also convinced that these challenges can be overcome and that the "tide is changing" in terms of finding broader institutional and funding support for agroecological work that is transdisciplinary, participatory, and action oriented.

1.5 EXAMPLES OF AGROECOLOGICAL INITIATIVES SEEKING A TRANSDISCIPLINARY, PARTICIPATORY, AND ACTION-ORIENTED APPROACH

In this section, we present two distinct case studies of agroecological initiatives that apply transdisciplinarity and PAR. We emphasize advances, challenges, and lessons learned so that they can be useful cases for others interested in pursuing similar approaches.

1.5.1 The Vermont Agricultural Resilience in a Changing Climate Initiative

The Vermont Agricultural Resilience in a Changing Climate Initiative (VAR) has explicitly incorporated transdisciplinary agroecology and PAR to address the effects and potential responses to climate change by Vermont farmers and other agri-food actors. Vermont, a state located in the northeastern United States, has a long and continuing agricultural history that has recently been strengthened by an energetic interest and support for alternative agri-food systems (e.g., the Vermont Farm to Plate initiative). VAR is inclusive of multiple partners, including researchers with a wide range of foci, a professional advisory committee that includes farmers and other collaborators, and farmers who cultivate a wide range of products. Its research approach is to work with a diversity of actors to identify on-farm management practices or climate change best management practices (CCBMPs) that could (1) best help farmers adapt to climate change now and in the future; (2) provide information on how farmers can contribute to greenhouse gas (GHG) emissions mitigation; (3) work with outreach professionals to deliver information about these practices to a broad community of farmers and other professionals; (4) assess the future needs related to climate change of a diversity of actors in the Vermont agri-food system; and (5) create and utilize tools to inform policy and governance that are specifically related to climate change and agriculture. In this section, we focus on objectives 1–3 of the broader VAR initiative as a way to illustrate the opportunities, challenges, and lessons of applying a transdisciplinary and PAR-oriented approach to this particular project.

1.5.2 Application of a Transdisciplinary and PAR Approach

The VAR initiative brought together eight faculty members from the following University of Vermont (UVM) units: Plant and Soil Science Department; Community Development and Applied Economics Department; the Rubenstein School of Environment and Natural Resources; and the Center for Sustainable Agriculture and the Northwest Crops and Soils Program, both from UVM Extension. Figure 1.3 shows the VAR phases, activities, and actors as well as illustrates the focus on transdisciplinarity and participation of a diversity of actors, including researchers, farmers, extensionists, and private agricultural service providers. The figure shows the wide diversity of methods used for the research, such as a random mail survey of about 1000 farmers (with 76 responses), key informant interviews, and on-farm research.

Transdisciplinarity was viewed in the same vein as previously discussed in this chapter; as an approach seeking to integrate academic disciplines (i.e., agroecology, agricultural economics, ecology, and public policy) with nonacademic knowledge (farmer experience), to address a specific issue (the impacts of and responses to climate change by Vermont farmers).

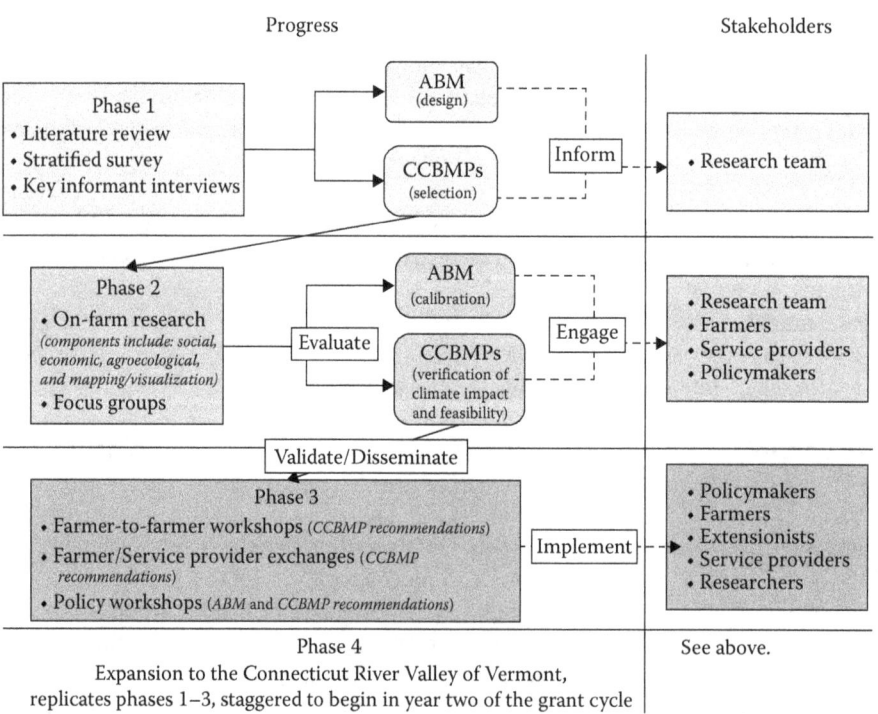

Figure 1.3 Planned phases and activities of the VAR Initiative. (From Méndez, V.E., et al., "Climate change adaptation and mitigation in the Lake Champlain Basin of Vermont," In Climate change adaptation and mitigation in the Lake Champlain Basin of Vermont," Grant proposal submitted to the University of Vermont Food Systems Initiative, 2012.)

1.5.3 Discussion of Selected Results

Although this project is still in progress, there are several key preliminary results worth mentioning that can be linked to the PAR and transdisciplinary approach. Rather than presenting the results themselves, which can be found in Schattman et al. (2015), we focus on describing the type of information that was collected:

- **Farmer perceptions and knowledge:** In staying true to both PAR and transdisciplinary tenets of including diverse voices and forms of knowledge, we sought to collect farmer perceptions and knowledge of management practices associated with climate change in four ways: (1) through written responses in a random mail survey; (2) analyzing text from farmer "Reports from the Field," which is an online newsletter managed by UVM Extension Professor Vern Grubinger posting farmer commentary on various issues related to vegetable and berry farming in Vermont (see http://www.uvm.edu/vtvegandberry/newsletter/datenavbar.htm); (3) through semistructured interviews with farmers and key informants (mostly extensionists and other service providers); and (4) by sharing and discussing with farmers landscape visualization posters with "before and after" scenarios of implementation of best management practices.
- **Establishment of an advisory committee:** This was done to keep our process accountable. The advisory committee consists of selected farmers, extensionists, staff from government and non-government organizations, and researchers. We periodically present project results and issues and request feedback, both in terms of potential impact for the groups they represent and general input in terms of the direction of the process.
- **On-farm research:** On going data collection with the participation of farmers on 12 farms, to produce results relevant to their farm contexts.
- **Farmer-to-farmer exchanges**: This phase is still pending, since it will take place once the on-farm research is concluded. It will include farmer-to-farmer knowledge exchanges related to the research on the 12 participating farms. We will invite farmers and service providers to attend these events in order to share our process. Although researchers may present some of their results, the idea is that these events will be led by farmer partners, with the research team taking a facilitating role.

1.5.4 Challenges, Opportunities, and Lessons

The experience of the VAR initiative has elucidated important opportunities, challenges, and lessons related to conducting transdisciplinary and PAR processes.

First, one of the most daunting challenges was keeping true to a meaningful transdisciplinary research endeavor. From the beginning, the complexity of bringing teammates together was evident. As the project progressed, each subteam went to work on their particular areas, and this occurred mostly in isolation from the rest. Although communications, full team meetings, and facilitation were kept frequent and consistent, there was little active interdisciplinary integration during the data collection process. Hence, those working on GHGs and costs of CCBMP implementation conducted their research without direct participation from other team members. At the current phase of the process, which focuses mostly on data analysis and developing publications, the team is intentionally seeking to do data integration, although it was mostly lacking during the data collection phase.

Second, even though the team was committed to engaging in a PAR process, we were limited in terms of farmer engagement in the following two ways: (1) the initial stages of design and fundraising for the project were led by researchers, since getting the team together required considerable effort and proposals had to be prepared with a deadline. However, we did engage extension partners as coinvestigators who work closely with farmers. (2) On-farm partners and farmers in the advisory committee have pointed out the need for the on-farm trials to be extended to several years. This is needed to account for climate change variability and better assess costs of maintaining adaptive practices over time. We were not able to secure funding for multiple year trials, but we are starting to develop proposals to do it in the next few years.

Third, leadership and facilitation are key components to move the process forward. In terms of leadership, it was evident that a strong presence from the project leader was necessary frequently and consistently. However, given the administrative demands of the project, shared or coleadership may be desirable. This has to occur from the onset, as trying to implement it after the project had started proved difficult in our particular case. Similarly, strong and effective facilitation of meetings, communications, and networking is essential for the success of these types of processes. In particular, deepening the transdisciplinary approach requires collaborators to share and better understand their multiple fields, as well as receive timely and synthetized information that they can use.

1.5.5 Food Security and Sovereignty with Smallholder Coffee Cooperatives and Farmers in Nicaragua

A team including researchers and staff affiliated with the nonprofit CAN, a faculty member at Santa Clara University, staff and farmers affiliated with the Promoter of Cooperative Development in the Segovias (PRODECOOP), a coffee-exporting cooperative union, and The Center for Information and Innovation within the Association for Development in northern Nicaragua (CII-ASDENIC), a local Nicaraguan NGO, designed the "Food Security and Sovereignty in the Segovias" project using the principles of transdisciplinarity, agroecology, and PAR. The project sought to address the persistence of seasonal hunger among the smallholder coffee producers of northern Nicaragua and develop options for reducing this situation. The relationships of trust and solidarity that established the initial context that would develop into a long-term partnership to eliminate hunger and create more sustainable food systems in the Segovias, started with research focused on the impacts of Fair Trade and organic coffee production in northern Nicaragua (Bacon 2005). This initiative also benefitted from strategic advice and targeted research conducted by UVM faculty and a graduate student, as well as insight and funding from Green Mountain Coffee Roasting Company (GMCR, now Keurig Green Mountain). The partnership focused on the shared production of knowledge, local capacity building, and the design of strategies to reduce seasonal hunger, increase access to healthy food, and promote sustainable agriculture among more than 1500 families affiliated with the PRODECOOP cooperative.

The study area incorporates coffee producing areas of Estelí, Madriz, and Nueva Segovia, commonly known as the Segovias region of Northern Nicaragua. The nested case study focuses on a primary level cooperative, consisting of about 100 families living in close proximity in the highlands of Condega, Estelí. Elevations range from 700 to 1550 m above sea level. The rainy season last from May through October followed by a dry season. The vegetation consists primarily of tropical dry forests at lower altitudes and semihumid and mixed oak and pine forests at higher altitudes. Most farmers in the study area produce a combination of cash crops (coffee), subsistence crops (corn and beans), fruit trees, and occasionally tubers and vegetables.

1.5.6 Application of a Transdisciplinary and PAR Approach

This partnership used an approach that was started by linking University-based researchers and CAN with Nicaraguan rural development organizations and small-holder cooperatives; thus we called it as community-based participatory action research (CB-PAR). The goal was to create a partnership that develops a more democratic approach to knowledge production and community change (Fortmann 2008; Hacker 2013; Minkler and Wallerstein 2008). The aim of the CB-PAR is to link the farmers' local and experiential knowledge with agronomists' technical skills, university researchers' academic knowledge, and cross-case expertise. To encourage this dialogue among knowledge systems and create a shared vision, we used participatory facilitation techniques from the *Campesino-a-Compesino* movement (Holt-Gimenez 2006). Chris Bacon, the principal investigator (PI) and professor at Santa Clara University, created a committee in which each member was given meaningful roles and responsibilities to facilitate participation in the CB-PAR process. The local research team consisted of PRODECOOP's agronomists working from the cooperative's central offices in the city of Estelí and a network of 24 primary cooperative level promoters (farmers who receive a small monthly stipend from CAN and PRODECOOP to coordinate a wide range of activities). CII-ASDENIC staff with training in field research methods, information technologies, and local development coordinated the reception and initial data capture from surveys. CAN and CII-ASDENIC staff drew from researcher recommendations, GMCR monitoring guidelines (developed by UVM researchers), and PRODECOOP's interest to identify indicators for project monitoring and evaluation. Additional details of the case study can be found in Bacon et al. (2014).

1.5.7 Selected Results

The multiple results that continue emerging from this initiative can be grouped under three of the four phases (i.e., action, research, and reflection) in a CB-PAR cycle (see Figure 1.2). The fourth phase is sharing, which represents an ongoing process that actually occurs simultaneously with each phase. Table 1.2 summarizes several of the more significant results. The initial project diagnostic included focus groups, interviews, participant observation, and a large baseline household survey conducted with participating farmers and rural youth. Local NGO staff also participated in community-based monitoring processes. After more than a year of data entry, quality control, and processing, preliminary results were discussed with farmers and other stakeholders. This information was then used to create local food security action plans and strategies to invest the international development funds received by the project (see Table 1.2).

This project generated a range of opportunities for research, local investment, institutional change, and follow-up international development projects. Capacity building took place through farmer-to-farmer exchanges, in workshops, and by sending staff from both PRODECOOP and CII-ASDENIC to international agroecology short courses organized by the CAN network in Chiapas and Quintana Roo, Mexico, and Estelí, and Nicaragua (hosted by project members). Furthermore, farmers learned about sustainable agriculture technologies from exchanges with neighbors within their own region and through experimentation on their plots. Several of the action steps implemented by CAN and PRODECOOP in this cycle (see Table 1.2) include (1) the development of community-based seed banks after an exchange suggested by the project PI and organized by leadership within Nicaragua's *Campesino-a-Campesino* program (see Holt-Gimenez 2006 for more on the history of this social movement); (2) workshops focused on using surface soil from surrounding forests, as

Table 1.2 Summary of CB-PAR Cycle from 2009 to 2013 in Northern Nicaragua

Research

Participatory baseline and diagnostic study
Participatory monitoring
Farmer exchanges with *Campesino-a-Campesino*
Best practices inventoried

Reflection and Planning

Agronomists and local staff train in workshops and International Agroecology Short courses with CAN
Diagnostic study results inform the creation of 11 Food Security Action Plans
Project scales-up to reach 1500 farm families

Selected Actions

Farmers plant 18,000 fruit trees on their land
Pilot project to change local corn and bean food system
Pilot project on community-based seed banks

an inoculant of beneficial microorganisms, for improved soil fertility management (learned following an exchange at an organic coffee producers' cooperative in Honduras); and (3) the cooperatives piloted an institutional innovation focused on relocalizing the local corn and beans distribution systems by using cooperative funds and infrastructure to purchase corn from affiliated members and local markets when prices were low during the harvest time and redistributing it at accessible prices during the lean months (June, July, and August), when corn prices generally spike.

The final published results from the household survey, focus groups, and interviews conducted in 2010 contributed to a second iteration of the CB-PAR cycle. The statistical analysis of the surveys revealed several anticipated patterns, finding that farmers with higher incomes and larger farms generally lived through shorter periods of seasonal hunger. However, we also found that farmers with larger corn yield and more fruit trees reported shorter periods of seasonal hunger (Bacon et al. 2014), which contributed to CAN and PRODECOOP's decision to support the planting of an additional 18,000 fruit trees and launch a series of experiments focused on the agroecological management of corn fields or *milpas*.

1.5.8 Opportunities, Challenges, and Lessons

Many opportunities emerged out of the necessity for tangible strategies that leverage direct investment and training in support of farmer-led sustainable agricultural actions and cooperative-led sustainable community development efforts. These were supported through adherence to a CB-PAR and agroecological approach from the inception of the project. The idea to create a broader coalition working to end seasonal hunger in northern Nicaragua, increase access to healthy food, and facilitate the transition to more resilient and sustainable food systems, leveraged resources and goodwill far beyond the constraints of this initial project. There are too many examples to summarize here. However, several specifics include the way that agroecology has been incorporated as a core transversal organizational strategy and programmatic area within both PRODECOOP and CII-ASDENIC. After the initial local diagnostic, it became evident that more direct investment was also needed to increase access to clean drinking water, and this project has moved forward at a significant scale. Finally, the researchers shared the results widely among interested academics and other stakeholders.

Like all initiatives there were also challenges and lessons learned. The detailed vision building, organizational, and administrative work associated with forming this coalition among staff from two highly capable local organizations in northern Nicaragua, the US-based CAN, and at least one university, with collaborating scholars from two additional universities, was not fully accounted for in planning timelines and budgets. The work invested in creating this collective effort and vision can be diminished, when new opportunities for international development research and local project funding emerge. Local development NGOs and cooperative leaders and/or well-positioned academics can request external funding directly from donor agencies without first doing the hard work of negotiating an agreement and a strategy with the key stakeholders who have created and started to implement the changing vision of this project. Given the current incentive systems for international funding and academic advancement (Fox 2006), there is no question that such dominant structures encourage this approach; yet the principles of generating an authentic PAR approach are clear and ask for transformational change from all participants. Another challenge is coordinating the timing and agendas of multiple project partners. However, this was resolved among the core PAR team through regularly scheduled time together in Nicaragua, every year, and to some extent through periodic visits that brought the Nicaraguan team to the United States. Among the key lessons learned is that this approach holds potential for generating knowledge and social change, provided that there is a long-term commitment to a place and reflexivity among all participants. These dynamic challenges and opportunities could be useful in the consideration of the broader debates about the global processes of scaling-up agroecology.

1.6 SCALING AGROECOLOGY OUT: OPTIMIZING PRODUCTION AND DEMOCRATIZING ACCESS

Smallholder farmers (less than 2 ha) represent 90% of farms worldwide, producing a substantial amount of the world's food supply (IAASTD 2009). Paradoxically, many of these farmers form part of the 842 million people who suffer from food insecurity (Parmentier 2014). The food riots of 2008 engendered a global discussion about the failures of our corporate-controlled industrial food system and the best way to feed the world's growing population (see Altieri and Holt, in this volume, on "corporate food regimes"). agroecology has gained traction as a possible solution among both peasant movements and multilateral institutions (De Schutter 2010; Rosset and Toress 2013). On the basis of social and environmental sustainability, agroecology's multidimensional approach offers a set of design principles particularly suited to the context of vulnerable small- and medium-scale farmers. However, can agroecology scale-up to meet the demand? Or more specifically, how can agroecology revitalize the small-farm sector, creating dignified livelihoods and vibrant local food economies that result in food access for all sectors of society?

Agroecology's tremendous potential lies in its capacity to address the root causes of environmental degradation and hunger. It is a context-specific approach that optimizes productivity of agricultural resources while enhancing local control over these resources throughout the agricultural value chain. Agroecology's engagement with the local ecology and culture is its cornerstone, yet has led to misconceptions about its scalability. Rather than a set of context-specific technologies, agroecology uses a set of principles that can be locally disseminated and adapted to multiple contexts. Farmer-to-farmer exchanges, based on the methodology of *campesino-a-campesino*, have proven to be one of the most useful tools for sharing agroecological practices. This form of horizontal exchange can be referred to as *scaling-out* (Parmentier 2014). Scaling-out centers the scaling process (or an "agroecological transition") on democratizing knowledge and practice by empowering farmers to be the experts and teachers of traditional and innovative approaches. For us, scaling-out also emphasizes democratizing the governance of resources and decision-making about agricultural development. Although policy change at the national and international level will be a crucial part in achieving an agroecological transition, consolidation of resource control and/or decision-making at this level undermines agroecology's very capacity to change the agri-food system.

We have identified three priorities for scaling-out agroecology: (1) shifting normative understandings of agricultural development toward optimization rather than maximization; (2) developing participatory and transdisciplinary research agendas to consolidate evidence supporting agroecology; and (3) supporting smallholder farmer, labor, and consumer movements focused on challenging corporate control over productive resources and food supplies.

Agroecology increases productivity through optimization by diversifying production systems, generating a variety of agricultural products tailored to farmers' practices and needs, as well as natural ecosystem processes. Research shows that small diversified systems have higher yields than large industrialized monocultures, if all output and externalities are measured. Furthermore, diversified systems produce an array of additional benefits—increased nutrient recycling, reduced dependency on chemical inputs, more robust food supply for local consumption, stronger rural economies, and farmers connected through exchange networks (Altieri and Nicholls 2008; Gliessman 1998; Parmentier 2014). However, farmers, agronomists, and development agencies tend to measure the success of agricultural development mostly in terms of single crop yields or income. Indeed, the Green Revolution succeeded in vastly increasing the yields of basic grains and increasing their availability at a lower cost to consumers (Evenson and Gollin 2003). This narrow notion of productivity externalizes social and environmental consequences, such as increased inequality among farmers, rising debt, rural–urban migration, environmental contamination, and low dietary diversity (Parmentier 2014). To scale out agroecology, we need to shift normative understandings of productivity away from maximization of yield and develop metrics for a multidimensional analysis of the benefits flowing from optimized agroecological production systems (Silici 2014).

Strong case study evidence demonstrates the multiple benefits of agroecology. However, further research is needed to consolidate the data and identify the most beneficial, context-specific agroecological strategies, and practices that contribute to scaling agroecology (Silici 2014). A participatory and transdisciplinary research agenda that includes the knowledge, experiences, and aspirations of the small-farm sector will best identify metrics for multidimensional analyses and identify barriers to adoption. Although agroecology may be particularly suited to the capacities of small farmers, investing labor and resources into an agroecological transition may not be feasible for farmers without secure land tenure, where disease or outmigration has vastly reduced available labor, or where farmers cannot afford to invest time in learning and experimentation (Silici 2014). Consolidated evidence regarding the challenges and opportunities of agroecological adoption can be a tool for advocating toward the creation of policies that shift greater resources toward agroecological research and development, while also incentivizing its implementation. To do this, it is necessary to shift policies that favor conventional chemical-intensive agriculture to those supporting agroecology (Altieri and Nichols 2008).

A shift in governance that incentivizes agroecology and jeopardizes the privileges of the corporate food regime will require political pressure (for specific examples of corporate influence over policies that shape the food system, see Parmentier 2014). Peasant movements, such as *"Via Campesina"* have embraced agroecology as a strategy to achieve food sovereignty (Rosset and Torres 2013). The Declaration of the International Forum for Agroecology (27 February 2015, Nyéléni, Mali) states that "Families, communities, collectives, organizations, and movements are the fertile soil in which agroecology flourishes. Collective self-organization and action are what make it possible to scale-up agroecology, build local food systems, and challenge corporate control of our food system." Scaling-out agroecology, as we understand it—as a science, practice, and movement—coincides with political demands for farmer control over productive resources and inclusive decision-making that values the perspectives of women, youth, and indigenous communities. The privileged scale of agroecology remains at the local level. To build an agri-food system based on agroecology, the adaptation and innovation of sustainable technologies must remain in the hands of many, not just a powerful few.

1.7 DISCUSSION OF THE CONTENTS OF THE EDITED VOLUME

This volume presents 14 diverse contributions that share a commitment to integrating transdisciplinarity, participatory, and/or action-oriented approaches within an agroecological framework. In this introductory chapter, we have sought to discuss the evolution of the field of agroecology and examine some of the contemporary debates that surround it. More specifically, we revisit the existence of different "agroecologies," a concept that we introduced two years ago (Méndez et al. 2013), as a key to understand how this dynamic field is constantly evolving. We also describe in-depth how we understand agroecology as a transdisciplinary, participatory, and action-oriented approach. To illustrate this, we present the opportunities and challenges of two case studies where we have tried to implement this particular agroecological approach. We finalize the chapter with a discussion of key issues surrounding the "scaling-out" of agroecology, a subject that is garnering increasing attention in research and policy circles. The subsequent seven chapters represent conceptual contributions that, in different ways, embrace the perspectives of transdisciplinary and participatory agroecology. These contributions are followed by six case studies that, in one way or another, have grappled with the integration of transdisciplinarity and PAR in agroecological work within different geographic and socioecological contexts.

Our introductory article is followed by a contribution from Steve Gliessman, which undertakes a historical analysis of his role in the development of agroecology, and his role in resisting industrialized agriculture. Through his work on traditional Mexican agriculture in the 1970s and 1980s, Gliessman examines the development of the "agroecosystem" concept, which drew from the work of Efraím Hernández Xolocotzi and the emergence of the field of agroecology. At this time, governments and international agencies were fully supporting the implementation of the Green Revolution in developing countries. The following chapter, by Sevilla-Guzmán and Woodgate, explores several social, political, and economic processes, such as agricultural modernization and environmentalism, as part of the foundations from where agroecology developed as both a "scientific discipline" and a transformative "agrarian social movement." The next piece by González de Molina proposes a stronger integration of political ecology into agroecology and the explicit development of a "political agroecology." González de Molina argues for the need to develop instruments and actions that interact with the political and institutional aspects of agroecological research and practice. In the following chapter, Francis and coauthors discuss the importance of involvement and reflection as key aspects of agroecological education. Their work is firmly grounded on a problem based, transdisciplinary agroecology program at the Norwegian University of Life Sciences (UMB), which engages students in real-world situations with rural communities. This article offers conceptual insights and an example of an innovative pedagogical model for agroecological teaching and learning. Subsequently, Vandermeer and Perfecto explore the potential for integration of ecological science with farmer knowledge to contribute to agroecological research and practice. Building on their experience analyzing ecological processes in agroecosystems, they argue that this integration could lead to the "generation of knowledge that is simultaneously deep and broad." This is followed by Holt-Giménez and Altieri's piece, which critiques what they term as the "new Green Revolution," discussed as a "re-ignition" of the previous green revolution with a slightly more environmental perspective, but replicating the socioeconomic and political approach that so severely damaged millions of smallholder farmers in the 1970s and 1980s. They emphasized the need for strengthening smallholders and their organizations as the backbone of alternative food systems, and caution of the danger of agroecology being coopted to strengthen the existing powers behind the new Green Revolution. The last of the conceptual chapters, by Francisco Rosado-May, introduces the reader to the interactions between agroecology and interculturality. The intercultural approach values and seeks out different forms of knowledge, links them together in a participatory process, and provides opportunity to generate new forms of action. As an example of an intercultural process, Rosado-May presents an in-depth historical perspective on the key role of agroecological work undertaken in Mexico and argues for the need to more explicitly integrate an intercultural perspective in current agroecological applications.

The following seven chapters of this volume present case studies, which were chosen for their alignment with agroecological, transdisciplinary, and participatory approaches. Chapter 9, by Guzmán-Casado and coauthors, discusses the need for an agroecological approach that goes beyond technological change. They propose PAR as the means to "collaborate with local communities and advance in the restructuring of physical flows, economies, and information that support local farming." These arguments are explored through an in-depth analysis of a case study with farmers in Andalusia, Spain. The authors conclude that despite some challenges associated with resources and longer time periods, PAR proved to be an adequate approach to foster an agroecological transition by farmers and other actors. The next chapter, by Fernandez and coauthors, undertakes a similar exercise as Rosado-Mayby analyzing the evolution of agroecology and its interactions with food sovereignty and urban agriculture in the United States. The authors find that although principles have been shared among the academic, agroecological perspective, and on-the-ground movements, an explicit collaboration is still elusive. The chapter concludes by providing specific recommendations for partnerships that can better integrate the strengths of agroecology as a participatory research approach and the experiences of alternative agri-food movements in the

United States. In Chapter 11, Heleba and coauthors share their in-depth and long-standing experience working closely with farmers through the UVMs Extension system. These highly respected Extension faculty and staff demonstrate the effectiveness and the need to have strong farmer involvement and collaboration in our efforts to more effectively implement a transition toward agroecology. Subsequently, Putnam and coauthors examine the application of the agroecological approach to implement a food security and sovereignty project in a coffee region of Nicaragua. Their findings support some of the themes discussed in this introductory chapter, as they demonstrate the efficacy of agroecology as a food security and sovereignty strategy, largely through farm diversification and farmer-to-farmer knowledge coproduction. The following chapter, by Gutiérrez and Ramírez, presents an example of a large international development initiative—the Mesoamerican Agroenvironmental Program (MAP) that sought to implement an integrated and multisectoral approach to "sustainable land management." MAP succeeded in incorporating a diversity of participatory and interdisciplinary approaches rarely seen in projects of this nature and stands as an example for similar future initiatives. The last chapter, by Alba Gonzalez Jacome, integrates perspectives from agroecology and ecological anthropology to analyze tropical agroforestry homegardens in Mexico. The author presents an in-depth historical analysis of this complex and traditional agroecosystem, which has inspired many agroecologists for its sustainability characteristics, and potential.

We selected the contributions in this edited volume as part of our efforts to support an engagement with an agroecology that is transdisciplinary, participatory, and action oriented. Both the conceptual and the empirical chapters provide insight into the opportunities, challenges, and lessons learned of applying this particular type of agroecology. We hope that this book will continue to encourage and advance critical and constructive "agroecological" debates, as well as provide inspiration for others seeking to embrace this challenging, yet promising, agroecological approach.

REFERENCES

Aeberhard, A. and S. Rist. "Transdisciplinary co-production of knowledge in the development of organic agriculture in Switzerland." *Ecological Economics* 68 no. 4 (2009): 1171–1181.

Altieri, M.A. *Agroecology: Principles and strategies for designing sustainable farming systems.* (2000). http://nature.berkeley.edu/~miguel-alt/principles_and_strategies. Accessed January 20, 2012.

Altieri, M.A. *Agroecology: The scientific basis of alternative agriculture.* Boulder, CO: Westview Press, 1987.

Altieri, M.A. and C.I. Nicholls. "Scaling up agroecological approaches for food sovereignty in Latin America." *Development* 51 no. 4 (2008): 472–480.

Altieri, M.A. and V.M. Toledo. "The agroecological revolution in Latin America: Rescuing nature, ensuring food sovereignty and empowering peasants." *Journal of Peasant Studies* 38 no. 3 (2011): 587–612.

Bacon, C. "Confronting the coffee crisis: Can fair trade, organic and specialty coffees reduce small-scale farmer vulnerability in northern Nicaragua?" *World Development* 33 no. 3 (2005): 497–511.

Bacon, C., V.E. Méndez, and M. Brown. "Participatory action research and support for community development and conservation: Examples from shade coffee landscapes of El Salvador and Nicaragua." Santa Cruz, CA: Center for Agroecology and Sustainable Food Systems (CASFS), University of California, 2005.

Bacon, C., W.A. Sundstrom, M.A. Flores-Gomez, et al. "Explaining the 'hungry farmer paradox': Smallholders and fair trade cooperatives navigate seasonality and change in Nicaragua's corn and coffee markets." *Global Environmental Change* 25 (2014): 133–149.

Bacon, C.M., V.E. Méndez, S.R. Gliessman, et al. (eds.). *Confronting the coffee crisis: Fair trade, sustainable livelihoods and ecosystems in Mexico and Central America.* Cambridge, MA: MIT Press, 2008.

Belsky, J.M. "Beyond the natural resource and environmental sociology divide: Insights from a transdisciplinary perspective." *Society & Natural Resources* 15 no. 3 (2002): 269–280.

Blaikie, P. and H. Brookfield. *Land degredation and society.* London: Longman Press, 1987.

Carruthers, D.V. "Indigenous ecology and the politics of linkage in Mexican social movements." *Third World Quarterly* 17 no. 5 (1996): 1007–1028.

Castellanet, C. and C.F. Jordan. *Participatory action research in natural resource management: A critique of the method based on five years' experience in the Transamazonica region of Brazil.* New York & Sussex: Taylor and Francis, 2002.

Cuéllar-Padilla, M. and Á. Calle-Collado. "Can we find solutions with people? Participatory action research with small organic producers in Andalusia." *Journal of Rural Studies* 27 no. 4 (2011): 372–383.

De Schutter, O. "Agroecology and the right to food." In *Agroecology and the right to food: United Nations Special Rapporteur on the Right to Food.* (2010) United Nations General Assembly. http://www.srfood. org/images/stories/pdf/officialreports/20110308_a-hrc-16-49_agroecology_en.pdf. Accessed April 1, 2015.

De Schutter, O. and G. Vanloqueren. "The new green revolution: How twenty-first-century science can feed the world." Solutions 2 no. 4 (2011): 33–44.

Evenson, R.E. and D. Gollin. "Assessing the impact of the green revolution, 1960 to 2000." *Science* 300 no. 5620 (2003): 758–762.

Fals-Borda, O. and M.A. Rahman (eds.). *Action and knowledge: Breaking the monopoly with participatory action-research.* New York, NY: Apex Press, 1991.

Fortmann, L. (ed.) *Participatory research in conservation and rural livelihoods: Doing science together.* Hoboken, NJ: Wiley-Blackwell, 2008.

Fox, J.A. "Empowerment and institutional change: Mapping 'virtuous circles' of state-society interaction." In *Power, rights, and poverty: Concepts and connections.* Alsop, R. (ed.). (Washington, DC: The World Bank, 2005), 68–92.

Fox, J.A. "Lessons from action research partnerships." *Development in Practice* 16 no. 1 (2006): 27–38.

Francis, C.A., G. Lieblein, T.A. Breland, et al. "Transdisciplinary research for a sustainable agriculture and food sector." *Agronomy Journal* 100 no. 3 (2008): 771–776.

Francis, C., G. Lieblein, S. Gliessman, et al. "Agroecology: The ecology of food systems." *Journal of Sustainable Agriculture* 22 no. 3 (2003): 99–118.

Freire, P. *Pedagogy of the oppressed.* London and New York: Continuum, 2000.

Gliessman, S.R. *Agroecology: Ecological processes in sustainable agriculture.* Ann Arbor, MI: Ann Arbor Press, 1998.

Gliessman, S.R. (ed.). *Agroecology: Researching the ecological basis for sustainable agriculture.* New York, NY: Springer-Verlag, 1990.

Gliessman, S.R. *Agroecology: The ecology of sustainable food systems.* 2nd Edition. Boca Raton, FL: CRC Press/Taylor & Francis, 2006.

Gliessman, S.R. *Agroecology: The ecology of sustainable food systems.* 3rd Edition. Boca Raton, FL: CRC Press/Taylor & Francis, 2015.

Gliessman, S.R. "Aspectos ecologicos de las practicas agricolas tradicionales en Tabasco, Mexico: Aplicaciones parala produccion." *Biotica* 5 (1980): 93–101.

Gliessman, S.R. "Nitrogen distribution in several traditional agroecosystems in the humid tropical lowlands of southeastern Mexico." *Plant and Soil* 67 (1982): 105–117.

Gliessman, S.R. (ed.). *Seminarios regionales sobre agroecosistemas con enfasis en el estudio de tecnologia agricola tradicional.* H. Cardenas, Tabasco: Colegio Superior de Agricultura Tropical, 1978.

Gliessman, S.R., R. Garcia-Espinosa, and M. Amador. "The ecological basis for the application of traditional agricultural technology in the management of tropical agro-ecosystems." *Agro-Ecosystems* 7 (1981): 173–185.

Godemann, J. "Knowledge integration: A key challenge for transdisciplinary cooperation." *Environmental Education Research* 14 no. 6 (2008): 625–641.

Goodman, D., M. DuPuis, and M.K. Goodman. *Alternative food networks: Knowledge, place and politics.* London: Routledge, 2011.

Greenwood, D.J. and M. Levin. *Introduction to action research: Social research for social change.* Thousand Oaks, CA: Sage Publications, 1998.

Grueso, L., C. Rosero, and A. Escobar. "The process of black community organizing in the Southern Pacific coast region of Colombia." In *Perspectives on Las Américas: A reader in culture, history, and representation.* Gutmann, M.C, Rodríguez, F.V., Stephen L, and Zavella P. (eds.). (Malden: Wiley-Blackwell, 2003).

Guzmán-Casado, G.I., M. González de Molina, and E. Sevilla-Guzmán. *Introducción a la agroecología como desarrollo rural sostenible.* Madrid: Ediciones Mundi-Prensa, 1999.

Guzmán-Casado, G.I. and A.M. Alonso-Mielgo. "La investigacion participativa en agroecologia: una herramienta para el desarrollo sustentable." *Ecosistemas (Spain)* 16 no. 1 (2007): 24–36.

Guzmán-Casado, G.I. and A.M. Alonso-Mielgo. "A comparison of energy use in conventional and organic olive oil production in Spain." *Agricultural Systems* 98 no. 3 (2008): 167–176.

Hacker, K. *Community-based participatory research.* Thousand Oaks, CA: SAGE Publications, 2013.

Hecht, S.B. "The evolution of agroecological thought." In *Agroecology: The science of sustainable agriculture.* Altieri, M.A. (ed.). (Boulder, CO: Westview Press, 1995), 1–20.

Holt-Gimenez, E. *Campesino a campesino: Voices from Latin America's farmer to farmer movement for sustainable agriculture.* Oakland, CA: Food First Books, 2006.

IAASTD. *Agriculture at a crossroads: Global report by the International Assessment of Agricultural Knowledge, Science and Technology for Development (IAASTD).* Washington, DC: Island Press, 2009.

Iles, A. and R. Marsh. "Nurturing diversified farming systems in industrialized countries: How public policy can contribute." *Ecology and Society* 17 no. 4 (2012): 42.

Kindon, S., R. Pain, and M. Kesby. "Introduction: Connecting people, participation and place." In *Participatory action research: origins, approaches and methods.* Kindon, S., Pain, R., and Kesby, M. (eds.). (Oxon: Routledge, 2007a), 1–7.

Kindon, S., R. Pain, and M. Kesby (eds.). *Participatory action research approaches and methods.* Oxon: Routledge, 2007b.

Kloppenburg, J. "Social theory and the reconstruction of agricultural science: Local knowledge for an alternative agriculture." *Rural Sociology* 56 no. 4 (1991): 519–548.

Kremen, C., A. Iles, and C.M. Bacon. "Diversified farming systems: an agroecological, systems-based alternative to modern industrial agriculture." *Ecology and Society* 17 no. 4 (2012): 44.

Martin, K. and J. Sauerborn. *Agroecology.* Dordrecht, the Netherlands: Springer, 2013.

Méndez, V.E. "Agroecology." In *Encyclopedia of Geography.* Warf, B. (ed.). (Thousand Oaks, CA: Sage Publications, 2010), 55–59.

Méndez, V.E., C. Adair, L. Berlin, et al. "Climate change adaptation and mitigation in the Lake Champlain Basin of Vermont." In *Climate change adaptation and mitigation in the Lake Champlain Basin of Vermont.* Grant proposal submitted to the University of Vermont Food Systems Initiative (2012).

Méndez, V.E., C.M. Bacon, and R. Cohen. "Agroecology as a transdisciplinary, participatory, and action-oriented approach". *Agroecology and Sustainable Food Systems* 37 no. 1 (2013): 3–18.

Minkler, M. and N. Wallerstein (eds.). *Community-based participatory research for health: From process to outcomes.* New York, NY: Jossey Bass, 2008.

Parmentier, S. *Scaling up agroecological approaches: What, why and how.* Belgium: Oxfam-Solidarity, 2014.

Patel, R. "The long green revolution." *Journal of Peasant Studies* 40 no. 1 (2013): 1–63.

Paulson, S., G. Lisa, and M. Watts. "Locating the political in political ecology: An introduction." *Human Organization* 62 (2003): 205–217.

Peet, R. and M.J. Watts (eds.). *Liberation ecologies: Environment, development, social movements.* London: Routledge, 2004.

Perfecto, I., J. Vandermeer, and A. Wright. *Nature's matrix: Linking agriculture, conservation and food sovereignty.* London, UK: Earthscan, 2009.

Rocheleau, D., B. Thomas-Slayter, and E. Wangari (eds.). *Feminist political ecology: Global issues and local experiences.* London: Routledge, 1996.

Rosset, P. and M.E. Torres. "La vía campesina and agroecology." In *La Vía Campesina's open book: Celebrating 20 years of struggle.* La Vía Campesina, 2013. http://viacampesina.org/downloads/pdf/openbooks/EN-12. pdf. Accessed April 11, 2015.

Schattman, R., V.E. Méndez, K. Westdjik, et al. "Vermont agricultural resilience in a changing climate: A transdisciplinary and participatory action research (PAR) process." In *Agroecology, ecosystems, and sustainability.* Benkeblia, N. (ed.). (Boca Raton, FL: CRC Press/Taylor and Francis, 2015), 326–343.

Sevilla-Guzmán, E. "Agroecología y agricultural ecológica: Hacia una 're'construcción de la soberanía alimentaria." *Agroecología (Spain)* 1 (2006a): 7–18.

Sevilla-Guzmán, E. *De la sociología rural a la agroecología.* Barcelona: Icaria Editorial, 2006b.

Silici, L. "Agroecology: What it is and what it has to offer." In *Agroecology: What it is and what it has to offer,* IIED Issue Paper. London: International Institute for Environment and Development (IIED), 2014.

Snapp, S. and B. Pound (eds.). *Agricultural systems: Agroecology and rural innovation for development.* Amsterdam: Academic Press, 2008.

Stringer, E.T. *Action research*. Thousand Oaks, CA: Sage Publications, 1999.

Thomas-Slayter, B., E. Wangari, and D. Rocheleau. "Feminist political ecology: Crosscutting themes, theoretical insights, policy implications." In *Feminist political ecology: Global issues and local experiences*. Rocheleau, D. (ed.). London: Routledge, 1996.

Tomich, T.P., Brodt, S, Ferris, H, et al. "Agroecology: A review from a global-change perspective." *Annual Review of Environment and Resources* 36 no. 1 (2011): 193–222.

Uphoff, N. (ed.). *Agroecological innovations: Increasing food production with participatory development*. London: Earthscan, 2002.

Vandermeer, J.H. *The ecology of agroecosystems*. Sudbury, MA: Jones & Bartlett Publishers, 2009.

Wezel, A., S. Bellon, T. Dore, et al. "Agroecology as a science, a movement and a practice. A review." *Agronomy for Sustainable Development* 29 no. 4 (2009): 503–515.

Wezel, A. and V. Soldat. "A quantitative and qualitative historical analysis of the scientific discipline of agroecology." *International Journal of Agricultural Sustainability* 7 no. 1 (2009): 3–18.

Whitmer, A, L. Ogden, J. Lawton, et al. "The engaged university: providing a platform for research that transforms society." *Frontiers in Ecology and the Environment* 8 no. 6 (2010): 314–321.

CHAPTER 2

Agroecology
Roots of Resistance to Industrialized Food Systems

Stephen R. Gliessman

CONTENTS

2.1 INTRODUCTION

One of the most complete definitions of agroecology today is the "ecology of the food system" (Francis et al. 2003; Gliessman 2014a). It has the explicit goal of transforming food systems toward sustainability, such that there is a balance between ecological soundness, economic viability, and social justice (Gliessman 2014a). However, to achieve this transformation, change is needed in all parts of the food system, from the seed and the soil to the table (Gliessman and Rosemeyer 2010). The two most important parts of the food system—those who grow the food and those who eat it—must be reconnected in a social movement that honors the deep relationship between culture and the environment that created agriculture in the first place. Our current globalized and industrialized food system is showing that it is not sustainable in any of the three aspects of sustainability (economic, social, or environmental; Gliessman 2104a). With a deep understanding of what a holistic, ecological view of the food system can be, the change needed to restore sustainability to food systems can occur.

2.2 CONCEPTUAL BACKGROUND: EXPLORING THE ROOTS

From the earliest appearance of the term agroecology, there has always been an emphasis on the relationship (or lack thereof) of the two fields of ecology and agronomy (see Gliessman 2014a, for a brief review of the history of agroecology). But from the beginning of its use, agroecology was divided between the agronomy of crop production and yields, and the ecology of crop distribution and plant or animal adaptation to the environment. The term most commonly used in these earlier times was crop ecology, with a very strong emphasis on developing technologies that allowed for adjusting or modifying the farm environment to meet the needs of the crop organism so that the highest yield could be obtained. Obviously, the array of machinery, fertilizers, pesticides, and other technological innovations that began to become available, especially after World War I, was the inputs used to modify the crop environment.

Interestingly, though, one of the first uses of the term agroecology was a response to the indiscriminant use of these inputs. Writing in the publication of the International Institute of Agriculture in Rome (a precursor to FAO) in 1930, Basil Bensin, a Russian agronomist, called attention to the need for international cooperation in agroecological investigation, and termed the science behind this investigation as agroecology (Bensin 1930). He observed that farmers were too often convinced by the organized advertising campaigns of the large companies who manufactured tractors, fertilizers, and seeds without really knowing if these inputs were appropriate for local conditions and meet the particular farmer needs. Advertisements, for example, claimed that a single tractor was "universal" and suitable for all kinds of soils, climates, and types of farming. By producing large quantities of a universal machine, the companies could compete more effectively on the market. But too often, Bensin observed, farmers experienced disappointment after having bought a farm machine advertised as universal. The same was true for seeds sold by the well-known seed and plant breeding companies of this time. Attracted by advertisements that claimed that these new seeds could succeed anywhere, farmers would order seeds produced in a place and under conditions very different from their farms, and too often they also required the machinery and fertilizers that are used to alter conditions to meet the needs of the new seeds. Local knowledge and experience was not included in the development of this new array of inputs. Farmers were being considered primarily as purchasers of production products and being taken advantage of in the process.

On the one hand, agroecology was seen by Bensin as a way to generate information through what he called "agroecological research" that would help farmers make better choices on what to purchase. Interestingly, though, he also stated the need "to regulate the purchase of fertilizers, machines, and seeds, so as to reduce the risk to the farmer." This can be interpreted as calling for some forms of resistance to the pressure being exerted by the corporations, a need that has only grown greater as the industrial model of agriculture has gained more and more dominance of our food system. But Bensin also saw agroecology as a multidisciplinary science, where all factors that have an influence on the development and success of a crop must be considered. For him, agroecological investigation needed to be grounded in botany, plant breeding, meteorology, climatology, soil science, and experimental agronomy—in some respects, grounded in knowledge of the entire ecosystem in which agriculture was occurring. He criticized the experimental agronomy of the time as being too focused on the yields obtained by the use of new inputs and practices, rather than a focus on the reasons and causes for the results obtained. In spite of its call for resistance, Bensin's agroecological proposal seems to have been reduced to crop ecology over the next several decades. The primary focus became meeting crop needs through environmental modification and agricultural inputs.

One of the best-known examples of agricultural ecology or crop ecology was the work of Azzi (1956). Building upon the same fields of agricultural meteorology and soil science as Bensin, Azzi proposed the field of agricultural ecology as a way to integrate all of the separate sciences that agronomy uses to understand how each affected the crop of interest. For him, agricultural ecology went beyond exploring the ecological characteristics of each species. It also provided a way to analyze yields differently as a way to discover what controls the complex relationships between plants, environment, and yield. Tischler (1965) further elaborated the need to understand the ecology of each of the components of the agricultural system from crop adaptations to insect management and land husbandry. These pioneers were creating a foundation for viewing agricultural systems as ecosystems but still emphasizing the crops, but not the people who grew them. The lack of a whole-system view of farming and agriculture, especially without any social component, may have been the main reason for the growth of a strong production emphasis, culminating eventually in the so-called Green Revolution of the 1960s.

It took work by ecologists rather than agronomists to finally formalize an ecosystem view of agriculture. One of the first to do this was Janzen (1973), starting with his paper on the concept of tropical agroecosystems. An ecologist who was very committed to protecting and preserving tropical forests, yet also very aware of the livelihood needs of local people in tropical regions, Janzen proposed what he called "sustained-yield tropical agroecosystems." These productive ecosystems, in his view, should be grounded in local ecological knowledge, adapted to local environments and cultures, and designed to meet local needs, first, rather than respond to the demands of export markets for commodity crops. By going against the thrust of the Green Revolution's focus on the market, Janzen was echoing Bensin's call more than four decades earlier for the need for agroecology, but with a special view toward the needs of people in the tropics. A bit later, a review by Loucks (1977) pointed out how the strengthening of our understanding of ecosystem structure and function in the 1960s, and which to a certain extent culminated with the classic work on ecosystem development by Odum (1969), brought us to a point where it was clear that agricultural ecosystems possessed similar characteristics to natural ecosystems. Agroecosystems and ecosystems differed, though, in the primary characteristic of continual removal of nutrients through harvest or loss through "leaks" in the agricultural ecosystem. This was due to the loss of interconnectedness and complexity in energy flow and nutrient cycles that characterizes the structure of industrialized agricultural systems. Loucks stressed the need for an agroecosystem approach not only for improving yield performance, but also to determine the long-term stability of such yield improvements and their impacts on ecosystems in the broader landscape in which the agroecosystems were located.

Loucks had participated in the preparation of a report sponsored by the newly formed International Association for Ecology (INTECOL) on the development of an international program for analysis of agroecosystems (INTECOL 1975). This report was commissioned by an ad hoc working group on agro-ecosystems that convened at the first International Congress of INTECOL in The Hague, The Netherlands, in September 1974. This also coincided with the publication of the first issue of the journal *Agro-Ecosystems*, which was designed as a forum to publish research that integrated the many fields of agriculture described by the early crop ecologists. The founding editor of the journal introduced the first issue with a call for research on the ecological interactions that occur in all human-managed ecosystems, from agriculture to forestry, and fisheries (Harper 1974). He emphasized the need for an ecosystem approach that recognized "that each part is a component of a whole and that at some point the whole itself must itself be a subject of study." The journal set in motion a broader, multidisciplinary view of agroecosystems intended to promote the understanding of the function and management of whole ecosystems, from the most extensive to the most intensive, and from the most natural to the most intensively altered by humans. The goal was to increase and maintain production in ways that were efficient, environmentally sound, and agronomically validated. Notably, the social science side of multidisciplinarity was not included, nor did the term "agroecology" appear.

In 1979, two books appeared that began to discuss the social component within the concept of the agroecosystem. The first was *Agriculture Ecology: An Analysis of World Food Production Systems* (Cox and Atkins 1979). Using a distinctly evolutionary approach, this book first located food production systems in an ecological and historical context, with agriculture seen as the result of a long process of coevolution between culture and environment. It is no coincidence that the cover drawing is a pre-Hispanic rendition of the remarkably productive system of raised beds and canals present in the Valley of Mexico when Cortez began his conquest. Considerable emphasis was placed on the value of local and traditional farming systems with a long history of experience, change, and adaptation, especially in developing countries where the highly mechanized and input-intensive production practices of the Green Revolution had not yet penetrated. By using an ecological lens through which agroecosystem dynamics could be investigated, present-day agriculture was examined for its strengths and weaknesses, with ecologically based alternatives proposed as needed. For example, the ecological impacts of cultivation, grazing, irrigation, and fertilization on the soil ecosystem were reviewed, with alternatives proposed that would help maintain a healthy, productive soil ecosystem. The negative impacts of the use of pesticides were also reviewed, along with positive alternatives such as biological control, crop rotations and diversification, sanitation, and new advances in chemical attractants, deterrents, and growth regulators. In all components of the agroecosystem, the book tried to look beyond the drive for yield increases at all costs, and instead presented a framework for increasing production without destroying agricultural lands or damaging global ecology. However, most importantly, the book emphasized the need to be aware of the cultural and economic contexts within which any change in agriculture occurs. By drawing attention to the weaknesses of single-crop economies, especially in the developing world, and to what were then recent sociopolitical "setbacks," resulting from the Green Revolution, the book makes a strong call for an agricultural ecology that will "reveal the ecological fitness of past and present agricultural systems as a basis for developing an ecologically sound approach to agriculture in the future" (Cox and Atkins 1979; Chapters 25 and 26).

The other relevant book was first produced as a text for students at the Tropical Agriculture Research and Education Center (CATIE) in Turrialba, Costa Rica (Hart 1979). Titled *Agroecosistemas: Conceptos Básicos* (Agroecosystems: Basic Concepts), the book was designed to give students of tropical agriculture an alternative to the technological focus imported to the tropics from mostly temperate parts of the developed world. Agronomy students were given a full training in the ecological concepts and principles that today form the foundation of agroecology. It provided an in-depth ecological content for understanding structure, function, relationships, and dynamics of agroecosystems from the individual plant or animal, from the farm to the region, and eventually, to the global food system. All components of the agroecosystem were viewed as subsystems, such as the soil, crops, weeds, pests, and diseases. By understanding the relationships between subsystems, a design for integrating them into a whole could be visualized. Perhaps the most important element of the book was that it began at the local level, with local farmers who had been living under a particular set of ecological, economic, and social conditions that had guided the development of their agroecosystems over time. Hart recognized this richness of knowledge and experience, and in fact, refers to these farmers as his "professors" for convincing him that there was much more to the agroecosystem than the yield of an individual crop plant or animal. These small-scale farmers, who were (and still are) the main food producers for people in the tropics and the rest of the developing world, were being forgotten by the Green Revolution. Both of these books became important components of the teaching and research programs in agroecology that are described subsequently.

2.3 THE ROOTS OF RESISTANCE IN MEXICO

By the late 1960s the Green Revolution had achieved a strong foothold in Mexico. The International Wheat and Maize Improvement Center (CIMMYT) was established in 1966 in the same rural town outside of Mexico City, where the National Autonomous Agricultural University of Chapingo (UACh) was located. The new "improved" high-yielding varieties of corn and wheat began to be introduced from CIMMYT to the school, agronomists, the extension system, seed outlets, and ultimately to farmers. Although CYMMIT's first efforts focused on improving crop varieties, the impacts of these new corn and wheat cultivars were more than just the introduction of new seeds. The seeds were the first step to displace a food system that was thousands of years old by what is known today as an industrialized agricultural system, characterized using high external inputs and fossil fuels, focused on export crops, and based on monoculture cropping. What was being displaced were diverse, low-external input, locally adapted farming systems such as the traditional intercrop of corn, beans, and squash. Despite their ability to deliver the promised dramatic increases in yields, these new Green

Revolution crops began having drastic negative impacts on rural and traditional farming systems. Mexico began to move from self-sufficiency in corn to being a net importer by the end of 1970. Food prices began to rise. Farms and their families began to abandon the rural areas they had lived in for generations. Agrobiodiversity began to decline. The reasons for these changes are many and complex, but at the beginning of the Green Revolution there was also a resistance movement taking root that was grounded in valuing the rich coevolutionary history and cultural memory of the local, indigenous, traditional farming systems of Mexico (Hernández Xolocotzi 1985, 1987; Gonzalez Jácome 2011).

The development of three programs occurred almost simultaneously in Mexico between 1974 and 1980. Together they formed both a resistance as well as an alternative to the Green Revolution. One of the most important actions was the work of agronomist and ethnobotanist Efraím Hernández Xolocotzi. In the 1950s and 1960s, he used his training in ethnobotany to lead extensive field collections of the immense agrobiodiversity present in the fields of local Mexican farmers. Through this research, he was able to see how this genetic richness was being used to create hybrid varieties that focused purely on raising yields and ignored the millennial coevolutionary processes that had led to the development of the systems in which these varieties had evolved. This realization motivated him to start another movement, which called attention to the strengths of traditional Mexican agriculture and the urgent need to keep it from being displaced, and which culminated in a national seminar in 1976 titled "Analysis of the Agroecosystems of Mexico" with its proceedings published in 1977 (Hernández Xolocotzi 1977). A key aspect of Hernández Xolocotzi's thinking is shown in Figure 2.1, where his conceptualization of an agroecosystem took the form of three axes that needed to be balanced for sustainability to occur. He argued that the Green Revolution ignored the ecological axis and emphasized introducing new inputs, practices, and technologies aimed at increasing yields to respond to market pressures and the dominant development thinking of the time. The socioeconomic axis was reduced to a purely economic one, and in the process, an entire culture of agriculture was being lost (Hernandez Xolocotzi 1985, 1987).

A second focus that was developing at this time in Mexico was called *agrobiología*. Its primary proponent was the ecologist and botanist Arturo Gomez-Pompa. He established the National Institute for Research on Biotic Resources (INIREB) which was headquartered in Xalapa, Veracruz. INIREB played an important role in drawing attention to the problem of deforestation in the tropics, especially in Mexico, and developed a range of alternatives grounded in biological and ecological knowledge linked with the traditional experience of local agricultural systems. In part, this effort was a form of resistance to the large-scale removal of tropical forests to install large internationally funded development projects using Green Revolution technology. His work on the reconstruction of different versions of wetland agriculture based on the model of raised fields or *chinampas* is a good example of pushing back on the industrialized agriculture model (Gomez-Pompa 1985). He termed his work as agrobiology.

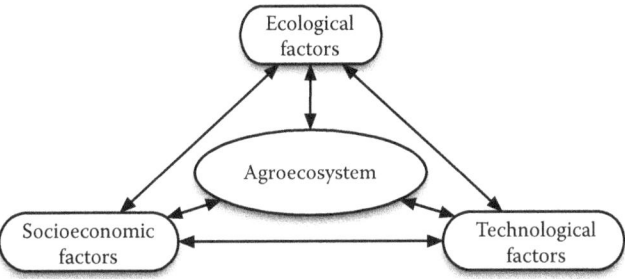

Figure 2.1   The factors influencing the coevolution of an agroecosystem. (Adapted from Gliessman, S.R. *Agroecology: The Ecology of Sustainable Food Systems. Second Edition.* Boca Raton, FL: CRC Press/Taylor & Francis Group, 2007.)

The third focus began in 1974 with the establishment of a small college of tropical agriculture near Cárdenas, in the state of Tabasco in southeastern Mexico (Colegio Superior de Agricultura Tropical [CSAT]). It was conveniently located in the middle of an immense International Development Bank (IDB)–funded project known as the Chontalpa Development Plan (*El Plan Chontalpa*), the first phase of which was a 90,000 ha clearing of tropical forest, draining of wetlands, moving of local communities to small housing villages located within the project, and the establishment of large-scale monoculture crops such as corn, beans, sugarcane, and improved pasture, all using Green Revolution technology. The region was to become the new granary of Mexico (Barkin 1978), with a primary focus on export crops, and CSAT was going to train the agronomists and test the technologies on its experimental fields to solve any problems that might arise. Due to several farsighted founding faculty in the Department of Ecology and the Department of Plant Pathology, as well as connections with Hernández Xolocotzi at the national school of agriculture, ecology courses formed part of the original curriculum at CSAT. But it soon became evident that ecology as a science separate from agriculture was not of interest to the students. To entice the students, most of whom were from the tropical regions of Mexico, ways of applying ecological concepts and principles to local agroecosystems had to occur. Soon, ecology courses began to morph into *agroecología*. International summer courses in agroecology were offered in 1978–1980, a master's degree program in tropical agroecology was begun in 1978, and research projects with the agroecosystem as the organizing concept and agroecology as the research process began as early as 1977 (Department of Ecology, CSAT, unpublished annual reports from 1978, 1979).

When an agroecological approach assessed the Green Revolution monocultures of corn, beans, rice, and sugarcane that were being grown on the experimental fields of CSAT, as well as the farmers' fields in the *Plan Chontalpa*, it quickly became obvious that they were not sustainable. This lack of sustainability was not just observed in the ecological realm, but also in relation to the social, economic, and cultural dimensions of these systems. As detailed in Barkin (1978), the social injustices and inequalities that the development project was bringing about were many. Farmers no longer grew the food they ate, planting decisions were made by the bank that funded the project, the farmers found it easier to contract salaried labor from outside the project area than do it themselves, and generations of local agroecological knowledge were being lost. It was at this time that the same ecologists who were now teaching *agroecología* at CSAT realized that there was another agriculture on the margins of the project, and in some cases, being practiced on unoccupied parcels within the project—traditional Maya agriculture.

A key event in the development of agroecology in Mexico was the organization of a regional seminar held at CSAT in March of 1978, with proceedings published that same year (Gliessman, 1978), with the title "Agroecosystems with an Emphasis on the Study of Traditional Agricultural Technology (TAT)." The seminar brought together Hernández Xolocotzi and his research group, the growing group of agroecologists at CSAT, persons or groups carrying out studies of TAT from around Mexico, as well as a large number of students and farmers. The agroecosystem focus was defined and applied to the richness of traditional farming systems all around Mexico, not just the lowland Maya region, and agroecology was presented as a way for these agroecosystem to be studied, preserved, improved, and expanded. In addition, a strong call made for all studies to include the full participation of farmers and their communities to reach the large number of rural cultures being rapidly marginalized by the Green Revolution.

For several years, intensive participatory surveys and research projects began to be carried out that demonstrated the strong combination of agroecological and cultural knowledge contained in the local Mexican systems. One of the most intensively studied local traditional farming systems was the intercrop of corn, beans, and squash, also widely known as the "3 sisters" (Amador 1980). With its roots in the prehispanic period, this polyculture cropping system that was common throughout Central America and Mexico was studied on local farms, using local seeds and practices, and most importantly, the local farmer knowledge (Figure 2.2).

Figure 2.2 The traditional corn–bean–squash intercrop system from Mesoamerica. Complex agroecological interactions are keys to the success of this cropping system.

On-farm research showed that corn yields could be stimulated as much as 50% beyond monoculture yields when planted with beans and squash on land that had a long history of only being managed using local traditional practices (Amador and Gliessman 1990). There was significant yield reduction for the other two associated crop species, but the total yields for the three crops together were higher than what would have been obtained in an equivalent area planted to monocultures of the three crops. As shown in Table 2.1, this comparison was made using the land equivalent ratio (LER), where a ratio greater than 1 indicates that an intercropping system shows distinct yield advantage in relation to monocultures of its component crops (Vandermeer 1989; Gliessman 2014b). Through a collaborative process that linked agroecological science with this local knowledge, the elucidation of a broad set of ecological processes and mechanisms began to be revealed, including the following:

- In a polyculture with corn, the beans nodulate more and are potentially more active in biological fixation of nitrogen (Boucher and Espinosa 1982).
- Fixed nitrogen may be made directly available to the corn through mycorrhizal fungi connections between root systems (Bethlenfalvay et al. 1991).
- Net gains of nitrogen in the soil were observed when the crops were associated, despite its removal during the harvest (Gliessman 1982).

Table 2.1 Yield of a Corn–Bean–Squash Polyculture Compared with Yields of the Same Crops Grown as Monocultures in Tabasco, Mexico

	Monoculture	Polyculture
Corn density (plants/ha)	66,000	50,000
Corn yield (kg/ha)[a]	1230	1720
Bean density (plants/ha)	100,000	40,000
Bean yield (kg/ha)[a]	610	110
Squash density (plants/ha)	7500	3330
Squash yield (kg/ha)[a]	430	80
Land equivalent ratio (LER)		1.77

Source: Data from Amador, M.F., *Comportamiento de Tres Especies (Maiz, Frijol, Calabaza) en Policultivos in la Chontalpa, Tabasco, Mexico.* Professional Thesis. Colegio Superior de Agricultura Tropical, Cárdenas, Tabasco, Mexico, 1980.

[a] Yields for corn and beans expressed as dried grain, squash as fresh fruits.

- The squash helps controlling weeds: the thick, broad, horizontal leaves block sunlight, preventing weed germination and growth, whereas leachates in rains washing the leaves contain allelopathic compounds that can inhibit the weeds (Gliessman 1983).
- Herbivorous insects are at a disadvantage in the intercrop system because food sources are less concentrated and more difficult to find in the mixture (Risch 1980).
- The presence of beneficial insects is promoted owing to such factors as the availability of more attractive microclimatic conditions and the presence of more diverse pollen and nectar sources (Letourneau 1986).

Interestingly, when the same varieties of corn, bean, and squash were simultaneously planted in the same way in a nearby soil that had at least 10 years of management history involving mechanical cultivation, synthetic chemical fertilizers, and modern pesticides, the yield advantages disappeared. Apparently, the positive interactions that occurred in the traditional farm field were inhibited by the alteration of the soil ecosystem that occurred with conventional inputs and practices. This result points to the important link between longer-term cultural practices and ecological conditions.

The corn–bean–squash intercrop is only one of many crop combinations that either exists or could be developed. By linking scientific inquiry with the long cultural experience of growing these unique cropping systems, we can better understand not only their persistence, but opportunities for improvement.

Another complex agroecosystem that received considerable attention from the researchers in Tabasco at this time was the tropical homegarden agroecosystem. This is an agroforestry system surrounding a household, with an overstory of perennial trees associated with a complex mixture of annual and perennial plants below the canopy, as well as small livestock such as chickens, turkeys, and pigs (Martínez Tirado 1980; Allison 1983). Homegarden agroecosystems are diverse agroecosystems that integrate ecological and cultural knowledge that has evolved over a longer period of time (Figure 2.3). This has proven to be the case, especially in the Maya lowlands of Mexico (Gonzalez-Jácome 2015). They embody many of the components of what we consider to demonstrate elements of sustainability. Agroecological studies of homegardens in Tabasco, Mexico, which began in the late 1970s, had explored these elements (Table 2.2). In relatively small areas, usually less than 1.0 ha, high diversity permitted the maintenance of gardens that in many aspects were similar to the local natural forest ecosystems (Allison 1983). The gardens studied had relatively high indices of diversity for cropping systems and had leaf area indices and cover levels that approximated the much more complex natural ecosystems of the surrounding regions (Ewel et al. 1982). Studies that have extended over time show that homegardens demonstrate both a capacity for resisting change as well as a remarkable

Figure 2.3 An example of a tropical homegarden agroecosystem in the lowlands of Tabasco, Mexico. The complex structure represents the coevolution between culture and environment.

Table 2.2   Characteristics of a Typical Homegarden Agroecosystem at Cupilco,
            Nacajuca, Mexico

Characteristics	Data from 1980 to 1982	Data from 2002 to 2005
Garden size	1.0 ha	1.0 ha
Useful species per garden	89	85
Diversity (Shannon index)	4.48	4.23
Leaf area index	4.5	–
Cover (%)	96.7	–
Light transmission (%)	21.5	–
Perennial species (%)	89.9	90.8
Tree species (%)	47.1	52.2
Ornamental plants (%)	21.3	15.4
Medicinal plants (%)	5.6	4.2
Mesoamerican species	58	52

Sources: Data from Ewel, J., F. Benedict, C. Berish, B. Brown, S.R. Gliessman, M. Amador, et al.
"Leaf Area, Light Transmission, Roots, and Leaf Damage in Nine Tropical Plant
Communities." *Agro-Ecosystems* 7 (1982): 305–326; Allison, J. An Ecological Analysis
of Homegardens (Huertos Familiares) in Two Mexican Villages. M.A. Thesis. University
of California, Santa Cruz, CA, 1983; Gliessman (unpublished data).

ability to adapt and evolve. Today these important agroecosystems provide an important foundation
for maintaining rural livelihoods while protecting local agrobiodiversity.

More specific agroecological studies that took place at this time delved into the structure and
function of traditional agroecosystems in the region (Garcia Espinosa 1978; Chacón and Gliessman
1982; Gliessman 1982), and development projects based on this knowledge were designed and
implemented in rural communities (Gliessman 1980; Gliessman et al. 1981). One of the most ambitious projects undertaken was a system of modular production units installed in rural communities,
where local knowledge and agroecological concepts were joined (Figure 2.4). The project integrated
a thorough understanding of the ecological processes functioning in traditional agroecosystems of
the Mayan farmers of the region, coupled with the development of production systems with which
the local farm communities could identify and understand. This established a foundation for local

Figure 2.4   A modular production unit in the community of Lazaro Cardenas, Tabasco, Mexico. Local tradi-
             tional farming knowledge is combined with agroecological concepts to design a more sustain-
             able local food system.

food systems which tried to provide a more varied and healthy diet, stability of production, reduced pest and disease problems, more efficient use of family labor, and the potential for intensive production despite the various limiting factors well recognized in tropical environments. As in the intercropped corn, bean, and squash system, as well as in the diverse homegarden systems, a remarkably large number of native and introduced plant species made up the repertoire of available elements in intensive vegetable, annual grain, and perennial cropping systems that were integrated into the modular units (Gliessman et al. 1981).

The modular production units were attempted to provide an alternative to the imposition of the monoculture, external-input-intensive, export-oriented Green Revolution systems, from outside the region that had little concern for the needs or desires of local farming communities. They were built on the considerable body of empirical, time-tested knowledge concerning the structure and management of traditional and local agroecosystems, as well as their rich local germplasm and agrobiodiversity. By linking this knowledge with agroecological understanding provided by the researchers collaborating with local farmers, a strong foundation for local sustainability was being built.

In all of the cases presented above, the local traditional agricultural knowledge was seen not only as a foundation for ecological sustainability, but also as a source of alternatives and opportunities for rural communities grounded in their own experience and culture. Agroecology was also seen as a way to pull modern agriculture back from its unsustainable track. As stated in one of the presentations at the TAT seminar: "Ecologists motivated by an agroecological approach are not blindly opposed to modern agriculture, but rather opposed to the blind practices associated with it" (Krishnamurthy et al. 1978).

Despite the fact that Hernandez X. died in 1991, INIREB was abandoned in the mid-1980s, and CSAT was closed by the government in 1985, the seeds planted during this time continue to grow as a movement. The Third International Congress of the Latin American Scientific Society of Agroecology (SOCLA), which was held in Oaxtepec, Mexico, in August of 2011, was attended by over 700 participants of which the majority was from Mexico. Agroecology, agroecosystems, and food systems were words that appeared on name tags of participants from universities, nonprofits, national, and international government programs, farmer organizations, extension personnel, and a most numerous number of students preparing to become the much needed change agents. A part of the closing declaration signed by the participants in the congress is a good way to consider just how deep the roots of resistance and the pathways for growing change have come:

"Agroecology must integrate science, technology, and practice, and movements for social change. We cannot let the artificial separation of these three areas be an excuse some may use to justify doing only the research or technology parts. Agroecology focuses on the entire food system, from the seed to the table. The ideal agroecologist is the one who does science, farms, and is committed to making sure social justice guides his or her action for change. We must help the people who grow the food and the people who eat the food reconnect in a relationship that benefits both. We must reestablish the food security, food sovereignty, and opportunity in rural communities throughout Latin America that has been severely damaged by the globalized food system. We must respect the different systems of knowledge that have coevolved for millennia under local ecologies and cultures. By doing this, we can avoid the eminent food crisis and establish a sustainable foundation for the food systems of the future" (Gliessman 2012).

2.4 FUTURE GROWTH

Reflecting on the growth of the agroecological movement since it put down its roots of resistance in the tropical lowlands of southeastern Mexico, one can see how the foundations for this book were formed. The agroecological approach to sustainable agriculture and food systems has been clearly enunciated for quite some time (Gliessman 1984). Today, it is active in multiple ways, from university degree programs, in farmer-to-farmer movements, and with consumer organizations. But like most other movements, the change is slow, and the roots of industrial agriculture are deep as well. Looking back at Hernández Xolocotzi's diagram of the agroecosystem, it is obvious that the social and ecological components of the food system must receive greater emphasis and support, or the strong link between market forces and the technology of production will continue to dominate. As agroecologist Carlos Guadarrama-Zugasti cautions, we must constantly maintain the interdisciplinary focus of agroecology so that its foundations of resistance are not captured or corrupted (Guadarrama-Zugasti 2007). The roots of resistance described in this book have penetrated deeply. Agroecologists at all levels of the food system, working in all three parts of agroecology—integrating science, practice, and participatory action for change—now have the responsibility to see that they flourish.

2.5 ACKNOWLEDGMENTS

This chapter is dedicated to Dr. Roberto Garcia Espinosa, plant pathologist and agroecologist, who was one of my main partners in the agroecological resistance that developed at CSAT in the latter half of the 1970s. He passed away shortly after completing his monumental work on agroecology and root diseases in agricultural crops, in large part developed in the intercultural environment of farmers' fields in the tropical lowlands of Tabasco and the classrooms and laboratories of CSAT (Garcia Espinosa 2010).

REFERENCES

Allison, J. An ecological analysis of homegardens (Huertos Familiares) in two Mexican villages. M.A. Thesis. University of California, Santa Cruz, CA, 1983.

Amador, M.F. *Comportamiento de Tres Especies (Maiz, Frijol, Calabaza) en Policultivos in la Chontalpa, Tabasco, Mexico.* Professional Thesis. Colegio Superior de Agricultura Tropical, Cárdenas, Tabasco, Mexico, 1980.

Amador, M.F., and S.R. Gliessman. "An ecological approach to reducing external inputs through the use of intercropping." In *Agroecology: Researching the Ecological Basis for Sustainable Agriculture. Ecological Studies 78.* Gliessman, S.R. (Ed.). (New York: Springer-Verlag, 1990), 146–159.

Azzi, G. *Agricultural Ecology.* London: Constable Press, 1956.

Barkin, D. *Desarrollo Regional y Reorganización Campesina: La Chontalpa como Reflejo del Problema Agropecuario Mexicano.* Mexico: Centro de Ecodesarrollo, Editorial Nueva Imagen, 1978.

Bensin, B.M. "Possibilities for International Cooperation in Agroecology Investigation. International Review of Agriculture. Part I. Monthly bulletin of agricultural science and practice." *The Institute for Agriculture, Rome* 21 (1930): 277–284.

Bethlenfalvay, G.J., M.G. Reyes-Solis, S.B. Camal, and R. Ferrera-Cerrato. "Nutrient transfer between the root zones of soybeans and maize plants connected by a common mycorrhizal inoculum." *Physiologia Plantarum* 82 (1991): 423–432.

Boucher, D., and J. Espinosa. "Cropping systems and growth and nodulation responses of eans to nitrogen in Tabasco, Mexico." *Tropical Agriculture* 59 (1982): 279–282.

Chacón, J.C., and S.R. Gliessman. "Use of the 'non-weed' concept in traditional tropical agroecosystems of Southeastern Mexico. *Agro-Ecosystems* 8 (1982): 1–11.

Cox, G.W., and M.D. Atkins. *Agricultural Ecology: A Analysis of World Food Production Systems.* San Francisco, CA: W.H. Freeman and Company, 1979.

Ewel, J., F. Benedict, C. Berish, B. Brown, S.R. Gliessman, M. Amador, et al. "Leaf area, light transmission, roots, and leaf damage in nine tropical plant communities." *Agro-Ecosystems* 7 (1982): 305–326.

Francis, C., G. Lieblein, S. Gliessman, T.A. Breland, N. Creamer, R. Harwood, et al. "Agroecology: The ecology of food systems." *Journal of Sustainable Agriculture* 22 (2003): 99–118.

Garcia Espinosa, R. "Reflexiones sobre el papel de los sistemas de cultivo en la incidencía de los fitopatógenos del suelo en el trópico húmedo." In *Seminarios Regionales sobre Agroecosistemas con Enfasis en el Estudio de Tecnologia Agricola Tradicional.* S.R. Gliessman (Ed.). (Tabasco, Mexico: Colegio Superior de Agricultura Tropical. H. Cardenas, 1978), 127–131.

Garcia Espinosa, R. *Agroecología y Enfermedades de la Raíz en Cultivos Agrícolas.* Montecillos, Mexico: Editorial de Colegion de Postgraduados, 2010.

Gliessman, S.R. (Ed.). *Seminarios Regionales sobre Agroecosistemas con Enfasis en el Estudio de Tecnologia Agricola Tradicional.* Tabasco, Mexico: Colegio Superior de Agricultura Tropical, H. Cardenas, 1978.

Gliessman, S.R. "Aspectos ecológicos de las prácticas agrícolas tradicionales en Tabasco, México: Aplicaciones para la producción." *Biotica* 5 (1980): 93–101.

Gliessman, S.R. "Nitrogen distribution in several traditional agroecosystems in the humid tropical lowlands of Southeastern Mexico." *Plant and Soil* 67 (1982): 105–117.

Gliessman, S.R. "Allelpathic interactions in crop-weed mixtures: Applications for weed management." *Journal of Chemical Ecology* 9 (1983): 991–999.

Gliessman, S.R. "An Agroecological Approach to Sustainable Agriculture." In *Meeting the Expectations of the Land: Essays in Sustainable Agriculture and Stewardship.* W. Jackson, W. Berry, and B. Coleman (Eds.). (San Francisco, CA:North Point Press, 1984), 160–171.

Gliessman, S.R. *Agroecology: The Ecology of Sustainable Food Systems. Second Edition.* Boca Raton, FL: CRC Press/Taylor & Francis Group, 2007.

Gliessman, S.R. "A voice for sustainability from Latin America." *Journal of Sustainable Agriculture* 36 (2012): 1–2.

Gliessman, S.R. *Agroecology: The Ecology of Sustainable Food Systems. Third Edition.* Boca Raton, FL: CRC Press/Taylor & Francis Group, 2014a.

Gliessman, S.R. *Field and Laboratory Investigations in Agroecology. Third Edition.* Boca Raton, FL: CRC Press/Taylor & Francis Group, 2014b.

Gliessman, S.R., R. Garcia Espinosa, and M. Amador. "The Ecological Basis for the Application of Traditional Agricultural Technology in the Management of Tropical Agro-Ecosystems." *Agro-Ecosystems* 7 (1981): 173–185.

Gliessman, S.R., and M.E. Rosemeyer (Eds.). *The Conversion to Sustainable Agriculture: Principles, Processes, and Practices.* Boca Raton, FL: CRC Press/Taylor & Francis Group, 2010.

Gomez-Pompa, A. *Los Recursos Bióticos de México (Reflexiones).* Xalapa, Veracruz, Mexico: Instituto Nacional de Investigaciones sobre Recursos Bióticos, Editorial Alhambra Mexicana, 1985.

Gonzalez Jácome, A. Analysis of Tropical Homegardens through an Agroecology and Anthropological Ecology Perspective. 2015. Guadarrama-Zugasti, C. "Agroecología en el siglo XXI: Confrontando viejos y nuevos paradigmas de produccón agrícola." *Revista Brasileira de Agroecología* 7 (2007): 204–207.

Gonzalez Jácome, A. *Historias Varias: Un Viaje en el Tiempo con los Agricultores Mexicanos.* Distrito Federal, México: Universidad Iberoamericana, México, 2011.

Harper, J.L. "The need for a focus on agro-ecosystems." *Agroecosystems* 1 (1974):1–12.

Hart, R.D. *Agroecosistemas: Conceptos Básicos.* Turrialba, Costa Rica: Centro Agronómico Tropical de Investigación y Enseñanza, 1979.

Hernández Xolocotzi, E. (Ed.). *Agroecosistemas de México: Contribuciones a la Enseñanza, Investigación, y Divulgación Agricola.* Chapingo, México: Colegio de Postgraduados, 1977.

Hernández Xolocotzi, E. Xolocotzia. *Obras de Efraím Hernández Xolocotzi. Tomo 1. Revista de Geografia Agrícola.* Texcoco, México: Universidad Autónoma de Chapingo, 1985.

Hernández Xolocotzi, E. Xolocotzia. *Obras de Efraím Hernández Xolocotzi. Tomo 2. Revista de Geografia Agrícola.* Texcoco, México: Universidad Autónoma de Chapingo, 1987.

International Association for Ecology (INTECOL). *Report on an International Programme for Analysis of Agro-Ecosystems.* Wegeningen, The Netherlands: INTECOL Working Group on Agro-Ecosystems, 1975.

Janzen, D.H. "Tropical agroecosystems." *Science* 182 (1973): 1212–1219.

Krishnamurthy, L., R. Garcia Espinosa., and S.R. Gliessman. "El impacto del hombre al cambiar las propiedades funcionales de los agroecosistemas tradicionales y modernos." In *Seminarios Regionales sobre Agroecosistemas con Énfasisen el Estudio de Tecnología Agrícola Tradicional.* S.R. Gliessman (Ed.). (Tabasco, México: Colegio Superior de Agricultura Tropical. H. Cárdenas, 1978), 108–115.

Letourneau, D.J. 1986. "Associational resistance in squash monoculture and polycultures in tropical Mexico." *Environmental Entomology* 15 (1986): 285–292.

Loucks, O.L."Emergence of research on agro-ecosystems." *Annual Review of Ecology and Systematics* 8 (1977): 173–192.

Martínez Tirado, J.E. *Caracteristicas Generales de los Huertos Familiares en la Sabana de Huimanguillo.* Professional Thesis. Tabasco, México: Colegio Superior de Agricultura Tropical, Cárdenas, 1980.

Odum, E.P. "The strategy of ecosystem development." *Science* 164 (1969): 262–270.

Risch, S. "The population dynamics of several herbivorous beetles in a tropical agroecosystem: The effect of intercropping corn, beans, and squash in Costa Rica." *Journal of Applied Ecology* 17 (1980): 593–612.

Tischler, W. *Agrarökologie.* Jena: Fischer Verlag, 1965.

Vandermeer, J. *The Ecology of Intercropping.* New York: Cambridge University Press, 1989.

CHAPTER 3

Transformative Agroecology
Foundations in Agricultural Practice, Agrarian Social Thought, and Sociological Theory

Eduardo Sevilla Guzmán and Graham Woodgate

CONTENTS

3.1 INTRODUCTION

Wezel et al. (2009) claim that, from the first scientific use of the term "agroecological" in the late 1920s, agroecology has developed through the intersection of agronomy and ecology. Subsequently, following the emergence of the word "agroecosystem" in the 1970s, the focus of research was extended from the field level eventually to encompass entire food systems. By the 1990s, the word "agroecology" was also being used in reference to agrarian social movements and environmentally friendly agricultural practices. For Wezel et al. this broadening of the scope of agroecology is a source of confusion and a cause for concern. We take issue with their position on two grounds. First, by focusing on "agroecology" and cognate terminology they miss the broader historical context of peasant agricultures and agrarian social thought from which the science of agroecology has emerged.

Similarly, they take no account of the numerous agrarian countermovements that have accompanied the development of capitalist agriculture and its associated food regimes (McMichael 2009). The restricted focus of their analysis leads to the second problem—flawed conclusions and recommendations. As Eric Wolf noted in his introduction to *Europe and the People Without History*, "the world of humankind constitutes a manifold, a totality of interconnected processes, and inquiries that disassemble this totality into bits and then fail to reassemble it falsify reality" (1982: 3). If we focus on what is signified rather than the signifier, a very different story comes into view where the science emerges from practice and in support of movements for agrarian justice and food sovereignty.

In our contribution to this volume, we do not seek to undermine agroecological science in its efforts to identify and develop more environmentally benign forms of agricultural production but rather to take issue with recent moves to construct and promote an apolitical natural science of agroecology (c.f. Tomich et al. 2011; Wezel et al. 2009). Our intellectual endeavour is to contribute to "transformative agroecology" (c.f. Méndez et al. 2013), which begins with the identification and analysis of situated or place-based agricultures and the productive/ecological, socioeconomic, and sociocultural/political milieux that influence what, how, where and by whom food is produced and consumed. Inclusion of the socioeconomic dimension suggests consideration of issues of social and economic equity within agrifood systems, while attention to the sociocultural/political milieu leads us to focus on the cultural and political contexts and agendas associated with the movement toward more sustainable agrifood systems and food sovereignty. All three dimensions build from critiques of what McMichael (2009) calls the "food from nowhere" regime of corporate agriculture and global food security, which we consider to be ecologically, economically, and politically bankrupt.

In order to explain the social foundations of transformative agroecology, this chapter is structured into three parts. In the first, we consider some of the basic characteristics of preindustrial food production and consumption and early concerns over the impacts of the development of capitalism in the countryside. This leads us into a discussion of intellectual engagement with these issues and with the countermovements responding to capitalist development in agriculture. The second part reviews some of the more important conceptual contributions that sociology and more recently environmental sociology have made to our understanding of the social and socioenvironmental relationships that are integral to a transformative agroecological endeavour. To facilitate this enterprise, Table 3.1 shows a schematic of our view of the historical pathway of social thought and theory that has led to the emergence of contemporary, transformative agroecology. The chapter concludes by illustrating some of the more recent history of solidarity among farmers, scientists, and social movements in pursuit of agroecological sustainability and food sovereignty.

3.2 FROM LOCAL AGRICULTURE TO A GLOBAL CORPORATE FOOD REGIME

The earliest evidence of agriculture has been found in the "Fertile Crescent," which runs north from the Nile plain in Upper Egypt, along the southern and eastern Mediterranean before following the Tigris and Euphrates on their journey south to the Persian Gulf. It was there that wheat and barley were first developed from wild grasses, and goats, sheep, and cattle were domesticated. According to Zeder (2008: 11597), "initial steps toward plant and animal domestication in the Eastern Mediterranean can now be pushed back to the 12th millennium cal B.P." In China, there are archeological indications of millet and pig farming between 7 and 8,000 years ago, while in the Indian subcontinent rice, bananas, tea, chickens, pigs, and buffalo were domesticated. Animal traction and the first ploughs appear to have been developed by Mediterranean farmers, while millet, sorghum, sesame, and coffee are all crops that were domesticated in a region occupied today by Eritrea, Ethiopia, and Somalia. The Americas have endowed us with another raft of familiar agricultural crops such as maize, climbing beans, squash, cotton, chilli, avocado, cocoa, and vanilla that were first domesticated in Mesoamerica, while the vast area covered by South America, from

Table 3.1 From Narodnism to Agroecology

Marx, Marxism, Narodnism and Anarchism (1850–1900)	
Narodnism "The backward march," "Uniting with the people"	A. Herzen, N. Chernishevski, P. Lavrov, A. Mikailov
Classical Anarchism: "mutual support as the motor of history," "the peasantry as revolutionary agents"	P. Kropotkin, N. Bakunin
Capitalism produces an "irreparable rift in the interdependent process of socioecological metabolism"	K. Marx
Marxism: "differentiation of the peasantry"	V.I. Lenin, K. Kautsky
Neo-Narodnism and Heterodox Marxism (1900–1940)	
Vertical cooperation	N. Bukharin
Social agronomy	A. Chayanov
Dependency and Underdevelopment (1940–1980)	
Center-periphery/world economy	A. Gunder Frank, I. Wallerstein
Internal colonialism	A. Gorz, P. González Casanova
Ethno development	G. Bonfil Batalla; R. Stavenhagen
Peasant Studies (1940–1990)	
Moral economy	K. Polanyi;
Cultural ecology	E. Wolf, K. Wittfogel, S. Mintz
Marxist neo-Narodnism	T. Shanin
Peasant technologies	A. Palerm; E. Hernández Xolocotzi
Postdevelopment (1980 to Present)	
Development and environment as historically produced discourse	Arturo Escobar
Co-motion rather than promotion	Gustavo Esteva
Environmental and Agrarian Social Theory and Agroecology (1980 to Present)	
Origins of agroecology in Marxist and libertarian social thought	E. Sevilla Guzmán
Marx's ecology (esp. the "metabolic rift")	J.B. Foster
Food regimes	H. Freidmann and P. McMichael
Coevolution	R.B. Norgaard
Conjoint constitution	W. Freudenberg et al.
Ecological debt	J. Martínez Alier, A. Simm
Food sovereignty	La Vía Campesina
Historical socioecological transition	M. González de Molina
Socioecological agency	D. Manuel Navarette and C. Buzinde

the Andes down through the Amazon Basin, offer proof of the early cultivation of potatoes, tomatoes, cotton, sweet potatoes and numerous other roots and tubers, as well as peanuts and pineapples (Toledo and Barrera-Bassols 2008). Since the advent of agriculture, the domestication, the breeding, and the production of crops and livestock, and the processing, distribution, and consumption of agricultural products have been accompanied and conditioned by place-based cultural learning and the establishment of a broad range of social institutions.

3.2.1 Agroecology as Agricultural Practice

Farmers have always experimented so that the diversity of agroecosystems and agricultures that have developed in different parts of the world over the last 10,000 years can best be conceptualized as "works in progress." Yet, for all this diversity, there are some basic characteristics that are central to agricultural sustainability. "We had long desired to stand face-to-face with Chinese and Japanese farmers; to walk through their fields and to learn by seeing some of their methods, appliances, and practices which centuries of ... experience have led ... [them] to adopt. We desired to learn how it is possible after 20 and perhaps 30 or even 40 centuries, for their soils to be made to produce sufficiently for the maintenance of such dense populations as are now living in these ... countries" (King 1911: 2). This short extract from the introduction to F.H. King's seminal text, "Farmers of Forty Centuries," captures two of the central premises of agroecological practice. First, sustainable agricultures have been developed around the world as a result of centuries of experience living in and with nature. Second, the fundamental basis of all such agricultures is a living and healthy soil. Similar observations were made in another foundational volume, Howard's (1940) *An Agricultural Testament*, which resulted from his experiences as Imperial Economic Botanist to the Government of India during the first quarter of the twentieth century. Although Howard had been sent to India to introduce Western agricultural practices, he came to the conclusion that he had more to learn from the Indian farmers than he could possibly teach them. Echoing King (1911), Howard (1940) considered a healthy soil to be the basis of healthy crops and livestock, nourishing food, and the well-being of human populations. As Barthel al. (2013: 1142) put it, sustainable agricultures have been "maintained through a mosaic of management practices that ... coevolved in relation to local environmental fluctuations, and ... [have been] carried forward by both biophysical and social features ... including: genotypes, artefacts, written accounts, as well as embodied rituals, art, oral traditions and self-organized systems of rules."

Traditional agricultural practices based around soil health and crop and noncrop biodiversity display a number of properties that are central to their long-term viability. First, preindustrial agricultures based on human and draught animal power are highly efficient net producers of energy. Manual agriculture can produce as many as 30 calories of energy for each calorie of energy invested (Wilken 1987), whereas the use of animal traction tends to reduce this ratio, farming with draught animals can still produce net outputs of 10 or more calories of useful biomass for each calorie of energy input (Martinez Alier 2011). These figures compare very favorably with modern industrial agriculture, which typically requires 10–15 calories of fossil fuel energy to produce just one calorie of food (McMichael 2009). As well as being net producers of energy, preindustrial polycultural agroecosystems also outperform industrial input-intensive monocultures in terms of total food productivity per unit area. Mesoamerican corn, beans, and squash intercropping can produce almost twice as much food per hectare as industrial maize monocultures and twice as much organic residue for composting and turning back into the soil, obviating the need for synthetic amendments (Altieri and Toledo 2011).

Another important characteristic of low input, biodiverse, polycultural systems that integrate annual and perennial crops is that, once established, they tend to be carbon sinks rather than carbon emitters and thus have the potential for climate change mitigation (IAASTD 2009; UNCTAD 2013). The industrial food system, on the other hand, is estimated to be responsible for more than 20% of greenhouse gas emissions (McMichael et al. 2007, cited in McMichael 2009). In addition to climate change mitigation, diversified agroecosystems are more resilient to the increasingly severe and frequent extreme weather events that are associated with global warming. A survey of more than a thousand farms in Central America reported by Holt-Giménez (2001) demonstrated that following the ravages of Hurricane Mitch in 1998, farms with biodiverse agroecosystems suffered significantly lower economic costs and recovered more rapidly than those where monocropping was prevalent, reflecting the inherent risk mitigating character of agroecological production.

As diversity confers resilience, in combination with traditional, place-based farmers' profound understanding of local ecological and cultural resources and relationships, it also imparts adaptability. Of course the coevolution of distinct agricultures has not been a smooth or linear process. Individual farming families, agrarian communities, and indeed entire civilizations have disappeared as a result of some irresistible environmental or social force. We will now turn our attention to some of the social and environmental concerns that have accompanied the development of capitalist agriculture.

3.2.2 Growing Concern over the Impacts of the Development of Capitalism in the Countryside

Various claims have been made about the origins of the political/economic systems we label "capitalist," although there is no such debate surrounding the centrality of property rights to the functioning of capitalism in whatever form. The tensions imposed by the establishment and (re)structuring of property rights in land have occasioned social critique and action since at least the sixteenth century, when the widespread enclosure of the open fields and commons as private sheep pastures in England denied ordinary people access to land and restricted their ability to feed themselves and their families. One of the earliest critiques of enclosure can be found in Moore's 1516 novel *Utopia*: "Your sheep ... which are usually so tame and so cheaply fed, begin now ... to be so greedy ... that they devour human beings themselves and devastate and depopulate fields, houses, and towns" (cited in Melville 1994: 6). In the seventeenth century, the problem of access to land became more acute, leading to numerous localized revolts and the coming together of the dispossessed into direct action movements to level the ditches and fences of the enclosures and invade and cultivate the land. Thus, they challenged the most fundamental element of the emerging capitalist economy—private property. "When men take to buying and selling the land ... they restrain other fellow creatures from seeking nourishment from Mother Earth ... so that he that had no land was to work for those ... that called the Land theirs; and thereby some are lifted up into the chair of tyranny and others trod under the footstool of misery, as if the Earth were made for a few and not for all" (Winstanley 1649, cited in Berens 2007 (originally published 1906): 70).

At the end of the eighteenth century as industrialization was beginning to revolutionize production, while it was a commonly held view among enlightenment thinkers that society was changing for the better and firmly set on a broadly upward and improving trajectory, others were less sanguine about the prospects for "perfecting society." Among them was Thomas Malthus, who questioned the ability of agricultural production to keep up with exponential human population growth. In his essay on the principles of population (1998 [1798]: 4) Malthus argued that because "population ... increases in a geometrical ratio [while s]ubsistence increases only in an arithmetical ratio ... the power of population is indefinitely greater than the power in the earth to produce subsistence for man". As Britain's industrial revolution gathered pace in the nineteenth century, however, agricultural production accelerated and Malthus's dismal predictions failed to materialize. Nonetheless, the means of achieving such elevated crop yields attracted criticism from a different quarter. In his text on agricultural chemistry, the German chemist von Leibig denounced Britain's success, pointing out that yield increases depended on imported nutrients, while none of the organic residues from food consumed in urban centers was recycled back to the soil (Foster 2000). Marx, who was particularly critical of Malthus's thinking, undertook a systematic analysis of von Leibig's work, leading him to one of the central concepts of his critique of industrial agriculture. As the nineteenth century progressed, agricultural production increasingly incorporated new mechanical, mineral and chemical technologies, protectionist measures aimed at encouraging cereal production were repealed, and labor was shed. Thus, as Britain was transformed from agrarian to industrial society, capitalist agriculture provoked "an irreparable rift in the interdependent process of *socioecological metabolism*" (Marx 1981: 949 emphasis added).

Kautsky's *Agrarian Question* (1899) employed the notion of "metabolic rift" (c.f. Foster 2000) in an analysis of what he characterized as the exploitation of the countryside by the cities. The agrarian question itself, however, concerned the fate of the Russian peasantry in the face of the development of capitalism and sought to respond to a debate that had been established in the second half of the nineteenth century between the Narodniks and Russian Marxists (see Table 3.1) following the emancipation of Russian peasants in 1861. Prior to the abolition of serfdom, the organization of agricultural production in Russia revolved around landed estates. The landlords provided their serfs with access to pasture and forests for fodder and fuel wood, and two adjacent strips of land:

one to be cultivated for their own subsistence needs and one to produce crops for the estate. On their emancipation, the peasants were allowed to retain their subsistence plots and this land formed the basis of village communes. However, the erstwhile landlords retained ownership of the adjacent strips and all of their pastures and forests. Thus, the peasants were denied access to vital fodder and fuel resources. The landlords sought to tax the peasants for access to estate pastures and in response some communes turned over some of their agricultural plots to grass but the landlords (or Kulaks) reacted by imposing tolls for each animal that crossed their agricultural strips on their way to the village pastures. In order to pay for access to the forest and acquire vital supplies of firewood, the peasants had no option but to cultivate crops for the Kulaks.

Inspired by the publication of Chernishevsky's novel, *What Is to Be Done?* (1863), the Narodniks viewed the peasants as a revolutionary force capable, with appropriate leadership from the intelligentsia, of overthrowing the tsars and building a form of socialism based around the village commune and developing cooperative forms of agricultural production utilising the resources of the old feudal estates. In early 1874, rural unrest between the peasants and Kulaks spread to the cities and prompted members of the Narodnik intelligentsia to head for the countryside to galvanize the peasantry and convince them of their revolutionary potential. Hence, Narodnik: the Russian term "narodniki" having the literal meaning of "going to the people." Despite the ultimate failure of the Narodnik movement—it met with little enthusiasm among the peasants and was harshly repressed by the tsarist police—in some important ways Narodnism prefigures the agroecological approach in its identification of the peasantry as a revolutionary force, its focus on the economics of solidarity, and the importance it attached to "going to the people."

The Marxists, however, believed that it was necessary for capitalism to develop before socialism could emerge, and thus claimed that the peasantry would have to disappear. In the same year that Kautsky published his *Agrarian Question*, Lenin published *The Development of Capitalism in Russia* (1986 [1899]), which begins with a chapter on "The Theoretical Mistakes of the Narodnik Economists." The subsequent chapter, entitled "The Differentiation of the Peasantry" described how the development of capitalism necessitated the polarization of the peasantry into small-scale capitalist farmers and associated rural proletarian classes. The idea that peasant modes of production were incapable of withstanding the development of capitalism was challenged by another Russian commentator, Alexander Chayanov, who developed what he called "social agronomy"—a form of natural resource management based on the social institutions and knowledge of peasant households and society. In his theory of the peasant economy, Chayanov (1989) explained how it was possible for peasant modes of production to continue to exist alongside capitalism due to the character of the peasant household as both unit of production and consumption and an alternative economic logic of balancing household consumption needs with the necessary amount of labor required to achieve them. Thus, in terms of their engagement with the peasantry and support for their role in the "sustainable development of agricultural production," we might consider both the Narodniks and Chayanov as a proto-agroecologists.

Following the death of Lenin in 1924, Stalin had Chayanov sentenced to the labor camps for his "antirevolutionary" ideas and set about modernizing Soviet agriculture through forced collectivization: a process met by fierce, but ultimately futile, peasant resistance. Pitirim Sorokin, a fugitive from the Russian Revolution, took up residence in the United States, where together with Carl Zimmerman and Charles Galpin he produced the three-volume *Systematic Source Book in Rural Sociology* (1930–1932). Within the framework of the United States Land Grant Colleges, rural sociology was employed to analyze and understand rural life and inform policies of agrarian modernization (Christy and Williamson 1992; Sevilla Guzmán and Woodgate 1997). In the second half of the twentieth century, the influence of rural sociology was felt in both the United States and Europe. In the United States, the Division of Farm Population and Rural Life was established within the Department of Agriculture and, under the leadership of Galpin, generated sociological understanding of the farm sector in order to influence New Deal policies and partially ameliorate the worst social impacts of industrialization on disadvantaged sectors of the rural economy. In post-War Europe the Common Agricultural Policy, directed chiefly at achieving food self-sufficiency, also included social payments aimed at maintaining vibrant rural communities.

The success of petroleum-based agricultural technologies in the advanced capitalist economies led to international efforts to increase global food production by promoting agricultural modernization in the South. Of particular relevance to this endeavour were the constituent institutions of the Consultative Group on International Agricultural Research, such as the Centre for the Improvement of Maize and Wheat in Mexico and the International Rice Research Institute in the Philippines. In concert with national agricultural development programmes and funding from the Rockefeller Foundation these international institutions drove what became known as the Green Revolution, which extended technological packages of hybrid seeds, synthetic fertilizers, and chemical pesticides across the Third World. Under optimum conditions, industrial technologies returned remarkable increases in production. The extension of industrial agriculture technologies to Third World nations through the institutions and policies of the Green Revolution during the third quarter of the twentieth century contributed to the reproduction of First World development models of a modern agriculture sector generating surplus capital for industrial development. At the same time, "agribusiness elaborated transnational linkages between national farm sectors, which were subdivided into a series of specialised agricultures linked by global supply chains … [and] … a 'new international division of labor' in agriculture began to form around transnational commodity complexes" (McMichael 2009: 141).

The extension of what McMichael (2009) calls the "corporate food regime" has concentrated land ownership through dispossession in the search for economies of scale, marginalizing smallholder agriculture and provoking large scale rural to urban migration. It has also transformed food from the most basic of human rights into a homogeneous set of globally traded commodities. Meanwhile, mechanization and the application of industrial inputs have degraded soils to the point where they function as little more than inert substrates with little or no inherent productive capacity. Furthermore, "the rate of biodiversity loss due to … chemically intensive monocultures is extraordinary. …Entire habitats and [the] wild species associated with them … have been lost or are on extinction trajectories … and it is now well established that the current loss of biodiversity in agroecosystems also erodes fundamental ecosystem services that underlie the resilience of production, such as soil fertility, pollination, and natural pest control" (Barthel et al. 2013: 1145).

These great transformations in the socioeconomic and productive/ecological dimensions of agroecosystems reverberate through the sociocultural/political dimension. In order to understand the dynamic interplay of social and ecological factors, we need to establish a broad conceptual framework. Thus, before moving to our assessment of the emergence and progress of transformative agroecology and its engagement with the politics of the food sovereignty, the next part of this chapter reviews some of the more important conceptual contributions that sociology and more recently environmental sociology have to offer.

3.3 FROM SOCIOLOGY TO ENVIRONMENTAL SOCIOLOGY

The central lesson that sociology teaches us is that, however, we conceive of a particular phenomenon our descriptions of it are always social because our perceptions are colored or filtered by our social context, its culture, its economics, and its politics. Social contexts are structured by institutions that carry and instantiate cultural, economic and political rules, and resources for social life. What environmental sociology reminds us is that we are all members of a biological population whose institutions and the behaviors they select for impact on nature to such an extent as to change the conditions of our existence. Thus, when we observe the world around us, it is like looking into a mirror in two important ways: what we see is what we have constructed from within our social context, and its materiality—its ecology—has been impacted by our activities. It is clearly impossible to explicate the entire canon of sociological theory in a few short pages, but in the sections that comprise this part of our chapter we will try to give a flavor of some basic sociological concepts and the key ideas that have contributed to our approach to transformative agroecology.

3.3.1 Sociology: Competing Visions of Society

Since Comte first set out his positivist sociology in the 1830s, with its analysis of "social dynamics" and "social statics" (Comte 1838), two key issues have characterized sociological enquiry: how social change occurs and how social order is maintained. With great simplification, we can say that the answer to the first question is that social change is brought about by social action, which can be understood by focusing on how individuals relate to each other and the world around them. This is the realm of interpretive or constructionist sociology, which provides us with ways of understanding how what people believe and the meanings and values they attach to things have discernible consequences in terms of their attitudes and behaviors. The answer to the second question is that order is maintained by social norms and institutions (structures) that define options for human behavior. Structuralism focuses our attention on social aggregates, such as the family, state, or market. Among his many contributions to sociological thought, Marx pointed out that social structures tend to favor the interests of elite classes—feudal lords or capitalist entrepreneurs, over the interests of the masses—the peasantry or proletariat, and thereby constrain progress toward more egalitarian societies. For Marx, social change required the active intervention of enlightened social actors in the form of "class struggle" and, thus, he adopted a position somewhere in between constructivism and structuralism, pointing out that while people make history they do not do so under conditions of their own choosing.

The twentieth century witnessed a proliferation of sociological approaches. Some set out largely structural explanations of society, while others proffered constructionist interpretations focused on human agency—the capacity of individuals to act independently of structure. Although for much of the twentieth century most sociology could be characterized by divisions related to the *structure/agency debate*, it is important to note that until the later decades of the century most sociology also shared a common assumption: human society represents an exceptional case in nature because humans have developed culture. According to this view, human culture changes more rapidly than nature's biology and thus progress can continue unchecked because, ultimately, all social problems can be resolved through cultural adaptation. In other words, human culture exempts societies from the ecological structures that shape the natural world. The agrarian modernization theories developed by rural sociologists in the second half of the twentieth and the policies that they informed were firmly embedded in this "enlightenment" thinking.

3.3.2 Agroecology and Development Theory: From Modernization and Dependency to the Rediscovery of Peasant Studies

What Catton and Dunlap (1978) labeled the "*human exemptionalist paradigm*" underlies most theories of development. While modernization theory cast underdevelopment as an original condition of "backward peasant farmers," in contrast, another broad theoretical orientation, which we can loosely term "dependency theory" (Table 3.1), claimed that it was an active process generated by structural inequalities between the advanced capitalist economies of the First World and the peripheral nations of the Third. For the more radical dependency theorists, such as Gunder Frank and Wallerstein, the greatest winners of "development" were the core nations, who enjoyed cheap food supplies imported from the periphery and the rapid growth and eventual transnationalization of their agricultural input industries and commodity trading corporations, which became key institutions of the corporate food regime.

While Green Revolution programs and methods increased commercial agricultural production and trade, the industrialization of agriculture also had the effect of robbing people of their identities and negating local knowledge and institutions. As we have already indicated, industrial agriculture also degraded soil structure and fertility and eroded agrobiodiversity. In short, capitalist agricultural industrialization represented a new form of colonialism in which modernization impoverished everything that did not follow the norms and rules that modernity dictated. These exploitative relations operated within as well as between nations as described by González Casanova and Andre Gorz (Table 3.1) in the concept of "internal colonialism." González Casanova (1965) had used the term to refer to the situation in Mexico in the 1960s. One of the first Southern nations to implement Green Revolution policies, Mexico was also among the first places where peasant technologies and institutions were studied and presented as valid alternatives to industrial agriculture (c.f. Xolocotzi, Table 3.1).

Some of the most important contributions of peasant studies to contemporary agroecology emanate from the works of Theodor Shanin. These include his research into the history of the agrarian question and the debate among the orthodox Marxists and Narodniki in nineteenth-century Russia, which we have already mentioned (also see Sevilla Guzmàn 2011). In particular, his rediscovery of the works of Chayanov have provided us with profound insight into the multiple levels of analysis that can and should be applied to peasant societies and their management of natural resources (c.f. van der Ploeg 2013). In Latin America in the 1970s and 1980s, the agrarian question was reignited by the rediscovery of Chayanov and the body of work developed by peasant studies. A fierce debate ensued between descampesinistas who, like Lenin and Kautsky, foresaw the eventual disappearance of the peasantry (campesinado) and those who concurred with Chayanov that the peasantry could continue to reproduce themselves at the margins of the capitalist economy: the campesinistas.

Despite the negative impacts of modernization on peasant agriculture and social organization, campesinistas such as Angel Palerm held that while peasants might participate in the market economy to generate cash, rather than the simple logic of capitalism, peasant life is organized through membership of kinship groups and participation in the community, by access to land through institutions other than private property, and by reciprocity and solidarity. The relevance of peasant studies to transformative agroecology is significant and well summarized in the following short quote from Angel Palerm's last book, *Anthropology and Marxism* (1980: 197 [our translation]): "The future of the organization of agricultural production appears to depend on a new technology based on the intelligent management of ... [natural] resources by means of human labor, utilizing minimal capital, land, and fossil energy. This model ... has its prototype in peasant farming systems."

As Palerm suggests, and innumerable studies of peasant communities and their use of natural resources confirm, the sustainability of peasant agriculture depends on distinctive social relationships as well as ecological processes, and these relationships and processes differ markedly from those associated with capitalist production. The peasant economy is a "moral economy" and while peasants may interact with commercial markets, as Polanyi (1944) claimed, the negative impacts of economic incorporation foster moral indignation and resistance. While ecology and agronomy may reveal important ecological and agronomic features of agricultural sustainability, in order to understand adequately the socioenvironmental relationships that underpin sound agricultural practice and the agrarian social movements that have arisen in defence of the peasant way of life, we need to make recourse to the field of environmental sociology.

3.3.3 The "Crisis of Modernity" and the Birth of Environmental Sociology

Since the industrial revolution, labor productivity and economic growth have been enhanced by the extraction and processing of what initially seemed like an inexhaustible supply of fossil hydrocarbons into fuels and industrial chemicals. In the early 1970s, the United States passed peak oil production and oil-rich Arab nations imposing an embargo on supplies to the United States, as a way of registering their protest at US support for Israel during the Yom Kippur War. This led to a quadrupling of the price of crude, bringing an end to the post-War economic boom. Beginning with the Mexican debt crisis in 1982, one country after another moved from inflation to recession and stagflation and had to turn to the international financial institutions for help. In return for debt relief, the Third World nations had to restructure their economies by selling off national assets and inefficient state enterprises, opening up domestic enterprises to international competition, and cutting back severely on public spending. Structural adjustment required funds that had previously been devoted to fostering agricultural modernization and economic growth, and addressing the social issues associated with capitalist incorporation to be diverted to debt repayments. Thus, the 1970s and 1980s brought an end to the era of the "developmental state" and ushered in neoliberal reforms.

It was not only the failure of state-led development that provided cause for concern in the 1970s and 1980s. Since the early 1960s, worrying accounts of the negative environmental and human health impacts of fossil fuel driven industrialization began to gain public attention. In particular, Rachel Carson's (1962) seminal work *Silent Spring* brought to light the negative ecological effects of chemical pesticides in agriculture. The socioeconomic impacts of resource scarcity were also becoming a focus for attention. In the early 1970s, the Club of Rome commissioned an assessment of the future of humanity in the context of varying levels of natural resource availability, industrialization, agricultural productivity, population control, and environmental protection. The report of their computer modeling exercise, "Limits to Growth" (Meadows et al. 1972), cast serious doubt over the future of humanity, with two of the three scenarios generated raising the spectre of Malthus by predicting that human population growth would outstrip the planet's carrying capacity and lead to the collapse of civilization before the end of the twenty-first century. If industrial capitalism was to survive, Carson's (1962) predictions had to be avoided. Such was Carson's concern about the long-term impacts of agricultural toxins that in bringing her book to a close she was moved to compare the contemporary conjuncture with the position described in Frost's "The Road Not Taken":

> We stand now where two roads diverge. ... The road we have long been travelling is deceptively easy, a smooth superhighway on which we progress with great speed, but at its end lies disaster. The other fork of the road—the one less travelled by—offers our last, our only chance to reach a destination that assures the preservation of the earth (Carson 1962: 277).

To put it more succinctly, the promise of modernization had been transformed into the crisis of modernity, the intertwined socioeconomic and productive/ecological dimensions of which provided the context for the birth of environmental sociology.

At the same time as the crisis of modernity was coming to light, the validity of the structure/agency debate in sociology was brought into question. Aware of the limitations imposed by adopting positions that favored *either* structure *or* agency, social theorists sought to bring them together within an integrated social ontology. For the purposes of our coming discussion of environmental sociology, we shall briefly discuss Giddens' Structuration Theory (1984 inter alia), in which the focus falls on "social practices ordered across space and time" (2). From this starting point, human agency is understood as the capacity of knowledgeable individuals to intervene in situations and change the course of events. Echoing Marx, however, Giddens suggests that while people make society, they do not do so under conditions of their own choosing: the daily activities of people in society are enabled and constrained by the rules and resources of social structures (institutions). The intended and unintended consequences of social practice lead to the reproduction or reformulation of social structures and systems over time.

Environmental sociology portrays a growing consensus surrounding what Giddens (1984) termed the duality of structure, with erstwhile structuralists incorporating human agency and social discourse into their analytical frameworks and scholars from constructivist traditions seeking to understand how structures emerge and are changed by agency. Political ecology, for example, while having structuralist roots, incorporated a constructivist element during the 1990s, and began to investigate the ways in which nature is socially constructed in discourses such as "sustainable development" and "biodiversity conservation," considering language to be constitutive of reality, rather than simply reflecting it (Escobar 1996). Similarly, while Hannigan's (1995) foundational text *Environmental Sociology* was subtitled, *A Social Constructionist Perspective*, the second edition (2006) dropped the subtitle and proposed that social order and social change can occur simultaneously. Such integrated socioenvironmental theory provides transformative agroecology with ways of understanding both the social processes that maintain peasant agriculture and the emergence of agrarian social movements in opposition to the depredations of the corporate food regime. More importantly, it offers an array of ideas that help us to think about the dynamics of societies' relationships with nature.

3.3.4 Environmental Sociology: Conceptual Food for Agroecology Thought

Before we can consider more of the conceptual contributions that environmental sociology has on offer, we need to distinguish it from a simple sociology of the environment that focuses on what are perceived as external environmental issues. In 1978, Catton and Dunlap published a paper in *The American Sociologist* claiming that recognition of ecological limits to growth implied that the exceptional characteristics of the human species could no longer be viewed as exempting societies from ecological constraints, as classical social theory had implied. Following their critique of the "human exemptionalist paradigm" of conventional social theory, Catton and Dunlap proposed a "new ecological paradigm" and defined environmental sociology as "the study of interactions between the environment and society," stressing that human beings are biologically constituted and ecologically embedded as well as culturally constituted and socially embedded.

So how, sociologically, can we get a handle on nature? One way is to examine how it coevolves with social forms. Norgaard introduced the notion of coevolution and suggested that the coevolutionary worldview could generate the epistemological basis for agroecology (in Altieri 1987). He explains social and environmental change as the outcome of coevolution between social systems (values, knowledge, technologies, and forms of organization) and environmental systems (climate, soils, biodiversity, etc.). All of the subsystems of society and environment are interrelated. Changes within each subsystem may be deliberate or arise randomly. Whether they survive or fall by the wayside depends on how well they fit with the currently dominant characteristics of the other subsystems and, thus, whether the other systems exert positive or negative feedback pressures on changes as they arise. Norgaard's coevolutionary model of society–environment interaction is thus neither environmentally nor culturally deterministic (Redclift and Woodgate 1997).

A more recent contribution in the coevolutionary perspective comes from Carolan (2005), who seeks to understand how nature is "coevolving in accordance with broader sociocultural processes (such as capitalism, globalization, etc.)," that is, to "understand what nature is" and to take the discussion forward by making analytical distinctions between the social and the biophysical, "while leaving conceptual space for interaction" (395). This is facilitated by distinguishing among three natures. *Nature* (upper-case) is used to refer to the physical laws of nature, *nature* (lower-case) denotes the meeting of both biophysical and social phenomena, while *'nature'* (in inverted commas) is employed to refer to discursive constructions alone. Thus the nature (lower-case) of agroecosystems can be understood as the outcome of the interaction of Nature and 'nature.'

Much environmental sociology has tended to focus on environmental degradation, often involving critiques of capitalist industrialization and globalization. Of particular relevance to transformative agroecology is the idea of ecological debt. Established on the principle of "environmental justice," ecological debt is the debt accumulated by the countries of the North and owed to the countries of the South through the extraction of more than its fair share of natural resources and the occupation of more than its fair share of environmental space through the dumping of production wastes. Ecological debt is an aggregate measure that brings together carbon debt, biopiracy, waste export, and environmental liabilities, inter alia. The concept acts as a counterbalance to the external financial debt of less industrialized countries, which continues to exert economic pressure toward further exploitation and degradation of environments in the South and the social deprivation of the "bottom billion." Although the idea arose in social movement discourse around the time of the first Earth Summit in 1992, since then it has engendered academic attention. In 2002, Martinez Alier made ecological debt a central theme in his book *The Environmentalism of the Poor: A Study of Ecological Conflicts and Valuation* and, with publication of the first edition of Simms' (2005) book *Ecological Debt: Global Warming and the Wealth of Nations*, the concept became firmly cemented in the environmental social science lexicon.

Other branches of environmental sociology, especially ecological modernization (EM) theory, have developed close links with policymakers and focused on the ecological restructuring of modern society rather than its worst environmental excesses. This more optimistic view sees producers responding to market signals and instruments of policy by developing new, green technologies and, more generally, improving the energy and material efficiency of production processes. At the same time, however, Joseph Huber, whom many consider to be the founder of EM theory, has cautioned that industry's efforts to increase productive "efficiency," even when combined with a shift in consumer behavior away from excess and toward "sufficiency," are unlikely to form an adequate response to our current environmental and human predicament. For Huber (2000) a third discourse is required, which he calls "consistency," entailing positive action to bring our social metabolism back in line with nature's metabolism and repair the "metabolic rift" (Foster 2000) between production and consumption that Marx had problematized in the nineteenth century (see Section 3.2.2). Socioecological metabolic consistency requires moving forward to a situation in which social reproduction is achieved without recourse to fossil hydrocarbons and without further depletion of biodiversity. Long-term sustainability cannot be achieved on the basis of finite resources (petrochemicals and mineral fertilizers) and the liquidation of natural capital (depletion of soil organic matter and deforestation). Ultimately, the entropy produced by socioecological catabolism must be balanced by the anabolic potential of solar radiation and biological processes.

Socioecological metabolic consistency is a core principle of agroecology, but how do we theorize the transition to consistency? We can begin by returning to structuration theory, with its notion of social practices ordered across time–space and integrating it into the coevolutionary approach proposed by Norgaard. If people are both culturally and biologically constituted, then our actions are better defined as socioecological practices, embedded within socioecological systems, and enabled and constrained by socioecological structures. In a study of agricultural industrialization in nineteenth-century Europe, González de Molina (2010) characterizes change as "socioecological transition" driven by complex interactions among factors including climate fluctuations, pest infestations and disease epidemics, human population growth, social inequality, technological developments, institutional change, and competing ideas about nature (also see Woodgate et al. 2005). These interactions impact on metabolic processes within a hierarchy of scales from the local to the global.

This characterization fits within and adds to coevolutionary theory and reflects Freudenberg et al.'s (1995) concept of "conjoint constitution" in which all phenomena, whether at first apparently social or natural, are in fact products of both social construction and ecological agency. Understanding nature as an active participant in historical processes of change is a basic tenet of environmental sociology and is central to the agroecological perspective, it is a fact which is now undeniable in the face of accelerated global warming and biodiversity decline. Indeed, globalization studies and work on climate change have begun to add credence to the view that ecological time is being "compressed." The pace of industrial developments in the nineteenth and twentieth centuries created the illusion of a timeless natural world, and the environment as a passive stage for the play of life, the most esthetically pleasing aspects of which could be preserved for all time. Yet, as the Anthropocene (c.f. Crutzen and Stoermer 2000) has unfolded in the late twentieth and early twenty-first centuries it appears that nature's time is accelerating. Ecologists and natural resource managers are revising their views of environmental change, suggesting that it may occur not in small incremental steps, but through major, relatively rapid regime shifts. Our eco-illogical past, it appears, is catching up with our individualized, social present and threatening our collective, future survival.

As agricultural practice, transdisciplinary science, and agrarian/environmental social movement, transformative agroecology involves diverse responses to the social cum environmental crisis of modernity. In order to understand these responses, a recent contribution to environmental social theory proves helpful—the concept of "socioecological agency." Manuel Navarette and Buzinde (in Redclift and Woodgate 2010: 141) suggest that overcoming the global environmental crisis requires the mutual cocreation of material and social structures to be mediated by a "self-reflexive, or transcendental form of agency enacted by individuals in their interaction not only with society and the environment, but also with themselves: with their inner worlds." Manuel Navarrete and Buzinde contend that such a transformation is "likely to emerge from a radical realization about the reciprocity and double directionality that exists between humanity and the planet as a whole" (ibidem).

3.4 AGROECOLOGY TODAY AND THE ROAD AHEAD

We noted at the beginning of this chapter that the constitution of agroecology as agricultural practice, transformative science, and agrarian social movements in pursuit of food sovereignty has led some commentators to claim that "these varied meanings ... cause confusion" (Wezel et al. 2009: 503). In the contemporary knowledge economy, while the legitimacy and importance of agrarian/environmental social movements may be grudgingly accepted, a premium is placed on systematizing agroecology as the science of sustainable food and fiber production in the context of accelerating environmental change (c.f. Tomich et al. 2011). For the contributors to this volume, however, the assumption that the science of agroecology can be separated from its practice and politics is deeply problematic. For us, there are no quick technological fixes to what we conceive of as the interrelated, agrarian, biodiversity, climate, cultural, economic, energy, food, and political crises of the present conjuncture. The farmers, researchers, and social activists of transformative agroecology demonstrate socioecological agency in practice, through their work together developing institutions and strategies in pursuit of the politics of food sovereignty. Transformative agroecology rejects the so-called "post-political" (Swyngedouw 2009) consensus of "sustainable development" and "global food security" promoted by the corporate food regime and implicit in the construction of agroecology-as-natural science, and adopts instead more collective, bio- and ecocentric positions.

As McMichael (2007) has argued, the recent return of peasant movements and politics (see also Peréz Vitoria 2005; van der Ploeg 2009) recasts development in at least four key senses: drawing on insight from the more radical dependency theories, poverty is viewed as the result of unsustainable development rather than an original condition. Place-based agricultures and agroecosystems are constructed as global goods that must be defended from enclosure and incorporation within global commodity markets. Individualization is challenged, the politics of solidarity are reclaimed, and a plurality of perspectives is adopted, making room for other rationalities beyond the narrow, economistic perspective of neoliberalism. Continuing in the tradition of the Narodniks of nineteenth-century Russia, peasant studies scholars in the 1970s and 1980s, and postdevelopment protagonists in the 1990s, transformative agroecology engages with peasant struggles. The remainder of this chapter sets out a little of the history of this recent engagement in order to illustrate some of the transformative agroecological institutions that have been built on the foundations of agricultural practice, agrarian social thought, and sociological theory.

3.4.1 Agroecologists: Going to the People

In the mid-1970s, following several years working in commercial enterprises in Costa Rica and Mexico, Stephen Gliessman took up a post as agricultural ecologist at the *Colegio Superior de Agricultura Tropical* in Tobasco, which had been established "to train the agronomists and test [Green Revolution] technologies on its experimental fields" (Gliessman 2013: 26). During his time in Central America, Gliessman had been intrigued by the agricultural practices of his peasant neighbors and, as an ecologist, it became clear to him that rather than trying to override natural processes the local peasant farmers worked with them. He took these insights to Tobasco, where he delivered what was probably the first university course in agroecology: "International summer courses in agroecology were offered in 1978–1980, a master's degree program in agroecology was begun in 1978, and research projects with the agroecosystem as the organizing concept and agroecology as the research process began as early as 1977" (ibidem). In 1981, Gliessman moved back to the United States and a post at the University of California, Santa Cruz, where he established the first agroecology program in the United States and set about building a team of colleagues and students who have subsequently established strong and enduring links with agroecological social movements throughout the United States. More recently, in 2008, Gliessman took on the editorship of the *Journal of Sustainable Agriculture*, engineering its transformation into *Agroecology and Sustainable Food Systems* from the beginning of 2013, the first issue of which ("Agroecology and the Transformation of Agri-Food Systems: Transdisciplinary and Participatory Perspectives") provided the inspiration for this book.

During the 1980s a multitude of development nongovernmental organizations (NGO) sprang up throughout Latin America as IMF-imposed structural adjustment programs forced states to close down development programmes and cut back on public spending. Toward the end of the 1980s, NGOs from Chile, Brazil, Argentina, Bolivia, Colombia, Ecuador, Paraguay, and Peru joined forces to form the Latin American Consortium on Agroecology and Development (CLADES). CLADES's technical advisor was Miguel Altieri, an agroecologist from University of California, Berkeley. Together with the likes of Peter Rossett and Clara Nichols, Altieri developed the Consortium's relationships with rural social movements and development NGOs, providing them with agroecological advice and training. Since 1991 CLADES has published *Agroecología y Desarrollo* (a journal dedicated to making agroecological knowledge and experience available to institutions working to promote ecologically and culturally relevant development practice) and provided a forum for debating the institutional challenges of sustainability (www.clades.cl).

Following the 1975 International Working Party for Peasant Studies at the University of Manchester, UK, where he had met and been encouraged by Teodor Shanin, Angel Palerm, Joan Martinez Alier, and Eric Wolf, Eduardo Sevilla-Guzman returned to Spain where, in 1978, he founded the Institute of Sociology and Peasant Studies (ISEC) at the University of Cordoba. ISEC became involved with the Andalusian landless workers movement (SOC), supporting SOC members as they occupied and began to cultivate abandoned haciendas in Andalusia, using agroecological techniques that they had learned from the peasant farmers who lived and worked around the old haciendas. The relationship between ISEC and SOC led to further important linkages with Latin American agrarian social movements, and these relationships made a significant contribution to the development of the militant perspective that characterizes agroecological research and teaching at ISEC to this day (Sevilla Guzmán and Martinez Alier 2006).

Interactions among UC Santa Cruz, CLADES, and ISEC led to the establishment of the first doctoral program in agroecology at ISEC in 1991, followed shortly after by a postgraduate programme at the International University of Andalucia, both of which continue to be offered today. Most of the contributors to this book have lectured or studied on these programs, and the personal and institutional relationships that have developed through this long period of interaction and cooperation have facilitated the training and diffusion of transformative agroecology practitioners, social movement activists, academics, and state functionaries throughout Europe, the Americas, and beyond. These transformative agroecological actors have contributed to the establishment and work of numerous associations such as the Brazilian Agroecology Association, the national umbrella group *Articulação Nacional de Agroecologia*, and the Latin American Agroecology Movement, (MAELA) many of which come together in SOCLA, the Agroecology Scientific Society of Latin America. At the same time, as it has been institutionalized within academic establishments and associations, agroecology has also become embedded in small-scale farmers' organizations, while the politics of food sovereignty have been adopted and pursued by both producer and consumer-led social movements.

3.4.2 The Movement of the People toward Agroecology and Food Sovereignty

In Brazil, the landless workers movement (MST—www.mst.org.br), like the Diggers and Levellers of England, came together in protest against the concentration of land in the hands of the few. Since 1984, the MST has led more than 2,500 land occupations, settling at least 350,000 families on somewhere in the region of 10 million hectares of land (www.mstbrazil.org). They promote agroecological methods among their members and in 2006 established the Latin American School of Agroecology on MST land in the State of Paraná. They also run an agroecological seed network to facilitate food sovereignty. In Europe, the Campaign for Seed Sovereignty (www.seed-sovereignty.org) represents the interests of more than 30 national and subnational organizations of small farmers and growers from nations of the European Union, united in their struggle against EU legislation aimed at the standardization and concentration of the seed market in the hands of a small number of seed industry corporations. Movements to defend traditional agriculture and advance food sovereignty include food consumers as well as producers and have the capacity to mobilize vast numbers of people in opposition to the institutions of the corporate food regime. On 25th May, 2013, some two million people took part in hundreds of rallies across more than 50 countries in protest against the corporate seed giant Monsanto. "March Against Monsanto" protesters call attention to the dangers posed by corporate control of the food system, genetically modified food, and the food giants who produce it.

Many national and regional agrarian organizations, confederations, and social movements are members of the peasant and small-farmer International, La Vía Campesina (LVC) (www.viacampesina.org). In 20 years, LVC has grown to encompass around 150 local and national organizations in 70 countries, representing about 200 million small-scale farmers in their struggle to "defend community-based agroecological farming as a cornerstone in the construction of food sovereignty" (www.viacampesina.org). Martínez Torres and Rosset (2010) trace the historical development of LVC from the early coalescence of numerous peasant and small-farmer organizations and confederations in Latin America. Established as a global social movement in 1993, during the 1990s the movement's leaders gained access to international policy fora, rejecting NGO representation and creating space for authentic peasant voices to be heard. In the twenty-first century LVC has taken on a global leadership role for agrarian struggles and positioned itself in opposition to the corporate food regime and its neoliberal discourse of "sustainable development" and "food security." In short, "peasants and family farmers have been able to build a structured, representative, and legitimate movement, with a common identity, that links social struggles on five continents" (op. cit.: 150).

LVC originally defined "food sovereignty" in its Tlaxcala Declaration as "the right of each nation to maintain and develop its own capacity to produce its basic foods respecting cultural and productive diversity" (La Vía Campesina 1996). At the local level, LVC works with member organizations to facilitate agroecological knowledge exchange and to share and develop the agroecological approach to food sovereignty. The movement has also established the Paulo Freire Latin American Institute of Agroecology (IALA) in Venezuela and built teams of agroecology trainers who organize continental scale encounters in the Americas, Asia, and Africa. In the face of global capital's relentless pursuit of profit through land grabbing, displacement of small-scale producers, and the patenting of seeds, knowledge, and technologies developed over generations of farming practice, the second Americas continental encounter in 2011 issued a declaration: "Agroecology is Ours and is Not For Sale." Peasant agriculture is part of the solution to the current crisis of the system. In this context, we reaffirm that "indigenous, peasant and family farm agroecology [can] feed the world and cool the planet" (La Vía Campesina 2011).

3.5 CONCLUSION

Vía Campesina's declaration is an unequivocal statement proclaiming the indivisibility of agroecology as science, movement, and practice. Today, agroecologists, whether farmers, scientists, or social movement activists (and many individuals operate in all three of these overlapping spheres of activity), are working together to defend rural communities and agroecological cultures against the negative impacts of capitalist industrialization. Reflection and dialogue within the social movements and between university-based agroecologists and peasant movement leaders are of key importance to the vitality and coevolution of an integrated, transformative agroecology. While their struggle is a global one, human experience of the negative impacts of capitalism in the countryside remains place based, and the local values, knowledge, institutions, and cultures of socioecologically situated people must be core elements in the construction of ecological sustainability and social justice. If the science of agroecology is separated from the place-based agricultural practices and agrarian social thought and movements with which it has grown up, we would argue that its transformative potential will be lost and agroecology will become just another instrumental discipline in the continuing saga of capitalism's struggle to overcome its own internal contradictions.

ACKNOWLEDGMENT

The authors are particularly grateful to Chris Bacon for his helpful comments and suggestions for improving this chapter.

REFERENCES

Altieri, M.A., and V.M. Toledo. "The agroecological revolution in Latin America: Rescuing nature, ensuring food sovereignty and empowering peasants." *Journal of Peasant Studies* 38 (2011): 587–612.

Barthel, S., C. Crumley, and U. Svedin. "Bio-cultural refugia—Safeguarding diversity of practices for food security and biodiversity." *Global Environmental Change* 24 no.5 (2013): 1142–1152.

Berens, L.H. *The Digger Movement in the Days of the Commonwealth.* London: The Merlin Press, 2007 [1906 was the original year of publication].

Carolan, M.S. "Society, biology and ecology: Bringing nature back into sociology's disciplinary narrative through critical realism." *Organization and Environment* 18 no. 4 (2005): 393–421.

Carson, R. *Silent Spring.* Boston, MA: Houghton Mifflin, 1962.

Catton, W., and R. Dunlap. "Environmental sociology: A new paradigm." *The American Sociologist* 13 (1978): 41–49.

Chayanov, A.V. *The Peasant Economy: Collected Works.* Moscow, Russia: Ekonomika, 1989.

Christy, R.D., and L. Williamson (Eds.). *A Century of Service: Land-Grant Colleges and Universities, 1890–1990.* New Brunswick, NJ: Transaction, 1992.

Comte, A. *Cours de Philosophie Positive.* Paris: Société positiviste, 1838.

Crutzen, P.I., and E.F. Stoermer. "The anthropocene." *IGBP Newsletter* 41 (2000): 12–13.

Escobar, A. "Constructing nature: Elements for a post-structuralist political ecology." In *Liberation Ecologies* (London: Routledge, 1996), 46–68.

Foster, J.B. *Marx's Ecology: Materialism and Nature*. New York, NY: Monthly Review Press, 2000.

Freudenburg, W., S. Frickel, and R. Gramling. "Beyond the nature society divide—Learning to think about a mountain." *Sociological Forum* 10 (1995): 361–392.

Friedmann, H. "From colonialism to green capitalism: Social movements and the emergence of food regimes." In *New Directions in the Sociology of Global Development* (Oxford, UK: Elsevier, 2005), 229–267.

Giddens, A. *The Constitution of Society: Outline of the Theory of Structuration*. Cambridge: Polity Press, 1984.

Gliessman, S.R. "Agroecology: Growing the roots of resistance." *Agroecology and Sustainable Food Systems* 37 no. 1 (2013): 19–31.

González Casanova, P. "Internal colonialism and national development." *Studies in Comparative International Development*. 1 (1965): 27–37.

González de Molina, M. "A guide to studying the socio-ecological transition in European agriculture." *Sociedad Española de Historia Agraria. DT-SEHA* 10 no. 06 (2010).

Hannigan, J. *Environmental Sociology: A Social Constructionist Perspective*. London: Routledge, 1995.

Hannigan, J. *Environmental Sociology*, 2nd ed. (New York, NY: Routledge, 2006).

Hernández Xolocotzi, E. Xolocotzia: *Obras de Efraím Hernández Xolocotzi*. Tomo I y Tomo II. (Universidad Autónoma Chapingo: Chapingo, Mexico, 1985).

Holt-Giménez, E. "Measuring farms agroecological resistance to Hurricane Mitch." *LEISA* 17 (2001): 18–20.

Howard, A. *An Agricultural Testament*. London: Oxford University Press, 1940.

IAASTD (International Assessment of Agricultural Knowledge, Science and Technology for Development). *Agriculture at a Crossroads: A Synthesis of the Global and Sub-Global IAASTD Reports*. Washington, DC: Island Press, 2009.

Kautsky, K. *The Agrarian Question*. London and Winchester, MA: Zwan Publications, 1988 [1899].

King, F.H. *Farmers of Forty Centuries or Permanent Agriculture in China, Korea and Japan*. Madison, WI: Democrat Publishing Co., 1911. https://archive.org/stream/farmersoffortyce00kinguoft#page/n7/mode/2up (Accessed July 3, 2013).

Huber, J. "Towards industrial ecology: sustainable development as a concept of ecological modernisation." *Journal of Environmental Policy and Planning* 2 (2000): 269–285.

La Vía Campesina. "*American Continental Encounter of Agroecology Trainers in La Via Campesina*." 2011. http://viacampesina.org/en/index.php/mainissues-mainmenu-27/sustainable-peasants-agriculturemainmenu42/1083-finaldeclaration-of-the-2nd-continental-encounter-of-agroecology-trainers-in-laviacampesina (Accessed September 11, 2012).

La Vía Campesina. "*Tlaxcala Declaration of the Vía Campesina (Declaration of the Second International Conference of Vía Campesina, Tlaxcala, Mexico)*." 1996. http://viacampesina.org/en/index.php/our-conferences-mainmenu-28/2-tlaxcala-1996-mainmenu-48 (Accessed March 1, 2014).

Lenin, V.I. "The development of capitalism in Russia." In *V. I. Lenin Collected Works*, Vol. 3, 4th ed. (Moscow: Progress Publishers, 1986), 21–608.

Malthus, T. *An Essay on the Principle of Population, as it Affects the Future Improvement of Society*. Electronic Scholarly Publishing Project, 1998 [1798]. http://www.esp.org/books/malthus/population/malthus.pdf (Accessed October 2, 2014).

Martinez Alier, J. *The Environmentalism of the Poor: A Study of Ecological Conflicts and Valuation*. Cheltenham, UK: Edward Elgar, 2002.

Martinez Alier, J. "The EROI of agriculture and its use by the via Campesina." *The Journal of Peasant Studies* 38 no. 1 (2011): 145–160.

Martínez Torres, M.E., and P. Rosset. "La Via Campesina: The birth and evolution of a transnational social movement." *Journal of Peasant Studies* 37 (2010): 149–175.

Marx, K. *Capital*. Vol. III. Harmondsworth, UK: Penguin, 1981.

McMichael, P. "A food regime genealogy." *Journal of Peasant Studies* 36 (2009): 139–169.

McMichael, P. "*Reframing development: Global Peasant Movements and the New Agrarian Question*." *Revista NERA* 10 no. 10 (2007): 57–71. http://www2.fct.unesp.br/nera/revistas/10/mcMichael.pdf (Accessed July 14, 2013).

McMichael A.J., J.W. Powles, C.D. Butler, and R. Uauy. "*Food, livestock production, energy, climate change, and health*." Lancet 6 (2007):1253-63.

Meadows, D.H., D.L. Meadows, J. Randers, and W.W. Behrens III. *Limits to Growth*. New York: New American Library, 1972.

Melville, E.G.K. *A Plague of Sheep: The Environmental Consequences of the Conquest of Mexico*. Cambridge, New York, and Melbourne: Cambridge University Press, 1994.

Méndez, V.E., C.M. Bacon, and R. Cohen. "Agroecology as a transdisciplinary, participatory, and action-oriented approach." *Agroecology and Sustainable Food Systems* 37 no.1 (2013): 3–18.

Norgaard, R.B. "The epistemological basis of agroecology." In *Agroecology: The Scientific Basis of Alternative Agriculture* (Boulder, CO: Westview Press, 1987), 21–27.

Palerm, A. *Antropologiay Marxismo*. Mexico: Editorial Nueva Imagen, 1980.

Peréz Vitoria, S. *Les Paysanssont de Retour*. Paris: ActesSud, 2005.

Polanyi, K. *The Great Transformation: The Political and Economic Origins of Our Time*. Boston: Beacon Press, 1957 [1944].

Redclift, M., and G. Woodgate (Eds.). *The International Handbook of Environmental Sociology*. Cheltenham, UK and Northampton, MA: Edward Elgar, 1997.

Redclift, M., and G. Woodgate (Eds.). *The International Handbook of Environmental Sociology*, 2nd ed. (Cheltenham, UK and Northampton, MA: Edward Elgar, 2010).

Sevilla Guzmán, E. *Sobre los Orígenes de la Agroecología en el Pensamiento Marxista y Libertario*. Cochabamba, Bolivia: AGRUCO-Plural Editores, 2011.

Sevilla Guzmán, E., and J. Martínez Alier. "New rural social movements and agroecology." In *Handbook of Rural Studies* (London: Sage, 2006), 472–482.

Sevilla Guzmán, E., and G, Woodgate. "Sustainable rural development: From industrial agriculture to agroecology." In *The International Handbook of Environmental Sociology* (Cheltenham, UK: Edward Elgar, 1997), 93–94.

Simms, A. *Ecological Debt: The Health of the Planet and the Wealth of Nations*. London: Pluto, 2005.

Sorokin, P., C. Zimmerman, and C. Galpin. *A Systematic Source Book in Rural Sociology*. Minneapolis: University of Minnesota Press, 1930–1932.

Swyngedouw, E. "The antinomies of the post political city: In search of a democratic politics of environmental production." *International Journal of Urban and Regional Research* 33 no. 3 (2009): 601–620.

Toledo, V.M., and N. Barrera-Bassols. *La Memorial Biocultural*. Barcelona: Icaria, 2008.

Tomich, T.P., S. Brodt, H. Ferris, et al. "Agroecology: A review from a global-change perspective." *Annual Review of Environmental Resources* 36 (2011) 193–222.

United Nations Conference on Trade and Development (UNCTAD). *Wake Up Before It Is Too Late: Make Agriculture Truly Sustainable Now for Food Security in a Changing Climate*. Geneva, Switzerland: UN Press, 2013.

van der Ploeg, J.D. *Peasants and the Art of Farming, a Chayanovian Manifesto* (Agrarian Change and Peasant Studies Series). Halifax and Winnipeg: Fernwood Publishing, 2013.

van der Ploeg, J.D. *The New Peasantries: Struggles for Autonomy and Sustainability in an Era of Empire and Globalization*. London: Earthscan, 2009.

Wezel, A., S. Bellon, T. Doré, C. Francis, D. Vallod, and C. David. "Agroecology as a science, a movement and a practice. A review." *Agronomy for Sustainable Development* 29 (2009): 503–515.

Wilken, G.C. *Good Farmers: Traditional Agricultural Resource Management in Mexico and Guatemala*. Berkeley, CA: University of California Press, 1987.

Woodgate, G., B. Ambrose-Oji, R. Fernandez Durán, G. Guzmán, and E. Sevilla Guzmán. "Alternative food and agriculture networks: An agroecological perspective on responses to economic globalisation and the 'new' agrarian question." In *New Developments in Environmental Sociology* (Cheltenham, UK and Northampton, MA: Edward Elgar, 2005), 586–612.

Zeder, M.A. "Domestication and early agriculture in the Mediterranean Basin: Origins, diffusion, and impact." *PNAS* 105 no. 33 (2008): 11597–11604.

Political Agroecology
An Essential Tool to Promote Agrarian Sustainability

Manuel González de Molina

CONTENTS

4.1 INTRODUCTION

Agroecology has developed a large arsenal of technical solutions to facilitate the transition of conventional farms to organic farming and to ensure the viability of the farm. However, the more macro aspects, especially the political aspects, have remained in the background. This has led to insufficient development of aspects that transcend the farm or local community scale of the agroecological focus, especially political and institutional aspects, which regulate social relationships in agriculture.

The link between politics and agroecology is not new. Many authors have demanded the need for socioeconomic structural reforms in order to be able to achieve sustainable agrarian systems (Buttel 1997, 2003; Rosset 2003; Levins 2006; Holt-Giménez 2006; Perfecto et al. 2009; Altieri and Toledo 2011). But this link between agroecology and politics has not received the attention it deserves by agroecologists. On the one hand, the agroecological movement is characterized by the scarcity of political proposals that go beyond the local sphere. The majority of agroecological experiences linked to nongovernmental organizations, academic institutions, and to a significantly lesser extent public governments continues to be in their majority experiences on farms or, in the remaining cases, community experiences, where the research, participatory action, and design of sustainable rural development strategies have been the favored tools. Meanwhile, the existing institutional framework, established by the corporate food regime (see Holt and Altieri in this volume), imposes a straitjacket that prevents the generalization of such experiences and hinders their continuity.

On the other hand, the agroecological movement reveals, especially in the academic field, the increasing influence of a current of agroecology that we could call scientistic or *technocratic*. This current considers agroecology almost exclusively as a scientific discipline, producing useful knowledge and technology for sustainable agriculture (Wezel et al. 2009; Tomich et al. 2011). It promotes technological solutions rather than institutional or social change solutions for the problems considered today by the global food system, based on what Pretty et al. call *sustainable intensification* (2011, 455). The term "agroecological intensification" has been used as well in an attempt to coopt agroecology into the dominant food system.

Although every scientific and social practice is political by nature, both trends deny politics, although for opposing reasons. The result is, on the one hand, a lack of efficiency and stability of agroecological experiences that barely reach the required size and expanse of land; and on the other hand, the spread of the false idea that technological innovation alone, without substantial social and economic change, will achieve more sustainable agriculture. The first leads to inefficiency, the second to inactivity, and both sever the possibilities of agroecology being an alternative to the ecological crisis in the countryside.

Agroecology has a practical dimension, which is inseparable from the scientific one. Agroecology cannot be limited to pointing out unsustainable factors in agroecosystems, followed by proposing management approaches and routes to their implementation that will restore these factors to a sustainable state. As stated by Gliessman (2011, 347), it is also a powerful tool to achieve change in the food system, in other words, a massive redesign of the economic structures that govern our food systems. This practical dimension of agroecology requires politics, that is, the disciplines responsible for designing and implementing institutions that make agrarian sustainability possible. In spite of this, agroecology is not yet equipped with the analytical instruments and criteria required to define strategies that could guide the change. Most agroecological experiences are still, with a few exceptions, local and uncoordinated. Agroecology is still closely bound to the scope of the farmer, the farm, and the local community. However, the participation of agroecologists in local and even national government is becoming increasingly widespread. Politics must develop within the heart of agroecology to provide agroecologists with instruments for analysis and sociopolitical intervention that would allow them to go beyond local experiences, encouraging their generalization and the essential changes in the food system at a higher territorial scale. Otherwise, experiences will be condemned to be "islands of success" amid a sea of privation, poverty, and environmental degradation (Altieri and Rosset 2010). This chapter examines the need to overcome this shortcoming and discusses some of the reasons that endorse this need.

4.2 THE DYNAMICS OF AGROECOSYSTEMS: THE PLACE OF POLITICS

The changing dynamics of agroecosystems make the need for power and politics more comprehensible. The search for sustainability implies a change in their dynamics that can come only from social agents by means of institutional mediation. It is this process of creation and establishment that political agroecology deals with.

As with social metabolism as a whole,* the dynamics of agroecosystems are a product of the relationship between the two poles of all socioecological relations—the population and the resources available to them. There are many factors that make up each of the two poles in this relationship and also many variables that alter them. In terms of resources, changes in the quality and quantity of environmental resources and the services offered by agroecosystems are determined

---

* On social metabolism theory and methodology, see Ayres and Simonis (1994); Fischer-Kowalski and Haberl (1997); Fischer-Kowalski (2003); Fischer-Kowalski and Hüttler (1999); Giampietro et al. (2012); Toledo and González de Molina (2007); and González de Molina and Toledo (2011, 2014).

by the dynamics of nature itself, dynamics with a long-term temporal dimension, but which do not preclude sudden changes. Similarly, the quantity and quality of the goods and services offered by agroecosystems can be modified by interferences caused by the population itself (i.e., society).

In terms of the population, factors that can alter the relationship with resources are not limited solely to the number of farmers living off agroecosystems. This pole of the relationship must be understood in a broad sense, encompassing not only the human population, but also its levels of consumption and the ease or difficulty with which it can access resources to satisfy them. These three aspects are institutionally conditioned.

For example, social inequality or territorial imbalance can induce changes in agroecosystems. From a physical point of view, it entails the unequal assignation of energy, materials, water, and environmental services. Pressure on the resources of agroecosystems can increase if part of the population is deprived of the wealth generated by their appropriation and transformation. The appropriation by one social group through exploitation mechanisms or the forced transfer of income can reduce the amount of biomass available to meet the endo- and exosomatic needs of the rest of the rural population or may increase social demand over the requirements of most of the population, increasing pressure on the agroecosystems. From the perspective of the *internal equity* of agroecosystems, an unequal distribution of natural resources usually puts pressure on increasing the productive effort.

Over the last century, farmers have been exposed to a new form of inequality, which has constituted the most powerful lever for productive intensification and the breaking apart of agroecosystems. We are referring here to the growing inequality that has been generated first in the national market and then in the global market in terms of distributing revenue between the agrarian sector and the other productive sectors or between different territories (unequal exchange), which we could term *external inequality* (Guzmán et. al. 2000). The global profitability of farming activity has been progressively declining since the start of the 20th century as a consequence of the unequal relationship of exchange between the agrarian sector, the industrial, and the services sector. Between 1900 and 1998, the cumulative effect was a decline of 62% in that relationship (Zanias 2005; Eisenmenger et al. 2007, 183). This loss of profitability has fostered processes of crop intensification to compensate for the decline in farming revenue. This process has made farmers more dependent on the market and on new technologies to achieve a minimum income threshold; in other words, more dependent on the agro-industrial complex as a whole.*

Social inequality, therefore, from an environmental perspective, constitutes an "ecosystemic pathology," a permanent source of instability and a powerful stimulus for conflict and socioecological change. This perspective is fundamental in our analysis, since it takes the concept of equity to the terrain of its effects on sustainability (Guzmán et al. 2000, 102). There are numerous cases, both historic and contemporary, in which poverty and the inaccessibility of resources lead to environmental degradation, deforestation and forest clearing, crop cultivation on steep slopes, overgrazing, the use of agrochemicals, and so on. However, in the opposite direction, the struggle for subsistence has often become a struggle to conserve resources and agrarian sustainability (Guha and Martínez-Alier 1997).

Undoubtedly, the relationship between population and resources can be altered by technological innovation. Certain technologies can increase the carrying capacity of an agroecosystem beyond its possibilities, increasing metabolic efficiency in the use of energy and materials. However, their adoption and even the very process of innovation depend on institutional arrangements, among them political power, and can be stimulated or not by public policies. Similarly, a rural community can increase the carrying capacity of its territory by importing resources from other countries or regions through *economic exchange*. This is, therefore, a very significant factor when explaining

---

* See, for example, part III of the FAO report for 1993: "Agricultural trade: entering a new era?"(http://www.fao.org/es/esa /es/pubs_sofa.htm).

the dynamics of agroecosystems. The market has been the vehicle through which the subsidization of energy and materials required to maintain the continued growth of agrarian production has circulated in industrialized countries. However, as Polanyi (1989) pointed out some time ago, the market is merely a power relation, at times conflictive, which must be regulated by political power.

Therefore, the decisions that emanate from the state are undoubtedly important in terms of explaining the dynamics of agroecosystems. We are referring here to the set of *stable power relations* (regulation and legal norms) or specific power relations (decisions or public policies), which aim to reproduce both the metabolism between nature and society, and the forms in which it is organized and, therefore, the ways in which energy and materials flow within agroecosystems. Influenced by the other components of change mentioned previously, this factor in turn has a decisive influence on them and, therefore, on the dynamics of agroecosystems. In this respect, the design of public policies that create a favorable institutional framework for the development of sustainable agroecosystems is fundamental. This is a task that falls within the scope of political agroecology; also to make its application possible through political action and the participation in state institutions. We shall later examine this decisive aspect for the development of agroecological experiences in a greater depth.

The scientist or technocratic streams of agroecology strip socioecological change of any collective dimension of human action. However, the distribution of political power and natural resources often gives rise to conflict. Conflicts between social groups and between territories constitute a potential source of socioecological change, and consequently, should be considered when it comes to studying the evolutionary dynamic of agroecosystems. For example, currently, environmentalist protests are helping to internalize environmental costs and, even if they do not abruptly change the approaches taken to management, they do lessen their harmful effects and pave the way toward agrarian sustainability (Guha and Martínez-Alier 1997).

Agroecology should pay particularly close attention to conflicts in which there are implicit or explicit motivations for changes in the agroecosystem status quo. These kinds of conflicts, which might have very diverse motivations and manifestations, could be classed as environmental conflicts. The resolution of such conflicts has historically been a source of modification or conservation for agroecosystems (González de Molina et al. 2009). For example, the protection afforded by many rural communities to natural resources against attempts to overexploit them by companies or the state itself has managed to save natural resources from overexploitation or deterioration. We could say, therefore, that environmental conflict can contribute to increasing the sustainability of an agroecosystem or agrarian metabolism as a whole or to decreasing it. This consideration of environmental conflict as a motor for socioecological change gives social movements a key role in the struggle for agrarian sustainability. In this respect, political agroecology is also a science of collective action in favor of sustainability—a philosophy of action.

4.3 POWER, POLITICS, AND SOCIAL METABOLISM

Accordingly, the organization and management of agroecosystems are not merely a technical or material question, but also a question of power, since it permeates social relationships in general and these determine and are determined by the environment. The ecological basis of this statement is very simple: while genetic instructions regulate the endosomatic and exosomatic metabolism of different species, establishing a stable, evolutionary framework to guide energy and material flows, the human species does not have equivalent biological support to establish a lasting organization for the energy and material flows, which make up its exosomatic metabolism. Despite the fact that human beings are social species, their behavior is not genetically regulated. To establish this kind of instruction and to make it permanent, the human species uses conventions or institutions, which guide those flows and give them meaning. But, by their very nature, social institutions are no more than relationships of power. Power is not only employed by the state or public institutions; it is also present in each social relationship or set of social relationships (family, sex, consumption, media, property, market, etc.). However, in contemporary societies such as ours, whose territorial scope is much wider than the local scale, institutionalized power in civilian organizations and, above all, in the state are of prime importance in the material organization of our metabolism with nature and, therefore, in the sustainability of our relationship with nature.

This, obviously, can be extended to agricultural systems. The organization and management of agroecosystems are usually seen from a material perspective. However, a rigorous socioecological approach requires us to afford equal relevance to those intangible mechanisms and instances with which and within which agroecosystems function. Agroecosystems are artificialized ecosystems that shape a particular subsystem within the general metabolism between society and nature and, therefore, they are the product of socioecological relations established within. For example, the use of crop change is a decision that often has socioeconomic roots and, at the same time, environmental consequences. These kinds of socioecological relations are part of social relations, in general, in which power and conflict are present. Consequently, from the simplest societies, technologically speaking, the specific assembly of each agroecosystem responds to different types of institutions, forms of knowledge, worldviews, rules, norms and agreements, technological knowledge, means of communication and governance, and forms of ownership (González de Molina and Toledo 2011, 2014). The sustainability of an agroecosystem is not only the result of a series of physical and biological properties, but also the reflection of power relations. Therefore, agroecology must have the theory required to deal with politics.

The sustainability of an agroecosystem is not, then, solely the result of a set of physical and biological properties but, rather, the reflection of certain power relationships; and so, "policy is the central programming function of a type of ecosystem, the social ecosystem" (Garrido 1993, 10). In this regard, the quest for the sustainability of agricultural ecosystems requires political agroecology, which is a new way of organizing agroecosystems and agricultural metabolism in general. In the same way the political power articulates the different subsystems of a socioenvironmental system, political agroecology should undertake the specific articulation of the different subsystems of an agroecosystem through the organization of energy, material, and information flows. Its articulation, programming, and functional orientation are the task of political agroecology, bringing continuity and order to the evolution of the agroecosystem.

4.4 POLITICAL ECOLOGY AND AGROECOLOGY

Of the two most common meanings of the term "politics" is "an art of domination or an art of integration," and we are especially interested in the latter—politics as "governability" (Foucault 1991), that is, the control and governance of a social group are settled in a specific territory. From this perspective, the fundamental objective of politics is to provide public goods by collective action (Colomer 2009). Considering the provision of said public goods is out of the citizens' reach individually, a coordinated effort is required, whether through voluntary or coercive means, or whether via collective action or public government institutions, which execute public policies. For example, sustainability is a public good that citizens cannot achieve individually. To achieve it, collective action, public policies, or a combination of both is required. Political ecology focuses on the study of this concept.

There is no agreement as to what political ecology is (Peterson 2000; Blaikie 2008, 766–767). The term gives rise to many meanings and ways of understanding its object. But all of them have in common their political economy approach to natural resources and their preferential application to developing countries (Blaikie 2008, 767). Our interpretation is the same as that of Gezon and Paulson (2005), for whom "the control and use of natural resources, and consequently, the course of environmental change" are shaped by "the multifaceted relations of politics and power and the cultural constructions of the environment." In this sense, political ecology combines political and ecological processes in the analysis of environmental change and it could be understood also as "the politics of environmental change" (Nigren and Rikoon 2008, 767). Paraphrasing Blaikie and Brookfield (1987), we could say that "political ecology [is] an approach for studying ecological and social change" (Blaikie and Broofield 1987), but *together*. In other words, political ecology is an approach for studying socioecological change in political terms.

In this regard, political agroecology would be the application of political ecology to the field of agroecology or the close association between these (Toledo 1999; Forsyth 2008). Using Paulson et al. (2003, 209) and Walker (2007, 208) we could say that political agroecology should "develop ways to *apply* the methods and findings [from political ecology research] in addressing" socioecological change in agroecosystems.

But political agroecology is not only a research subject, it has another practical dimension closely linked and considered as a central goal-acheiving agrarian sustainability. Many agroecologists are involved in a "'popular political ecology' that ties research directly to activist efforts to improve human well-being and environmental sustainability through various forms of local, grassroots activism, and organization" (Walker 2007, 364). In this respect, political agroecology should develop in two directions: as an ideology which, in competition with others, is dedicated to dissemination and turning the organization of agroecosystems based on an ecological and sustainable paradigm into the dominant system (Garrido 1993) as well as a disciplinary field responsible for designing and producing actions, institutions, and regulations aimed at achieving agrarian sustainability.

Political agroecology is based on the fact that agrarian sustainability cannot be achieved using only technological (agronomical or environmental) measures, which help to redesign agroecosystems in a sustainable manner. Without a profound change in the institutional framework in force it will not be possible for successful agroecological experiences to spread and for the ecological crisis in the field to be combated effectively. Consequently, political agroecology examines the most suitable way to participate in these moments and to use those tools that render institutional change possible. Such a change, in a world still organized around nation-states, is possible only through political mediation. In democratic systems, for example, it implies-collective action through social movement, electoral political participation, the game of alliance between different social forces to build government majorities, and so on. In other words, it calls for the creation of essentially political strategies. The design of institutions that favor the achievement of agrarian sustainability (Ostrom 1990, 2001, 2009) and the way to organize agroecological movements in such a way that they can be implemented comprise precisely the two main objectives of political agroecology.

Political agroecology is, therefore, more than a specific proposal for a program. For example, the demand for food sovereignty, promoted by *VíaCampesina* and other social movements, is *a* specific proposal for a program that *can* emerge from applying political agroecology to the current conditions of the global food system. But, like any specific proposal for a program, it can change depending on the social and political scale and context to which it is applied. Political agroecology is responsible for establishing it and, as a new branch of agroecology, it is not a political proposal or program to get agrarian sustainability.

Political agroecology employs the concept of autonomy, an attribute of sustainability (Altieri 1995; Gliessman 1997), which besides, has its roots in agroecological epistemology. It is this use of the concept of autonomy that leads the agroecological political discourse to demand food sovereignty, as a current expression of this attribute, that is, as the best way to strengthen the autonomy of the agroecosystems and those who manage them. In other socioenvironmental and political contexts, the principle of autonomy can have other specific dimensions. Political agroecology is not a new alternative term for food sovereignty. It seeks to produce knowledge that renders agroecology and food sovereignty something that can be practiced, exploiting the knowledge accumulated by political ecology, and the experience of social movements and green political parties.

4.5 THE SCALES THAT MAKE AGROECOLOGY "POLITICAL"

The process of agroecological transition, from a spatial point of view, takes place at different scales which, although interrelated, displays different characteristics. At least five can be identified such as crop, farm, community or village, national, and global. Although other intermediate levels can be considered (districts, watershed, regions, etc.) at more aggregated scales, agroecology has to deal with emerging properties that do not emerge at lower levels.

At the scale of individual crops and farms, agroecology has developed a complete arsenal of technical solutions that have made it possible to design sustainable systems. The next level at which the transition occurs corresponds to the organization of the agroecosystem. In this case, during the industrialization of agriculture, there has been a growing segregation in land uses and the losses of production and functional synergies generated by agro–silvo–pastoral integration. The result has been the loss of spatial heterogeneity. With it, flows of energy and materials, which tended to be local and closed (renewable), have become global, provided by fossil fuels. This is one of the most underdeveloped aspects of agroecology, which has led to the lack of focus on landscape agroecology. It is at this level that decisive aspects of the agroecological transition are revealed; for example, which territorial arrangements will be required for agriculture to be sustainable (Guzmán and González de Molina 2009; Guzmán et al. 2011). At a national or global level, the industrialization of agriculture has entailed the constitution of a global agrarian market and a single global food system, in which agroecosystems also tend to integrate in a specialized way.

At the scale of the crop or farm, changes in farmers' attitudes could be sufficient to drive a transition, as could changes in patterns of consumption, which can be achieved at an individual level when expressed by turning to the market or other institutions to procure food. But when talking about community and particularly the State and the world, political power and collective action are two properties that emerge and with them the need for political action (Zimmerer and Basset 2003; Swyngedouw 2004; McCarthy 2005; Paulson et al. 2003; Rangan and Christian 2009). Political agroecology appears then as an urgent need to which practically no attention has been paid thus far. Many essays have been published on peasant movements and food sovereignty (among the latter see Altieri 2009; Holt-Giménez 2006, 2011; Holt-Giménez and Patel 2009; Martínez-Torres and Rosset 2010; Perfecto et al. 2009; Petersen 2009), but there has not been a systematic and articulated reflection from agroecology.

A brief review of the attributes of agrarian sustainability should illustrate this need. First, it refers to productivity, which is often only considered at the scale of the individual crop or farming estate, without taking into account the interrelations which, from the perspective of land uses, take place at the level of the agroecosystem or in relation to nearby agroecosystems. The possibility of closing cycles and using locally generated energies depend on territorial planning and organization. These tasks fall to local government or the state and depend on public policies.

The same can be said of stability, referring to the capacity of an agroecosystem to maintain its productivity over time. As maintained by Altieri (1995), certain properties of an agroecosystem, such as climate conditions, have very prolonged cycles over time, and a farmer's capacity to influence them is fairly limited. However, a farmer can try to maintain and even increase the biological stability of an agroecosystem or a specific estate by improving practices such as irrigation or the integration of agriculture and livestock farming. These properties and practices, owing to their territorial impact and economic cost, exceed the scope of the community and are the responsibility of the state or its regional planning bodies. The formation of product pricing, inputs used, subsidies and incentives and, therefore, the economic stability of farming businesses is dependent on established decision-making and regulatory spheres that are often far removed from rural communities.

The resilience of an agroecosystem does not depend solely on its productive arrangements. State institutions, responsible for managing natural and socioeconomic disasters, can create favorable or adverse conditions for the recovery of the productive capacity of an agroecosystem. In this respect, there are institutions that favor the resilience of an agroecosystem more than others. In contrast to private or simply state property, communal forms of ownership, characteristic of traditional rural cultures, result in management approaches that adapt more easily to "surprises" or changes experienced by ecosystems (Holling et al. 1998; Holt-Giménez 2001). In this respect, agroecology must provide an analytic approach regarding forms of organization for decision-making and institutional design that increase the resilience of agroecosystems.

One of the attributes of sustainability considered fundamental by agroecology is social equity. Access to resources and the distribution of agrarian revenue are organized by institutions that, such as ownership or the market, can significantly condition the sustainability of an agroecosystem. The rules and regulations that ensure sufficient income for farmers are the responsibility of the state, just as an unequal distribution of property can also be modified by the political power of government actions such as agrarian reforms. It is also the responsibility of the state to reverse the sustained deterioration experienced through the relationship of exchange between food and agricultural raw materials, and the inputs and manufactured products consumed by agrarian businesses or farming families. It falls to the political institutions to establish the opportune regulations in markets that guarantee sufficient income for farmers or to establish the necessary compensations by means of subsidies and fiscal incentives that redress market imbalances. It is also the responsibility of political authorities to establish a fair assignation of resources for future generations. Political institutions must guarantee, by imposing regulations regarding management, the right of those who are not yet born to an agroecosystem in good physical and biological conditions.

Finally, the level of autonomy is an essential attribute of sustainability and is closely related to the internal capacity to supply the flows of energy and materials required for production. The current model of agriculture generates high external dependence through an unbalanced relationship of mercantile exchange that is damaging for farmers, especially for small farmers. The growing integration of farmers into the world market and the food system has stripped them of their decision-making capacity about the type of crops they grow, their management and guiding knowledge, or the final destination of production. Hence, the concept of food sovereignty has been proposed as an alternative to the classic concept of food security. In short, the mission of political agroecology is producing knowledge that makes possible the establishment of institutions and social movements favorable to the development of agrarian sustainability.

Similarly, at more aggregated scales of the transition, properties emerge regarding the relationship with other metabolic processes; for example, the link that has been established in recent decades between agrarian production, the processing and transformation of foods, transportation, distribution, and the ways in which foods are conserved, cooked, and finally consumed. This has compelled agroecologists to adopt a much broader vision, adopting an approach that focuses on the food system (Francis et al. 2003; González de Molina and Infante 2011), which also necessarily requires politics and collective action.

4.6 THE POLITICAL DIMENSION OF CHANGE

The planet is facing a serious structural food crisis caused by growing competition for the use of productive land, which is in limited supply and whose *per capita* availability is falling continuously, since it is highly dependent on a management model based on increasingly expensive and scarce fossil fuels and since it is highly vulnerable to the climate change currently under way. To all of this, we must add the environmental damage caused by the predominant model of chemical agriculture. This damage is diminishing—and will do so more seriously in the future—the capacity of agroecosystems to produce food and raw materials and to offer environmental services. The dietary habits of rich countries, with a propensity toward disproportionate consumption of meat and dairy products, are rerouting large amounts of cereals away from direct consumption and away from the most needy areas of the world, perpetuating hunger and malnutrition and inflating prices in such a way that these basic foodstuffs are inaccessible to the poorest. The global food system is currently incapable (Dixon et al. 2001, 2) of feeding the human population, although there is sufficient raw material available. It has made little progress in the eradication of rural poverty and is starting to show clear signs of exhaustion (FAO 2007a).

Furthermore, the functioning of the markets and the subordinate role played by agricultural activity in economic growth have caused an acute loss of profitability. According to the FAO, the real prices of major agrarian products have decreased by 50% since 1983 (FAO 2007a). This decline is the cause of abandonment in rich countries and hunger, rural depopulation, and poverty in poor countries. Paradoxically, prices have experienced a significant increase over the last three years. The sustained increase in the consumption of grain, the increase in the consumption of meat, particularly in Asia, the rising price of oil, and the scarcity of land, which has been highlighted by the expansion of agro-fuel crops, are expressions of the structural crisis in the world food system. In relation to the increasing scarcity, a dense speculative Web has also been weaved, which has accentuated inflationist tension even further.

The traditional production imbalances between countries, the unequal distribution of land, and the control of large agrifood multinationals and big banks over global agricultural markets turn food insecurity, endemic hunger, and the poverty of vast rural areas of the planet into structural characteristics of agriculture worldwide (McMichael 2009). At the same time, the technological model of intensive agriculture, which arose after the Second World War and gave rise to the so-called Green Revolution, is now exhausted. It cannot maintain growth in agricultural production or do so without degrading natural resources. Furthermore, it is physically and economically unhealthy. The technological model of intensive agriculture produces contaminated foodstuffs, causes degenerative illness, and does not guarantee farmers' income or the maintenance of jobs.

In view of the crisis, the two objectives of an agroecological strategy are to eradicate hunger and malnutrition and raise the income of farmers, especially in countries with a higher index of poverty, and to reduce or possibly eliminate environmental damage, all through the promotion of sustainable management approaches for agroecosystems. But how can this be achieved? The scientific and political consensus (FAO 2007b; De Schutter 2010) are that agroecological methods can significantly increase production and yield by combining new technologies with the development of agronomics, and local knowledge, and resources.

However, if eating habits do not change in rich countries—reducing the consumption of meat, eggs, and dairy products—and the demand generated by this diet continues to increase, both pressure on importing food from countries with food security and hunger problems will intensify. Hence, the advances that might be made run the risk of proving insufficient. In the West, the adoption of an agroecological approach should, therefore, give rise to a different strategy based on *degrowth* in their food systems (González de Molina and Infante 2011).

All of this entails a change in several dimensions. First, citizens must individually change their eating habits, especially in Western countries. But that is not enough; the implementation and multiplication of collective experiments in sustainable production and responsible consumption through the creation and strengthening of production and consumptions groups, associations of producers and consumers, and so on constitutes a second dimension that is essential. Throughout the planet, a good number of agroecological experiments have arisen, both rural and urban, for production and consumption, providing the avant-garde for a new food system.

Agroecology has focused on small farmers as subjects since they possess high agroecological potential. Among other reasons, they can be closer to rural rationality and practices that make the sustainable management of agroecosystems possible (Altieri and Toledo 2011). Traditional peasant communities have also developed highly efficient methods of cooperative and communal management of resources (Ostrom 1990, 2001), which are of great interest to agroecology. As a response to the consolidation of the "food regime," there has been for some time a *repeasantization*, which the current economic and financial crisis has accentuated. This process is characterized by the struggle for autonomy and decommoditization and by the quest for local solutions to global problems. It goes much further than a mere resistance movement (Ploeg 2008). "It is present in the unfolding of organic farming, just as it is the main driver of the many forms of endogenous rural development that we are witnessing in Europe and in the new forms of production that are being developed in the campamentos created by the Brazilian Movimento dos Sem Terra (MST) and in the inlands of Chiapas..." (Ploeg 2010). To all of this must be added the emergence throughout the world of food sovereignty movements with which alliances are being forged (Holt-Giménez 2011).

Other subjects also become relevant when dealing with food systems, especially consumers. No agroecological transition will be fully successful without a major alliance between producers and consumers. But to ensure the majority participation of these groups, a partnership is required in turn with the green movement in its broadest dimension. This will not be achieved without political and institutional mediation, in other words, without the development of public policies that drive the transition forward. Indeed, for these experiments to reverse the ecological crisis in the countryside, they must be expanded and achieve a sufficient quantitative and qualitative dimension. The development of public policies and the dynamic action of social movements are crucial in this task (Altieri and Toledo 2011). Within this context, the role of the state and social movements becomes fundamental, as does the decision-making process of democracy itself. This raises the question of how to achieve a strong presence in the government to promote public policies that favor rural sustainability, either alone or by partnering up with other social and political forces. The experiments developed in Andalusia, Spain (González de Molina 2009), and Brazil (Caporal and Petersen 2011) prove this. The debate around how to make this possible is one of the most urgent tasks facing agroecology.

4.7 THE MULTIDIMENSIONAL NATURE OF COLLECTIVE AGROECOLOGICAL ACTION

For this to be possible, it is essential to overcome the dichotomy which often exists between political–institutional action and the transformational practice of local agroecological experiences. The innovative nature of these experiences in production, distribution, and consumption make them the *avant* guard of an alternative food system. They are the foundations on which a more sustainable future will be built, but they are not sufficient, in themselves, to produce a change on a wider social scale or to guarantee their own survival. A simple sum of experiences does not produce the desired change on the immediately higher territorial scale. To believe that this is possible is idealistic and it is also a mechanical view, which is contrary to the very theory of agroecology. The step from the micro to the macro scale is possibly the main challenge facing political agroecology. It certainly requires us to overcome the rhetoric of spontaneity, which attributes the necessary strength to the virtues of agroecological experiences for them to multiply and replace the dominant development model. This kind of utopian thinking is unachievable, among other reasons because it replaces one type of determinism with another. History has already shown that the future is constructed in the present and that it depends on the correlation of social forces, in other words, through politics.

Structural and institutional change does not come about through the sum of individual or local changes, or through abrupt revolutions, which make change possible overnight. Human societies are complex systems whose evolutionary dynamics are not linear and, therefore, neither is the logic of social change. Local actions are effective agents for global change insofar as they involve strategic interaction between the local and the global by means of transformative collective action. The sum of the parts is not equal to the whole, and the emergent properties of all social systems and their recursive capacity finally condition the dynamics of the components (Morin [1999] 2007, 68).

To think of the relationship between global change and local change (either one or the other) or to express preference for one or the other is not, then, a rigorous position and, above all, it is ineffective. Local changes, since they are partial, are not enough to produce agroecological transformation in agricultural production and consumption since the generalization of the change is slowed and hindered by the institutional structure of the food system. Without changes in the institutional arrangements, local change will be reduced or confined and will take place in conditions in which they will be weaker in the face of attacks by the dominant system. But without local changes, global change is not possible, as it would not be supported by social forces and innovative experiences which drive collective action for systemic change and which build successful agrifood transformation programs. The function of local changes is to liberate spaces ruled by agribusiness but, above all, its function is to become an agent for change, a social movement which pursues global, universal aims through the replication of successful practical experiences.

Political agroecology is, then, in favor of multidimensional collective action, which works on different planes and on different territorial scales. Local agroecological experiences in production and consumption multiply their efficiency and chances of survival if they take place within the framework of multilevel collective action and cooperation. The reductionism of simple actions by individuals or small groups in isolation from social institutions (markets, family, etc.) and simple, exclusive institutional political intervention must be avoided.

4.8 AN ESSENTIAL INSTITUTIONAL CHANGE FOR PROGRESS IN AGROECOLOGICAL TRANSITION

As we have been proposing, agroecological transition at a higher scale than the farm or the local community requires decisive state intervention through a raft of actions and the corresponding provision of human, material, and financial resources. This is what we conventionally call "public policy," and they may be of different types: regulatory, constitutive, redistributive policies, and so on. They may take the form of programs, projects, laws, publicity campaigns, technological innovations, government subsidies, and so on. In some countries and at different scales of action, public policies have been implemented to promote organic and alternative agriculture, which *a priori* represent a very significant opportunity for progress toward agroecological transition.

However, these policies are developed in an institutional environment, which is conceived for the benefit of conventional agriculture and, therefore, they finally encourage technological modernization without encouraging the closure of the ecological cycles and the shortening of commercial circuits. That is to say, they end up as an obstacle to the structural coupling of agroecosystems with the territories where they are based. Public policies in support of organic agriculture, for example, are often oriented by market criteria and do not represent a real alternative to the configuration of the dominant food system.

This is the result of ignorance of the fact that the different scales or levels on which agroecological transition takes place are mutually dependent. The debate between collective community action and institutional and regulatory action should be at the heart of any agroecological strategy to advance the transition. In other words, local experiences depend on the limiting capacity of the institutional arrangements at the higher scales of social organization, as we saw in Section 4.5. It is, then, the task of the State, among other institutions, to create the conditions necessary so that the numerous agroecological experiences do not end up being isolated experiences. But in the same way that local action must be subjected to new types of social and institutional relationships, and there is no guarantee *per se* that public policies will really produce progress toward the transition if there is no promotion of an institutional change at the state level. It is the task of political agroecology to accompany public policies and collective agroecological action with changes in the institutional design to guarantee that there is real progress toward transition.

Otherwise, the different versions of sustainable agriculture run the risk of becoming a specific sector that does not question the current configuration of the food system or the power relationships that sustain it. With no change in the institutional environment, which is typical of market economies, public policies in support of sustainable agriculture inevitably lead to *conventionalization*. By this, we mean the process by which organic, alternative, or agroecological-based family agriculture becomes a version which is little different from conventional agriculture, reproducing the same story and sharing the same social, technical, and economic characteristics (Buck et al. 1997; Hall and Magyorody 2001; Darnhoffeer et al. 2010). If it is not counteracted by the conscious action of the administrations and the collective action of the agroecological social movements, the logic of conventional food markets pushes organic or alternative producers toward intensification (Guthman 2004).

Customer preferences in the marketplace could help to achieve sustainability, for example, by purchasing organic and fair trade food products. In fact, this market is one of the so-called green markets, which are a strong pillar of support in the struggle for sustainability. But the market alone is incapable of adequately valuing organic foodstuffs or their environmental functions. Market regulation directly influences price setting and facilitates or hinders responsible consumption. Without political regulation, it is not possible to guide the markets, including the green markets, toward the path of sustainability.

Food market forces, in which large-scale distributors play a predominant role vis-a-vis a fragmented agricultural sector, push organic farmers into conventionalization. As they are currently regulated, they are unable to force conventional agriculture to take on the *territorial cost of sustainability* (Guzmán Casado and González de Molina 2009; Guzmán Casado et al. 2011). In an effect, the pressure for prices that are perceived as being lower—a trend that conventional agriculture bears for structural reasons, but which also affects organic agriculture—stimulates a response, which leads to greater externalization of territorial costs (less rotation, fewer crops, high-response seeds, more plant health treatment, etc.) and, therefore, greater dependence on external inputs. In this way, farmers are encouraged to take a shortcut to obtain greater profits at the expense of sustainability. This trend is usually favored by a regulatory structure governing organic production that allows and even encourages this type of external solution.

An example of this can be seen in what has happened in recent years in Andalusia (southern Spain), a pioneering region in the design and implementation of organic agriculture promotion plans. Since the Greens left the regional government in 2007, the government has tended to consider organic foodstuffs to be simply another product, though of higher quality. It has declined to intervene in the sector through active regulation policies and has left it at the mercy of the "free" market. This has caused very significant imbalances from the production and distribution point of view and a marked tendency to become a specific kind of production whose purpose is to meet the demand of a segment of the population with a high disposable income (González de Molina 2012). Something similar was noted by Allen (2004) in the United States and by Levidow et al. (2013) in Europe, and it could also occur with respect to organic production in Brazil (Petersen et al. 2014). This risk threatens the experiences of peasants, which have arisen in response to the crisis in the food system, to which we have already referred. Without institutional change, this process of repeasantization runs the risk of becoming a process of reexploitation of peasants and of subordination to the demands of capital. Without institutional change, there is no guarantee that repeasantization will follow an unequivocally agroecological path. The crisis of the dominant food system has created conditions that are favorable to the maintenance of a peasant economy, but the continuity of the institutions that regulate it is what guarantees that peasant farms remain subordinate to their requirements and continue to be exploited and their surpluses appropriated. As shown by van der Ploeg (2010, 17), "specific expressions of peasant struggle (such as particular parts of organic production) have been embraced, whether not appropriated, by powerful actors such as Wal-Mart, Dole, and Starbucks." It is, then, essential to redesign the institutional operation of the industrial metabolism so that the process of repeasantization is more than just an antisystem resistance movement.

In the same way, without a regulatory change that creates the conditions for the collective management and use of natural resources, the closure of the biogeochemical cycles and the productive autonomy of farms will be difficult to achieve. The legislation which, in many countries, protects private ownership of the land against any decision, even state decisions, and prevents or restricts access by farmers to usable land is an objective obstacle to the achievement of agrarian sustainability. Therefore, without institutional change—promoted, for example, by means of agroecological agrarian reform, that is, which includes not only a plot of land but also all of the territorial resources that make agrarian sustainability possible—there will be no progress toward transition. Without a specific, agroecologically oriented institutional design, all organic productions remain in the hands of the markets, stimulated only by some public policies, which are often insufficient or even counterproductive.

But the institutional sphere is also a field of conflict between the forces that dominate the food systems and the social movements today, which are fighting for change. Without the strength of these movements, which are deeply rooted in local experiences, institutional change is improbable. Consequently, political agroecology should be oriented toward institutional change, and the implementation of public policies that reinforce the social movements that are fighting for sustainable agriculture and food justice (Holt-Giménez and Shattuck 2011). In this way, the recursive loops between institutions and the process of building institutions are strengthened.

4.9 INSTITUTIONAL DESIGN AND COGNITIVE FRAMEWORKS

For this, it is essential to have the appropriate instruments to generate "cognitive frameworks" and "institutional frameworks," which are favorable to agrarian sustainability. The cognitive frameworks prefigure the perception of individuals, largely unconsciously, and the institutions condition their behavior, restricting the set of possible alternatives. Political agroecology, as it generates its own cognitive and institutional frameworks, encourages a reflexive automation of the perceptions, ideas, and behavior of the players. The institutional design establishes the material and socioenvironmental conditions that drive the practical implementation of the ideological program proposed by the cognitive framework as being preferable or desirable for the agroecological transition. If these aspects are not contemplated, agroecological experiences will finally be abandoned to the dominant cognitive framework (neoliberal ideology) and the diktats of "self-regulated" markets (the institutional framework), hindering their development and pushing them toward conventionalization.

In this regard, political agroecology should offer both a precise diagnosis of the crisis in the food system and a theory of collective action and action by the political institutions to progress toward agroecological transition. We have already sketched out the former, and the latter could be based on the three cognitive principles put forward by Garrido (2011) for political ecology and which, obviously, political agroecology shares: "biomimesis," "neuromimesis," and "ethomimesis." "Biomimesis" is "imitating nature" when reconstructing human productive systems to make them compatible with the biosphere. All the epistemologies of agroecology consist of the creative application of the principle of "biomimesis" to the design of agroecosystems (see Gliessman 1997), the strengthening of the connective diversity and density of agroecosystems (rotation, genetic diversity, interactions, and synergies, etc.), the closure of the nutrient cycles, the efficient exploitation of waste, the reduction of external inputs, the use of solar energy, the energy exploitation of crops at a level of efficiency close to that of photosynthesis, and so on.

"Neuromimesis" consists of using the human brain as a cognitive social model, that is, to copy the devices, rules, and mechanisms that make the human cognitive system the most efficient and complex of all machines. Weighing only 1.3 kg and with energy consumption of some 400 cal, the brain holds over 100 billion nerve cells (neurons), which are able to develop a million synapses (connections between neurons) per second, with a total potential interneuronal connective density of 10^{14}, the greatest connective density in the known universe. In contrast, the supercomputer designed by IBM (BG), which simulates the activity of 10,000 neurons, consumes 100 kW. The human brain is capable of recognizing any of the trillions of sentences contained in any natural language in only tenths of a second. This is possible thanks to the enormous social and neuronal connective density, the capacity for symbolism and the synaptic plasticity, which the brain and human society use. The key to neurocognitive efficiency lies in the substitution of the consumption of materials and energy by the consumption of organization and information. The brain is a model organism from a thermo-dynamic point of view. The study of the human cognitive system, then, allows us to design efficient, sustainable organization and communication systems.

Finally, "ethomimesis" attempts to learn from our evolutionary history as a species and from the behavior of many other species, about the use of resources, and relationships with the natural environment. In this case, the ethological keys to efficiency are to be found in social cooperation (between individuals of the species), in thrift in the use of natural resources, and lavishness in the use of social relationships (e.g., in play). Most primitive communities and communities of the great apes manage natural resources as finite resources and treat social and emotional resources as infi-nite (play, celebration, duels, sexual rites, etc.).

We have already mentioned that the development and success of agroecological movements and experiences depend on optimal institutional design. It is a matter of progressing from acceptance of the existing institutions to an alternative institutional design, that is to say, to the production and/ or reform of the institutions by means of rules and regulations and a structured series of signals, stimuli, payments, and sanctions with the intended purpose of encouraging a desired type of agro-ecological behavior. All of this must take in the five basic elements of all institutional designs: the regulations, the institutions, the mechanisms or devices, the agents and the target functions, or ends. In this regard, the eight criteria proposed by Ostrom and her school are of particular interest for the design of efficient institutions for the management and conservation of common goods (Ostrom 1990, 2001). In the same way, when compared with competition, we know that cooperative behavior is the most efficient social or species behavior that has ever existed in our evolutionary history as a species (Bowles and Gintis 2013) and, therefore, it is the behavior which is most appropriate for the collective management of common goods (Garrido 2013). With all this material and these instru-ments, political agroecology can find the necessary "sustainability heuristic" (Klauer et al. 2013, 83) "situated at the interface between theory and practice: it collects from theory all the relevant facts and causal relations and processes this information in such a way that it can guide practical action. In doing so it provides some basis for bridging sustainability science and politics."

REFERENCES

Allen, P. *Together at the table: Sustainability and sustenance in the American agrifood system.* University Park, PA: Pennsylvania State University Press, 2004.

Altieri, M., and P. Rosset. "The potential of agroecology to combat hunger in the developing world." 2010. (www.agroeco.org.)

Altieri M., and V. Toledo. "The agroecological revolution in Latin America. Rescuing nature, ensuring food sovereignty and empowering peasants." *Journal of Peasant Studies* 38 (2011): 587–612.

Altieri, M.A. "El 'Estado del Arte' de la Agroecología y su Contribución al Desarrollo Rural en América Latina." In *Agricultura y Desarrollo Sostenible.* A. Cadenas Marín (Ed.). (Madrid: Ministerio de Agricultura, 1995), 151–203.

Altieri, M.A. "Agroecology, small farms and food sovereignty." *Monthly Review* 61 no. 3 (2009): 102–111.

Amate, J., and M. González de Molina. "'Sustainable de-growth' in agriculture and food: An agro-ecological perspective on Spain's agri-food system (2000)." *Journal of Cleaner Production* 38 (2013): 27–35.

Ayres, R.U. and U.E. Simonis. *Industrial metabolism: Restructuring for sustainable development.* New York, NY: United Nations University Press, 1994.

Blaikie, P. "Epilogue: Towards a future for political ecology that works." *Geoforum* 29 (2008): 765–772.

Blaikie, P., and H. Brookfield. *Land degradation and society*. London: Methuen, 1987.

Bowles, S., and H. Gintis. "¿Ha Pasado de Moda la Igualdad? El Homo Reciprocans y el Futuro de las Políticas Igualitaristas." In *Razones para el socialismo*. R. Gargarella, and F. Ovejero (Eds.). (Barcelona: Paidos, 2013).

Buck, D., C. Getz, and J. Guthman. "From farm to table: The organic vegetable commodity chain of northern California." *Sociologia Ruralis* 37 (1997): 3–20.

Buttel, F.H. "The politics and policies of sustainable agriculture: Some concluding remarks." *Society and Natural Resources* 10 no. 3 (1997): 341–344.

Buttel, F. *Envisioning the future development of farming in the USA: Agroecology between extinction and multifunctionality?* Wisconsin, USA: New Directions in Agroecology Research and Education, 2003. www.agroecology.wisc.edu/downloads/buttel.pdf.

Caporal, F.R., and P. Petersen. "Agroecologia e Políticas Públicas na América Latina: o Caso do Brasil." *Agroecología* 6 (2011): 1–16.

Colomer, J.M. *Ciencia de la Politica*. Barcelona: Editorial Ariel, 2009.

Darnhoffeer, I., T. Lindenthal, R. Bartel-Kratochvil, and W. Zollistsch. "Conventionalisation of organic farming practices: From structural criteria towards an assessment based on organic principles. A review." *Agronomy for Sustainable Development* 30 (2010): 67–81.

De Schutter, O. "Informe del Relator Especial sobre el Derecho a la Alimentación." *Nueva York, Naciones Unidas, Asamblea General, Consejo de Derechos Humanos* (2010): 1–24.

Dixon, J., A. Gulliver, and D. Gibbon. *Sistemas de Producción Agropecuaria y Pobreza. Cómo Mejorar los Medios de Subsistencia de los Pequeños Agricultores en un Mundo Cambiante*. Roma: FAO, 2001.

Eisenmenger, N., J. Ramos, and H. Schandl. "Transition in a contemporary context: patterns of development in a globalizing world." In *Socioecological transitions and global change. Trajectories of social metabolism and land use*. M. Fisher-Kowalski (Ed.). (Cheltenham, UK: Edward Elgar, 2007), 179–222.

FAO. *SOFA. The state of food and agriculture*. Rome: FAO, 2007a.

FAO. *Organic agriculture and food security*. Rome: FAO, 2007b.

Fischer-Kowalski, M. *On the History of Industrial Metabolism*. Vienna: Greenleaf Publishing, 2003.

Fischer-Kowalski, M., and H. Haberl. "Tons, joules, and money: Modes of production and their sustainability problems." *Society and Natural Resources* 10 (1997): 61–85.

Fischer-Kowalski, M., and W. Hüttler. "The intellectual history of material flow analysis, Part II, 1970–1998." *Journal of Industrial Ecology* 2 no. 4 (1999): 107–136.

Forsyth, T. "Political ecology and the epistemology of social justice." *Geoforum* 39 (2008): 756–764.

Foucault, M. "La Gubernamentalidad." In *Espacios de Poder*. R. Castel, J. Donzelot, M. Foucault, J.P. de Gaudamar, Cl. Grignon, and F. Muel (Eds.). (Barcelona: Ediciones de La Piqueta, 1991).

Francis, C.A., G. Lieblein, S.R. Gliessman, T.A. Breland, N. Creamer, R. Harwood, et al. "Agroecology: The ecology of food systems." *Journal of Sustainable Agriculture* 22 no. 3 (2003): 99–118.

Garrido, F. *Introducción a la Ecología Política*. Granada: Editorial Comares, 1993.

Garrido, F. "Aproximación a una Fundamentación Eológica de la Democracia." *Dilemata* 5 no. 12 (2013): 63–74.

Garrido Peña, F. "Ecología Política y Agroecología: Marcos Cognitivos y Diseño Institucional." *Agroecología* 6 (2011): 23–28.

Gezon, L., and S. Paulson. "Place, power, difference: Multiscale research at the dawn of the twenty-first century." In *Political ecology across spaces, scales and social groups*. S. Paulson and L. Gezon (Eds.). (New Brunswick, NJ: Rutgers University Press, 2005), 1–16.

Giampietro, M., K. Mayumi & A. H. Sorma. *The Metabolic Pattern of Society*. Routledge, 405 pp, 2012.

Gliessman, S.R. *Agroecology. Ecological processes in sustainable agriculture*. Chelsea, MI: Ann Arbor Press, 1997.

Gliessman, S.R. "Agroecology and food system change." *Journal of Sustainable Agriculture* 35 (2011): 345–349.

González de Molina, M. (Ed.). *El Desarrollo de la Agricultura Ecológica en Andalucía (2004-2007). Crónica de una Experiencia Agroecológica*. Barcelona: Icaria, 2009.

González de Molina, M. "Luces y sombras del Crecimiento de la Producción Ecológica en Andalucía durante el Ultimo Quinquenio (2007–2011)." *CUIDES* 9 (2012), 153–192.

González de Molina, M., D. Soto, A. Herrera, and A. Ortega. 2009. "Peasant protest as environmental protest. Some cases from the 18th to 20th century." *Global Environment* 4 (2009): 48–77.

González de Molina, M., and J. Infante. "Agroecología y Decrecimiento. Una Alternativa Sostenible a la Configuración Actual Sistema Agroalimentario Español." *Revista de Economía Crítica* 10 (2011): 113–137.

González de Molina, M., and V. Toledo. *Metabolismos, Naturaleza e Historia. Una Teoría de las Transformaciones Socio-Ecológicas.* Barcelona: Icaria, 2011.

González de Molina, M. and V. Toledo. *The social metabolism. A socio-ecological theory of historical change.* New York, NY: Springer, 2014.

Guha, R., and J. Martínez-Alier. *Varieties of environmentalism. Essays north and south.* London: Earthscan, 1997.

Guthman, J. *Agrarian dreams. The paradox of organic farming in California.* Berkeley: University of California Press, 2004.

Guzmán Casado, G.I., and M. González de Molina. "Preindustrial agriculture versus organic agriculture. The land cost of sustainability." *Land Use Policy* 26 (2009): 502–510.

Guzmán Casado, G.I., M. González de Molina, and A. Alonso Mielgo. "The land cost of agrarian sustainability. An assessment." *Land Use Policy* 28 (2011): 825–835.

Guzmán, G., M. González De Molina, and E. Sevilla. (Eds.). *Introducción a la Agroecología como Desarrollo Rural Sostenible.* Madrid: Ediciones Mundi-Prensa, 2000.

Hall, A., and V. Magyorody. "Organic farmers in Ontario: An examination of the conventionalization argument." *Sociologia Ruralis* 41 (2001): 399–422.

Holling, C.S., F. Berkes, and C. Folke. "Science, sustainability and resource management." In *Linking social and ecological systems.* F. Berkes (Ed.). (Cambridge: Cambridge University Press, 1998), 346–362.

Holt-Giménez, E. (Ed.). *Food movements unite! strategies to transform our food systems.* Oakland, CA: Food First Books, 2011.

Holt-Giménez, E. "Measuring farms' agroecological resistance to hurricane mitch." *LEISA* 17 (2001): 7–10.

Holt-Giménez, E. *Campesino a campesino: Voices from Latin America's farmer to farmer movement for sustainable agriculture.* Oakland, CA: Food First Books, 2006.

Holt-Giménez, E., and R. Patel. *Food rebellions: The real story of the world food crisis and what we can do about it.* Oxford, UK: Fahumu Books and Grassroots International, 2009.

Holt-Giménez E., and A. Shattuck. "Food crises, food regimes and food movements: rumblings of reform or tides of transformation?" *Journal of Peasant Studies* 38 (2011): 109–144.

Klauer, B., R. Manstetten, T. Petersen, and J. Schiller. "The art of long-term thinking: A bridge between sustainability science and politics." *Ecological Economics,* 93 (2013): 79–84.

Levins, R. "A whole-system view of agriculture, people, and the rest of nature." In *Agroecology and the struggle for food sovereignty in the Americas.* A. Cohn, J. Cook, M. Fernández, R. Reider, and C. Steward (Eds.). (Nottingham, UK: Russell Press, 2006), 34–49.

Levidow, L., M. Pimbert, P. Stassart, and V. Vanloqueren. "*Agroecology in Europe: Conforming or transforming the dominant agro-food regime?* Conference on 'Agroecology for Sustainable Food Systems in Europe: A Transformative Agenda' (2013) (http://www.ensser.org/increasing-public-information/agroecology-conference/).

Martınez-Torres, M.E., and P.M. Rosset. "La Vıa Campesina: The birth and evolution of a transnational social movement." *Journal of Peasant Studies* 37 no. 1 (2010): 149–75.

McCarthy, J. "Scale, sovereignty, and strategy in environmental governance." *Antipode* 37 (2005): 731–53.

McMichael, P. "A food regime analysis of the 'world food crisis.'" *Agriculture and Human Values* 26 (2009): 281–295

Morin, E. "La Epistemología de la Complejidad." In *El Paradigma Ecológico en las Ciencias Sociales.* F. Garrido, M. González de Molina, J.L. Serrano, and J.L. Solana (Eds.). (Barcelona: Editorial Icaria, 2007), 55–84.

Nigren, A., and S. Rikoon. "Political ecology revisited: Integration of politics and ecology does matter." *Society and Natural Resources* 21 (2008): 767–782.

Ostrom, E. *Governing the commons. The evolution of institutions of collective action.* New York, NY: Cambridge University Press, 1990.

Ostrom, E. "*Commons, institutional diversity of encyclopedia of biodiversity,*" Volume I. Waltham, MA: Academic Press, 2001.

Ostrom, E. "A general framework for analyzing sustainability of social-ecological systems." *Science* 325 no. 24 (2009): 419–422.

Paulson, S., L. Gezon, and M. Watts. "Locating the political in political ecology: An introduction." *Human Organization* 62 (2003): 205–217.

Perfecto, I., J. Vandermeer, and A. Wright. *Nature's matrix: Linking agriculture, conservation and food sovereignty.* London: Earthscan, 2009.

Petersen, P. *Agricultura Familiar Camponesa na Construção do Futuro.* Rio de Janeiro: AS-PTA, 2009.

Petersen, P., E. Mussoi, and F. Dal Soglio. "Institucionalización del Enfoque Agroecológico en Brasil: Avances y Desafíos." *Agroecología* 8 no. 2 (2014).

Peterson, G. "Political ecology and ecological resilience: An integration of human and ecological dynamics." *Ecological Economics* 35 (2000): 323–336.

Polanyi, K. *La Gran Transformación.* Madrid: La Piqueta, 1989 [1944].

Pretty, J., C. Toulmin, and S. Williams. "Sustainable intensification in african agriculture." *International Journal of Agricultural Sustainability* 9 no. 1 (2011): 5–24.

Rangan, H., and A.K. Christian. "What makes ecology 'political'? Rethinking 'scale' in political ecology" *Progress in Human Geography* 33 no. 1 (2009): 28–45.

Rosset, P. "Food Sovereignty: Global Rallying Cry of Farmer Movements." *Institute for Food and Development Policy Backgrounder* 9 no. 4 (2003): 1–4.

Swyngedouw, E. "Scaled geographies: Nature, place, and the politics of scale." In *Scale and geographic enquiry: Nature, society, and method.* E. Sheppard, and R. McMaster (Eds.). (Oxford: Blackwell, 2004), 129–153.

Toledo, V.M. "Las "Disciplinas Híbridas": 18 Enfoques Interdisciplinarios sobre Naturaleza y Sociedad." *Persona y Sociedad* 8 no. 1 (1999): 21–26.

Toledo, V.M., and M. González de Molina. "El Metabolismo Social: Las Relaciones entre la Sociedad y la Naturaleza." In *El Paradigma Ecológico en las Ciencias Sociales.* F. Garrido, M. González de Molina, and J.L. Serrano (Eds.). (Barcelona: Editorial Icaria, 2007), 85–113.

Tomich, T., S. Brodt, F. Ferris, R. Galt, W. Horwath, E. Kebreab, et al. Agroecology: A review from a global-change perspective. *Annual Review of Environment and Resources* 36 no. 15 (2011): 1–30.

van der Ploeg, J.D. *The new peasantries: Struggles for autonomy and sustainability in an era of empire and globalization.* London: Earthscan, 2008.

van der Ploeg, J.D. "The peasantries of the twenty-first century: The commoditisation debate revisited." *Journal of Peasant Studies* 37 no. 1 (2010): 1–30.

Walker, P.A. "Political ecology: Where is the politics?" *Progress in Human Geography* 31 no. 3 (2007): 363–369.

Wezel, A., S. Bellon, T. Doré, C. Francis, D. Vallod, and C. David. "Agroecology as a science, a movement and a practice. A review." *Agronomy for Sustainable Development* 29 (2009): 503–515.

Zanias, G.P. "Testing for trends in the terms of trade between primary commodities and manufactured goods." *Journal of Development Economics* 78 no. 1 (2005): 49–59.

Zimmerer, K., and T. Bassett. *Political ecology: An integrative approach to geography and environment-development studies.* New York, NY: Guilford, 2003.

CHAPTER 5

Learning Agroecology through Involvement and Reflection

Charles Francis, Edvin Østergaard, Anna Marie Nicolaysen,
Geir Lieblein, Tor Arvid Breland, and Suzanne Morse

CONTENTS

5.1 INTRODUCTION

5.1.1 Action Learning in Agroecology

On a sunny September morning in southern Norway, 25 students and two university teachers chatter in French, English, and Norwegian as they bicycle through rolling hills of forest interspersed with fields of ripening wheat and islands of rock outcrops hiding ancient Viking graves. The path is cool, rising, and falling through dark spruce forest and then emerging into a farmyard bordered by a white 18th-century farmhouse, a red dairy barn, and several smaller outbuildings, plus a shiny new combine near the barn. We learn that this is the Haugen Farm.

Solveig, Osmund, and their son, Einar, greet this polyglot group in clear and measured English, generously sharing their family story of cultivating these fields of glacial till and marine clays, fields farmed by many inhabitants over the last 2000 years. Theirs started first as a traditional mixed farm and after a period of mainly cereal monoculture, today has become rediversified with organic grain, green manure, and hay crops, plus Norwegian heritage breeds of red cattle and sheep. These same fields have passed through the generations guided by Viking law; just one son or (today) one daughter purchases the farm from the parents. Each generation brought a new view of how to farm these fields; each generation struggled to bring their vision to fruition under changing laws, policies, and economics, under the ever-present, watchful eyes of their parents, now living in a smaller house close by on the same farmstead.

The challenge for our student team is one of mutual understanding of what will it take for this farm to continue to persist and to flourish. How can the students quickly learn about the multifaceted nature of this farming system? How might the farmers and students, all overflowing with their own notions of how best to farm, learn together, and explore alternatives to today's practices that can help the farmers move closer to their visions for a sustainable farm now and into the future?

(The names of farmers and students are changed.)

We have begun this journey through the landscape of action learning by describing the involvement of students with farmers, recognizing the power of stories to build meaning and catalyze education (Bruner 1986, 2002; Neuhauser 1993; Rossiter 2002). We continue by providing reflections of students in response to this initial field experience as they join their instructors in the quest to become agroecologists.

One of the team, Kristin from Norway, gives this description of what happened in the team at this stage, written in her reflection document at the end of the course:

Through our first meeting with the farmer it became apparent to us that there were some clear issues that needed to be dealt with on the farm—and straight after the interview we in the group started discussing and deliberating how we could best identify and come up with possible solutions to these problems. At the same time, we were afraid: afraid of stepping on the toes of the farmer. This became one of the main discussions within the group: Should we tell the farmers that some of their practices were unsustainable? Should we give suggestions of alternative, more sustainable practices? How should we in those cases deliver that message to the farmer? Would they not be offended? The group was not in unison. Some were reluctant to tell the farmer that he should or ought to do anything at all, others found the practices of the farmer upsetting and wanted to tell him how he could do better. This is where our different backgrounds came to show: Some were focusing on the tilling practices, others were focusing on the economical aspects, some were concerned with the aspect of global sustainability and ethics (this was me), or the generational shift going on at the farm, or the nutrient cycles, and for some the management of the forest was a main concern. We were dispersed in focuses and knowledge. We found a compromise, I thought, by aiming to describe to them a completely different set of farm practices than the current ones, as an alternative for the farmers to draw inspiration from.

Another student, James from New Zealand, provided this description of what happened before the second meeting, which included a participatory visioning and action planning workshop with the farmer [Osmund] and his oldest son [Einar]:

We had fallen into the old paradigm of picking out problems of the farm we saw to then solve with hard knowledge. We felt that Osmund was over-tilling the soil. We recognized the structure of the soil to be quite poor proven by the farm's incessant issue with crabgrass, a bioindicator of poor soil structure. We had prepared our entire workshop to confront this issue with Osmund and Einar, thinking they would be receptive to us and change the farm accordingly. We set to research things the farm could do to improve its soil structure by simultaneously reducing Osmund's weed problems. It was his responsibility, we thought; our knowledge of the "truth" would merely enlighten him to the fact. We were wrong. He and his son almost entirely dismissed our findings at the workshop; instead the farm's Organic status was in danger due to Einar feeling it would not bring in enough money to support his future family he dreamt of having.

A lecture in class by Dr. Mette Vaarst about her experiences from working in Denmark and Uganda with farmer field schools (FFS) (Vaarst et al. 2007a, b) apparently made a big difference in guiding how the group then proceeded in their work and communicated to the farmer family:

Our professors provoked us during the first class. They called us "Agents of Change," and their aspirations for us were no less ambitious. Far from understanding the interrelated nature of systems, we were taught to act as a force, or so I thought; within the Holon, or at least, the farming systems of a Holon .... What came to pass during this journey, and I use the word "journey" knowing its sensational effect, was that acting as an agent of change does not mean acting as a force against a system, rather acting as an agent of understanding. A profound shift in the way I relate to the world, I was set to understand this Holon of ours through observation, reflection, and participation; not as an "aggressive" (or violent, referred to by M. Vaarst) force against what I think is wrong, or bad. My world view had not shifted, but my place within it had, which is nothing short of an enlightenment in my mind and it was helped by my journey on this course [....] This lecture was fundamental to my enlightened experience. Change must come from within; it does not come by force or hard knowledge alone. This lecture changed the way my group approached our work [....] Mette was right; one cannot just point out problems and solutions. It is an ineffective means for change, especially when the farmers did not even recognize the problem as a leading one. We completely changed the report and our research; it was wonderful that my group was so understanding and flexible to change ourselves. Instead, we wanted to increase [Haugen's] own network among other organic farmers. It is very difficult to stay motivated as an organic farmer when you are alone; the problems of the farm needed to be realized and solved by the farmers themselves. On reflecting about Vaarst et al.'s paper on FFS, one notices how farmers must learn from themselves and be motivated to learn, and change. It cannot be forced, an agent of change, as I now understand it, will only complement a farmer's motivation and process of change [....] I genuinely feel that my group began to rethink our role as agents of change after this [James].

The student [Kristin] from Norway describes the shift from focusing on mainly agronomic issues to dialoging and visioning with the farmers:

I think the new perspective she [Mette Vaarst] brought in made it obvious to us (I was not the only one she touched with her lecture) that we needed to redefine our role, objectives, and approaches on the farm. We shifted our focus onto extracting the problems the farmers themselves had identified, and channelized the energy in the workshop to try to get them to come up with solutions or ideas for how to solve the problems they were facing.

5.1.2 Education for Responsible Action

The situations and reflections described in the visit to the Haugen Farm are taken from the autumn course "Agroecology: Action Learning in Farming and Food Systems" (http://www.nmbu.no/course /pae302) in an Agroecology M.Sc. program at the Norwegian University of Life Sciences (http://www .nmbu.no/en/studies/study-options/master/master_of_science_in_agroecology). Experiences in reflective learning in agroecology short courses at our university evolved into a complete M.Sc. degree path based on what students and teachers learned in week-long doctoral courses in the mid-1990s (Lieblein et al. 2000). We observed early in this process that students learned best by starting with *observations* on the farm, an experience that included transect walks across the landscape (Francis et al. 2012) and in-depth interviews with farmers (Østergaard et al. 2013). Students in subsequent courses were immersed in the farms through participating in hands-on work as well as conversations with farmers.

In the current autumn semester course, each student team spends 4 months in *dialog* with farmers, other food system stakeholders, teachers, and within the project groups, but dialogue with stakeholders is a key to the group work process. On the basis of this comprehensive and intense activity, student teams identify themes that are relevant for the development of the farm and perform a *visioning* exercise to imagine a desired future state that represents a substantial improvement from the current situation. The ideas that emerge from the visioning exercise are presented and further developed in dialogue with the farmers. Perhaps unique to the program is a series of structured *reflection* sessions at each step of the process. A parallel study of the food system in a Norwegian community is conducted simultaneously, both enriched and complicated by the multiple goals and ambitions of multiple stakeholders connected to food, including consumers.

This reflective learning process is an integral and continuing element in the first semester of the 2-year M.Sc. program. The second and third semesters are dedicated to classes at Norwegian University of Life Sciences (NMBU) or at a cooperating university, as well as planning thesis research, which is conducted in the fourth semester. This program, which is intended to prepare students for responsible action (Lieblein et al. 2007), has been carefully evaluated and reported in the literature. Subsequent section describes learning goals, the process of developing team project documents both on the farm and in the local food system, and a final personal reflection by each student where they assess their own learning in the course (Francis et al. 2011, 2012; Lieblein et al. 2008, 2012; Østergaard et al. 2010). The program is described in detail in Section 4, where we expand on the course history, specific features of the concept, and the present 2-year study program with focus on the autumn semester as a case study in practical education. In the course, we are intent on connecting phenomena on the farm and in the community with the learning process, as described in Section 5.1.3.

5.1.3 Experiential and Reflective Learning in Agroecology Education

Because of the broad nature of agroecology, the learning process and the desired competencies of an agroecologist, we apply an experiential and holistic approach to reflective learning and teaching (Kolb 1984). Challenges facing decision makers in farming and food systems are most often complex. The entire process from farming to food on the table can be perceived as comprising human activity systems with multiple, often conflicting goals. They include biophysical as well as economic and social elements that interact within and across various hierarchical levels in space and time (Francis et al. 2003, 2004). Some key features of such situations are difficult to understand unless students become directly involved, and some properties do not surface until one tries to change the current situation. For example, the need for and power of observing a situation from multiple perspectives (Rickerl and Francis 2004) become immediately evident when a group of students from different parts of the world and with diverse educational backgrounds, ranging from natural sciences to social sciences, are given the task to understand, envision improvement, and facilitate action together with farmers and stakeholders in food networks. Furthermore, the story above shows that even after careful observation of the Haugen farm, some issues did not surface until the students, as part of the action learning process, had a visioning session with the farmer and his son regarding the desirable future situation of the farm.

Motivation is a key to learning, and we observe students respond to a stimulating learning environment. In our experience, most students become highly motivated and dedicated when they are given the task of dealing with a concrete, open-ended case, including people who appreciate participation and dialogue. The social and subject-related interactions within the student group working on a case add to this motivation gained in a learning environment with multiple stimuli. Lessons learned firsthand are more likely to be retained longer than knowledge gained from reading or from a lecture in the classroom, based on our student evaluations in the Norway program. Also, the likelihood that the knowledge and skills acquired will be used in later professional life is higher the more similar the students' learning environment is to their later working situation. Reports from graduates confirm our observations about how they retain and apply their academic and practical agroecology education in the M.Sc. program (Francis et al. 2011; Lieblein et al. 2000). We theorize that linking the students' experiences to relevant theory and reflecting on these links both deepens the students' understanding and enhances the transferability of newly acquired knowledge to other similar situations that they confront.

We consider the experiential approach a prerequisite for the development of competencies needed for agroecologists to contribute to responsible action (Lieblein et al. 2004; Østergaard et al. 2010). We focus on observation, participation/action, dialogue, and visioning with reflection imbedded as an integral part of each step. This thread on reflection continues through the education as well as through the chapter, a key educational tool that links our activities in the agroecology course to the five key competencies that will be valuable to students throughout their studies as well as future professional lives. The process is illustrated in Figure 5.1, where learning and action are seen as outcomes of fluxes between the five core activity areas.

We depict learning as a web because we have learned through our own observation and reflection with the student teams that the process is rarely linear and does not necessarily proceed step by step around the circle in a clockwise direction. Rather, the process is messy and complex. For example, after reflection on initial observation, a student team may decide quickly that more observation is needed and return to that previous step. Likewise, participation and dialogue may reveal gaps in facts or other parts of the farm context that require more observation rather than a move toward visioning and action. The process then reflects the dialogues among the student team, farmer or community

The Learning–Action Web

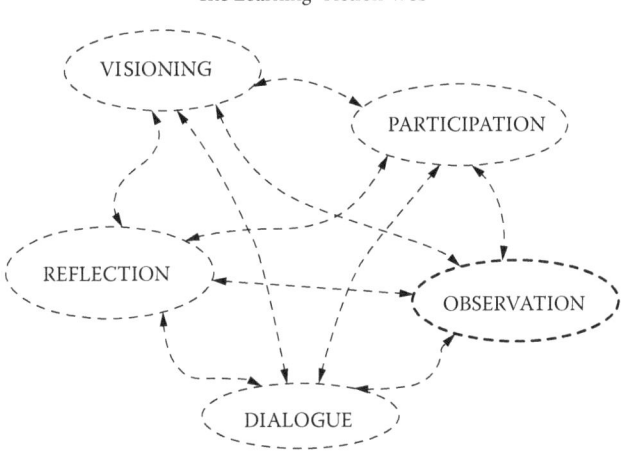

Figure 5.1 Reflective learning web with the steps in learning through involvement in agroecology farm and community student team projects.

stakeholders, and instructors, and the team adaptively manages learning and interactive activities as more of the current situation unfolds for them. The process is described in more detail in a later section on key agroecological competencies and another on the implementation of an open-ended case strategy that is appropriate to place and time, as well as to the students on each project team. The story continues with an exploration of why this educational approach in agroecology is relevant to learning how to achieve responsible action in working with clients and improving farming and food systems.

5.2 CONNECTING PHENOMENOLOGY AND LEARNING

From the beginning of our M.Sc. program in agroecology, an experiential approach to learning and teaching (Kolb 1984) has been of major importance. At a later stage, we realized that phenomenology could provide an additional conceptual foundation for our educational activities (Østergaard et al. 2010). The phenomenological approach to agroecology education draws on experiences from phenomenon-based science education. In both science education and agroecology education the intention is to connect phenomenology as a branch of philosophy with concrete applications in education. Phenomenon-based learning in science education has been developed on the foundations and applications of this philosophy by many teachers in different countries over the past several decades (Østergaard et al. 2007). The primary nature and value of phenomena in concrete, lived experiences has been an essential component also in applying phenomenology to agroecology education.

A core of phenomenology is to return "to *concrete*, lived human experience in all its richness" (Moran 2000, p. 5; italics in original). The intention of phenomenology is to provide a basis for open-mindedly exploring the diversity of lifeworld phenomena and human relationships to them. *Lifeworld* can be described as "the world of everyday experiences" (Østergaard et al. 2007). Husserl's dictum, "to go back to the 'things themselves'" (Husserl 1970a, p. 68), has for more than 100 years been a guideline for phenomenologists. This type of learning is "preconceptual" in the sense that no concepts are searched for; rather, gaining a rich, sensual expression of the phenomenon itself is the primary goal that leads to learning. Phenomenology forms a critique of the "theory-first dogma," which rules academia in the conventional classroom, the knowledge and theories behind what the students experience have become more scientifically correct than the experience itself. This situation easily creates a gap between the world of scientific knowledge-based explanation and students' experienced lifeworld. In contrast to theory-centric learning and teaching of traditional academia, our approach to agroecology is grounded in phenomenon- and experience-centric learning.

In his later philosophy, Husserl, the German founder of philosophical phenomenology, argues that the natural sciences have lost contact with the lifeworld. In his *Crisis in the European Sciences*, Husserl (1970b) maintained that the scientific culture of Europe has without substantial critique accepted the Cartesian dualism and its consequent objectivistic and naturalistic view of knowledge and learning of knowledge. Therefore, he argues, science is unable to consider how the researcher as a person participates in the constitution of scientific knowledge. Husserl raises the challenge that people pursuing natural sciences no longer relate science to everyday experiences, nor recognize that scientific knowledge in fact always presupposes the lifeworld as its ontological foundation (Dahlin 2001). In his account of Husserl's critique of science, Harvey (1989) defines the "ontological reversal" as an ontological position where abstract models from science are considered to be more real than reality itself. It means that what actually is secondary, ontologically speaking, becomes primary (Dahlin et al. 2009). The abstract, often mathematical, models are seen as truly describing the causes behind everyday experiences. From an ontological point of view, the understanding of what is *real* has been reversed.

According to the ontological reversal, personal and sensual experience is regarded as subordinate in relation to objective and conceptual understanding. This turn in the understanding of what is real has severe consequences for learning and teaching, especially when it comes to science. As long as abstract, scientific models are taken to represent the true causes behind our experiences in everyday situations that by nature are to be understood and explained, teachers tend to put less emphasis on students' perception and experience. The phenomenological critique of this ontological turn is explicitly expressed by Husserl and Heidegger and by science educators like Wagenschein (1990). This critique also forms the very basis for an ontological *rereversal*, where lifeworld phenomena are given back their ontological primacy.

A pedagogical implication of the ontological reversal is that teaching is planned "from the end" (Wagenschein 1990). This implies that concepts and theories are useful points of departure, but not the end products of an exploratory learning process rooted in lifeworld experiences. In conventional teaching approaches, phenomena are given a secondary significance, whereas theory of the phenomenon, relevant concepts, theories, and models to explain phenomena all are of primary importance. A lifeworld-based teaching starts ideally in open-minded sense experiences, in students' everyday, personal, and intuitive knowledge. From this foundation, the teacher designs a learning path toward theories, models, and abstract knowledge. This ensures that scientific concepts are rooted in experience, and not merely jumped into for convenience. Agroecology teaching planned "from the start" involves a primary focus on perceptual lifeworld experience and a secondary focus on cognitive activities in which these experiences are reflected or explained (Østergaard and Dahlin 2009). In conventional science classes today, we would have no problems of finding a pervading attitude, which Bo Dahlin (2001) refers to as "the primacy of conceptual cognition." Sense experience is reduced to a mere instrument for quasi-openly looking for what has already been defined. This is how the ontological reversal is carried out in practical teaching. Obviously, an ontological rereversal also implies giving experience and perception back its role in education. We could even say that our approach to agroecology transcends most current science education.

Phenomenology is relevant to inform new perspectives on teaching and learning as it is concerned with our being and acting in the world as prior to knowledge about the world. According to the French phenomenologist Maurice Merleau-Ponty (1992), our consciousness is not in the first place a matter of "I think" but of "I can." Merleau-Ponty's emphasis on embodied consciousness and the primacy of "I can" has led to a more explicit focus on concrete acting in concrete settings as primary to cognition and theorizing about those settings. As such, an action orientation in agroecology education and research is not an elective "add-on" issue, but is in the core of the conceptual foundation of our M.Sc. program.

The phenomenological perspective elaborated here is in accord with the works of John Dewey on learning and experience. Dewey urges educators to integrate students' prior knowledge with new knowledge through reflective experience. The relationship between experience knowledge and interaction with nature is developed in-depth in his *Art as Experience* (Dewey 2005). Here, aesthetic experience is regarded as "the result, the sign, and the reward of that interaction of organism and environment which, when it is carried to the full, is a transformation of interaction into participation and communication" (Dewey 2005, p. 22). He also goes further by claiming that this form of experience integrates person and environment. In the true experience, the viewer and the viewed are one. This concept of the *aesthetic* experience is related to phenomenology's emphasis on rich sense experiences. The word *aesthetics* is derived from the Greek words *aisthetikos*, "sensitive, perceptive" and *aisthanesthai* which means "to perceive (by the senses or by the mind), to feel." Etymologically speaking, an aesthetic experience is a precognitive, sensuous experience, an experience opened up for through sensuous perception (Østergaard 2015). As Dewey emphasizes, in the aesthetic experience, there is no distinction of self and object, "since it is esthetic in the degree in which organism and environment cooperate to institute an experience in which the two are so fully integrated that each disappears" (Dewey 2005, p. 259).

The phenomenological approach to agroecology education is related to other similar educational approaches. The introduction and acceptance of constructivism in education philosophy have provided a major shift of attention for the teacher, from the knowledge content and the task of transmitting this to the students, to a focus on the learning processes and "knowledge construction" in the students themselves. One branch of constructivism, sociocultural learning, argues that knowledge construction is inextricably connected to its cultural context (Cobern and Aikenhead 1998). However, whereas constructivism focuses on cognition and construction of knowledge, phenomenology more strongly emphasizes the precognitive phase, including the roles of sensing and feeling as different from purely conceptual cognition. The phenomenologist agrees with the constructivist that knowledge is created by the learning person and that we need to consider learning processes in cultural contexts. However, phenomenology tries to balance the predominance of abstract conceptual explanations by intentionally connecting theoretical knowledge to preconceptual experience in the world as the basis for cognitive understanding.

Another pedagogical direction of relevance to phenomenon-oriented science education is context-based learning, with its origin in the early 1980s. This approach to teaching is characterized by using a specific context as a starting point for developing understanding of scientific ideas (Bennett et al. 2005, p. 1522). Context-based learning is similar to the phenomenological approach as they both intend to bridge the gap between lifeworld experiences and scientific concepts. However, they differ substantially on one point, as phenomenon-oriented science teachers tend to regard lifeworld phenomena as the primary basis, and not mere illustrations, of scientific knowledge. Thus, in phenomenology one does not search for the suitable context for promoting students' understanding of specific, predefined science contents. Rather one seeks to develop students' ability to unfold and act in relation to lifeworld phenomena as primary sources to understand the theory connected to those phenomena.

A phenomenological approach to teaching and learning agroecology has several practical pedagogical implications. These include both tools for teaching and learning and the training of specific competencies needed by agroecology graduates (Lieblein et al. 2012). A valuable tool for phenomenon-oriented learning, as conceptualized and applied in the Norwegian agroecology program, is an open-ended case strategy for study of farms and communities (Francis et al. 2009) as described earlier. Such a strategy implies a double challenge. First, there is a need to return to the primacy of sense experiences as a necessary foundation for scientific learning, not merely a nice introduction to concept learning. Second, it is necessary to cultivate the skill of phenomenon unfolding. This includes practicing to be wide-eyed learners and approach learning in the field without preconceptions and with open curiosity about everything. In this process, the problem of "not enough openness" occurs when insufficient attention is invested on planning and ensuring the open inquiry phase. For agroecology students, the challenge is expressed both mentally and practically, when they give too little attention to an initial exploration of the case. If the students act as if they know what is important in a case, they will not be prepared to observe what is there, as strikingly evident from the story in Section 5.1.1 The experience itself and others involved in the process will be less likely to teach students what they need to know—and the complex picture the students are trying to construct will be correspondingly less complete. This problem should be discussed openly with students, emphasizing that a truly open-ended case study depends on an unbiased attitude toward inquiry. Students must be encouraged to immerse in unprejudiced experiences, which might serve as entrances into conceptual understanding (Østergaard 2015). At the same time, these experiences form the content of students' reflection. In the open-ended case study they have—at least to some degree—shared *common* experiences that are bases for common reflection sessions.

The shift to phenomenology in agroecology education also has important implications for the notion of transdisciplinarity. The shift from theory to our everyday knowing, acting, and experiencing as being of prime importance demands multiple points of view. The formation of disciplines that need to be linked somehow arises from a division of lifeworld into academic departments, which is a dominant organizational pattern in universities that should be reconsidered. Further, the lifeworld as the ontological foundation for scientific knowledge is "pre-scientific"—it is *not* divided into disciplines and thus, in a certain sense, *predisciplinary*. The direct experience comes before the (academic) division into disciplines. With a foundation in phenomenology, we argue that lived experience should provide the starting point for exploring the spaces between the theoretical disciplines.

In Heidegger's phenomenology, particular emphasis is placed on "the nonrepresentational aspect of human beings' involvement with the world" (Donnelly 1999, p. 934). These aspects relate to our involvement in and with the world that have not yet become representations. In the abovementioned open-ended case strategy, the students are encouraged to explore the interface between phenomena and the existing knowledge or conceptions about them. In this sense, learning has two dimensions: the *outer* dimension is connected to learning about the agricultural system's properties and aspects, whereas the *inner* dimension is connected to learning about oneself. It is our task as educators to design a setting that allows for students to have a grounded experience on which they can both build and reflect. This must start by facilitating the students to encounter with lifeworld phenomena as they are *before* they are turned into representations and examples of scientific, agroecological theory. This is not so much a question of choosing the right phenomena as allowing the chosen phenomena to speak on their own terms. Students can only conceptually grasp knowledge about lifeworld phenomena, whether this knowledge is allowed to be grounded in experience (Østergaard 2015). The inner and outer dimensions of learning could be described as a "dual learning ladder" (Lieblein et al. 2007).

5.3 KEY AGROECOLOGICAL COMPETENCIES

To become agroecologists, students need to acquire and practice certain key competencies that will be essential through their academic and field studies, thesis research, and subsequent activities in future professional positions. We consider these competencies to include skills in *observation*, in *participation*, in *dialoguing* with stakeholders and team colleagues, and in *visioning* possible future scenarios. Imbedded in each of these skills is an ability to *reflect* on one's own experience as well as the process of the team.

5.3.1 Becoming an Agroecologist

Due to the program's strong emphasis on students *becoming agroecologists*, not on merely understanding the concepts of agroecology (Østergaard et al. 2010), the question of skill training is crucial. Based on our own experiences and research (Lieblein et al. 2001) and supported by the diverse contributions within phenomenology, we have become aware of specific skills to be practiced when the student goal is to become an agroecologist. In the following sections, we enumerate these skills, illustrated by statements from agroecology students that were written in their reflection documents, and, which play a central role in the full-semester course in agroecology at NMBU.

5.3.2 Skills in Observation

One of the first class exercises we had this semester was a transect walk. We were paired with another student and sent for a 45-minute walk in silence to observe our surroundings. I found the exercise refreshing for numerous reasons. Sometimes it is nice to not have the pressure to make conversation with someone you do not know well and I appreciated the fact that I could take this time to focus on the new environment I was in .... I also think this activity helped me to be more aware of what it means to be a "good observer." This exercise of quiet observation was meant to encourage me to suspend judgment, to let myself be a sponge to understand the situation, and observe freely. If as a researcher, you go into situations with preexisting notions it makes it more difficult to see the system for what it really is, and certain qualities or elements may be left unnoticed (Karen from USA 2012).

Based on Husserl's phenomenology, we emphasize pure, unbiased skills of *observation*. This predisciplinary, preconceptual, nonjudgmental approach is important to allow for a rich, aesthetic experience. Suspension of judgment is further an important prerequisite for being able to deal with the whole of a situation, and not just some predefined part. Valuing the role of observation as an important source for learning and action, and not just as one that provides an illustration of what is already known, is another prerequisite for developing this skill. The skills of observation and reflection are interconnected and often difficult to separate. Through careful and precise observations the natural, agricultural, and cultural phenomena are unfolded in sense-perceptual experience. This is in accordance with Husserl's demand of letting phenomena "speak" on their own terms by, as far as possible, bracketing what one already believes or knows about them. The ability to reflectively observe is important in agroecology education since the students tend to regard scientific knowledge as something "given" and often believe that their task, as agroecologists in the field, is "only" to transmit this knowledge to the stakeholders. This kind of preunderstanding may hinder the process and potential of a phenomenon to experientially unfold itself, as happened in the student group work described in Section 5.1.1.

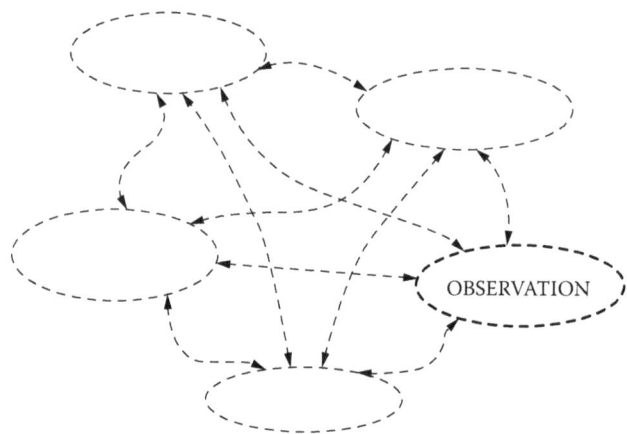

5.3.3 Skills in Reflection

When I eventually gathered my courage and decided to confront it [the challenge], it took me four hours in a row of reflecting, trying to apply it to my past experience, and making my own diagrams to finally feel like I had understood it. Now that I am familiar with it, I don't quite understand what was so complicated for me to grasp. However, I don't regret that it has been [challenging], for the process of trying to understand was extremely exciting and interesting, and gave material for reflection .... I have much left on my "reflection list" that I haven't included in here, finding it difficult to articulate it with the rest, for the moment. I am also conscious that I haven't talked about what I had learned on the content in the field. This is because I haven't yet had time to assimilate it [yet information] is still there, up in the air, disparate bits of information waiting to be processed. But all the reflection can't be done at once, it takes time, and I have to accept that. The process of writing this document has been like looking at myself in a mirror, scrutinizing every inch of myself. If you have tried it, when you spend a long time looking at yourself in a mirror, you end up not recognizing yourself anymore (same process as when you repeat a word until you don't know what it means anymore). This is encouraging me to stop here. The perspective of continuing the course now that this process has been fully activated is nonetheless very exciting, and I am looking forward to it (Natalie from France 2011).

This semester has been my first experience using this type of deep thinking and contemplation about the experiences I have had and I found it a challenge to do well. It is easy for me to just recall a situation, but much more difficult to contemplate how it made me feel, what I had learned from it, and how I could connect it to my existing knowledge and experiences. I have to say that in retrospect reflecting felt somewhat laborious at the beginning of the semester … always thinking! But of course I very much recognize the value in structured thought and now appreciate this concept and believe it is actually essential to learning. Reflection of this semester has taught me to look and think about situations from all scales and points of view, to look and understand without judgment, or to at least recognize my existing biases and how that may affect my decisions. I have found conscious observation and reflection also help me to relate situations by focusing on how I see it (my biases, my feelings, my observations), but this also helps me to remember the situation more clearly and focus on the information I find most pertinent. This in turn then allows me to apply my existing experiences and knowledge to my future experiences (Katherine from USA 2012).

From Heidegger's phenomenology and his explicit focus on our already being in the world and our preunderstanding as a prerequisite/condition for understanding and reflection, we draw the skill of *reflection*. In agroecology, reflection is characterized by a "Janus-quality" (after the Greek and Roman god, Janus), with one face looking outward into the world of food, agriculture, and the environment, and the other face looking in the opposite direction, into the inner world of the student. The challenge for the student is to value the importance of both perspectives and to cultivate the links between the two. Our task as educators is to provide a safe and encouraging learning environment, where students can explore ideas and learn to link their prior experience with new knowledge and skills, and to combine these into a capacity for visioning a desirable future.

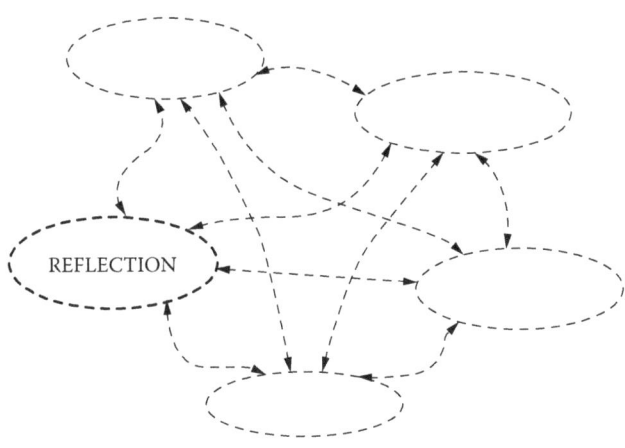

5.3.4 Skills in Participation

In the case of our work with Røis farm, we just began the participatory action research process, so I am doubtful of sustainable results. The best way to judge such a project from the farm point of view would be to observe changes made there. I do not think our moments of mutual liberation were significant or sustained enough to affect such change. Time will tell. This may be of secondary importance for students who can carry experiences and reflections forward to future work with agroecosystems, but it is of dubious value to the individual farmers, who come to be seen as resources to be used for our learning purposes …. I am worried that too much thinking about the complexity of nature takes us away from the experience of being immersed in the whole world. I think of this as the explosion of the mind and the death of the body (Dennis from Canada 2004).

We often tend to regard participation and action as being outside of the academic realm. From Merleau-Ponty's phenomenology with its focus on bodily lifeworld presence and acting as foundation for conceptions, we derive the skills of *participation* and involvement. The recognition of the value of participation and involvement for the learning process is a vital part of developing this skill.

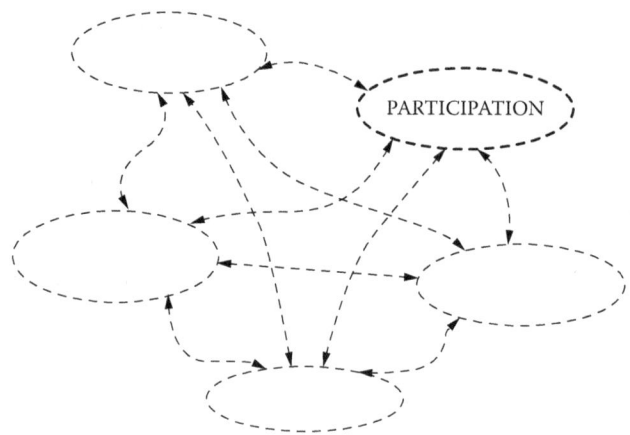

5.3.5 Skills in Dialoging

In some instances, I have learned to ease the chatter by taking a few minutes beforehand to focus on achieving mental clarity, much in the same way that Marjorie Parker describes the practice of listening. One of the primary guidelines for dialogue that she identified during her guest lecture was listening without thinking of a response (Parker 2013). I would practice clearing my head, taking deep breaths, and focusing only on the first three words I was planning to say, trusting that the next three would come at the right time. Also as Parker suggested, it takes a shift in mind-set from having to feel competent to allowing oneself to become vulnerable .... Unfortunately, our attention to this method of communication waned as the workload of the semester increased. It would have been good to start earlier, and we could have developed a system for incorporating it into our work, in order to make sure it was used. I found that as the semester progressed, and stress increased, some of the values we held initially were no longer so esteemed. Taking time for silence and dialogue becomes less appealing when a deadline looms. But to me, this is the crucial moment to apply the knowledge we have been acquiring. In order to reinforce your grasp of something—to make it habitual—you must attempt to apply it in novel situations (Alice from USA 2013).

We are most often trained to communicate in a polarized manner, in a debate or discussion format, where the main goal is to win, to show that your thinking and arguments are better than those of your opponent. Dialogue-based communication is found in the other end of the communication continuum. Here the aim is to explore a topic together and to create a space for collective learning. The described shift in mind-set represents an important prerequisite for developing the skill of dialogue. And, as for all skills, it needs to be practiced, not just talked about.

To communicate in a dialogic manner implies an ability to actively listen both to fellow students, to teachers, and to people students meet in the case studies, as well as the ability to express one's own experience and preunderstanding without forcing them on others. As teachers guiding the students' exploration, we try to be sensitive to their ways of expressing their experience and to ask them questions that can lead them on to new insights. It implies an open attitude towards seeing and promoting the students' activity in exploring phenomena.

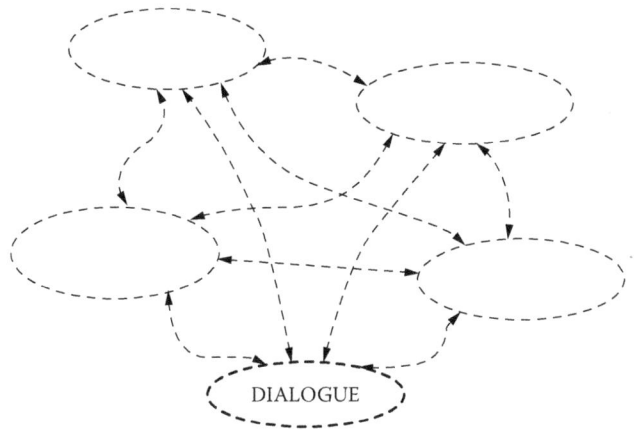

5.3.6 Skills in Visioning

Agroecology seems fairly future oriented. More than likely, our careers as agroecologists will be future oriented as well (Marcus from USA 2005).

As a reoccurring theme throughout this paper I chose to emphasize silence and the quieting of one's mind. I believe that before one can truly learn about her/himself, and grow as a person, one must be quiet and silent enough to allow time for reflection and self-exploration. Also, silence is an extremely important part in the visioning process to allow the mind to experience and test new ideas without the overpowering need to communicate these verbally or on paper. We must be quiet in order for the mind and the soul to reconnect and reflect on what it has experienced (Dennis from Canada 2005).

The visioning seminar was the main factor making PAE 303 very different from all the previous courses I have taken. Never before had I approached a situation from the perspective of what I wanted to see in the future. All other work had taken into account the current situation and then attempted to solve the problems to make the system better. The visioning seminar was particularly appropriate for a situation, which is hard to divide into its smaller parts. The emergent properties of a food system are simply too complex to reduce into its individual parts. Visioning allowed us to look at the whole system without becoming bogged down in unimportant details (Michael from USA 2005).

It is hard to accurately describe how we worked because of the visionary thinking seminar. Inspired is certainly not the right word, maybe directed is better. Before the visionary thinking seminar, we had little direction. We felt like we were treading water, but not getting anywhere with our work. While we were far from strongly confident after the seminar, we felt we had much better material to work with afterwards (Roger from USA 2005).

As a part of our work in Trondheim we worked with visioning as a tool. Through our work with visioning and sensing, the strength of the feedback received from the recipients of our work with the vision made me think. On one side we felt much empowered to work with visioning—the liberating and creative feeling of not being stuck in the web of complex reality. On the other we felt the positive feedback from those who listened to the results of our work. It has made me think a lot about how visioning can be a way to empower and engage people in change. To work with the task to bring people together and "let them loose" on their ideas for the future—seems like a fabulous way to bring ownership and desire to change. This will be one of the things to bring on into my future studies and work (Erik from Iceland 2012).

Visioning as a tool was completely new to me. In the beginning I felt quite critical about this. It was nice to use in a class of open-minded students, but it felt quite strange to use it with people who had never heard of it before. For me it helped not to think about the whole process too much, but to focus on the goal and the benefits. And it is a brilliant way to create a goal for yourself or a shared one together. Not emphasizing current problems or limitations in your planned outcome really helps to get to the core and to be creative. You don't want to solve one or two problems, you want the whole situation to move to a higher level and a better situation. I think this makes it an ideal tool to use in messy situations (Annika from Holland 2012).

Features of the past do not necessarily contain what is needed to deal with the challenges of the present and the future. It is therefore not sufficient to look at what we did last week or last year to find the solutions for what needs to be done tomorrow. Again we need to bring in a "Janus-approach" of not only looking into the past, but also looking into the future. We then open up for a fundamentally different kind of learning, a learning from the envisioned future (Scharmer and Kaufer 2013). This skill is perhaps the most important to develop in becoming an agroecologist.

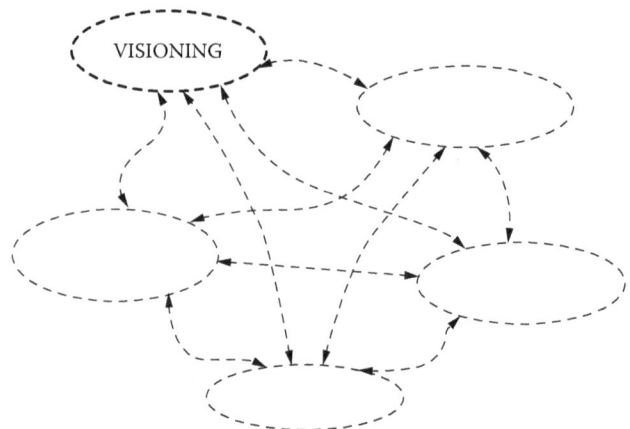

5.4 OPEN-ENDED LEARNING: AGROECOLOGY M.Sc. IN NORWAY

One practical example of how an educational program can help students attain these key competencies is the Agroecology M.Sc. program at the NMBU in Ås, Norway. As a foundation for this program, from 1995 to 1997, we hosted 3-week-long doctoral courses in Norway that focused on agroecology in (1) integrated farming practices, (2) holistic and integrated farming systems, and (3) farming and food systems (Lieblein et al. 2000). This publication summarized a dialogue and reflection over three days in January 1999 by a group of nine instructors and nine former students who had participated in the doctoral courses. Each person in the group was able to put the educational experiences into the current context of their continuing studies and teaching. Although some lectures were included in the doctoral short courses for the introduction of participatory study methods, preparation for farmer interviews (Østergaard et al. 2013), and dealing with qualitative data, we observed that the most effective learning took place while students were interacting with farmers and other stakeholders in the field and community.

From the doctoral short courses emerged a plan for the 2-year M.Sc. program, and an early step was to articulate clear general learning goals, as well as goals specific to knowledge, skills, and attitudes. These have been modified as we observe the students immersed in the learning process each year and reflect on their feedback through each semester. Goals include the following:

General learning goals we intend for students to attain:

- Ability to handle complexity and change
- Ability to link theory to real-life situations
- Communication and facilitation capacities
- Capacity for autonomous and lifelong learning

Knowledge goals:

- Knowledge of farming and food systems
- Agroecosystem/food system structure and functioning
- Methods for dealing with complex issues in food systems
- Systems analysis and assessment of sustainability
- Specific features of ecological agriculture/organic farming

Skills goals:

- Action competence: how to manage complexity and change
- Communication to bridge the gap between knowing and doing
- Team work and people management
- Critical thinking and problem evaluation
- Autonomous, lifelong learning by learning how to learn
- Tools for cooperative visioning toward a desirable future

Attitude goals:

- Clarify and incorporate ethical, personal, and cultural values
- Encourage open-mindedness, determination, and enthusiasm
- Model and reward inclusiveness, gender equity, and tolerance
- Stimulate concern for relevance and responsible action

We recognize that many of these goals are beyond the scope of most university courses, which focus primarily on knowledge and skills. In organizing the agroecology course and choosing learning methods, we strive to help students look beyond "what IS today" and learn to envision "what COULD BE a more desirable future." The goal is not to impose our own ethics or moral imperatives on others, but as instructors to each be transparent about our personal worldviews and to model how we apply these in a learning process that will lead to responsible action. It is up to each student to clarify their own values and to discover how these are intimately linked to decisions in career, family, and communication with others on the job and in personal life. One of our goals in the learning process is to put these issues on the table, to open up discussions, and to provide both a safe environment for students to grow as well as confront challenges that are similar to those they will face later in life. We have found that in agroecology this involves immersing in the context of the farm and the food system, communicating with empathy with others to gain better understanding, and emerging from the educational process as confident and compassionate learners who are ready to deal with the future. Within this philosophy and educational context, here are some of the learning methods that we have found valuable.

Our learning strategy builds on the open-ended case approach (Francis et al. 2009), a method where students identify and interview key farmers and other professionals in the food system. This includes people in processing, marketing (both wholesale and retail), government agencies involved in food and environment, nonprofit groups, and consumers. Learning is enhanced because students realize that the answers to current challenges are not known to instructors or to clients, unlike the typical decision case study where the challenge is to be clever enough to get a good grade by figuring out what the teachers already know. We have found this open-ended approach both stimulating and challenging, as in most classes students are told exactly what to do in courses and what will be on the exam. In agroecology, they are asked to be creative and seek their own path, using tools that we provide, as well as individual and team initiative.

Conventional case studies in agriculture have been reviewed by Simmons et al. (1992) and assembled by the ASA (2006). A comparison of the conventional decision case study and the open-ended case we use is shown in Table 5.1 (from Francis et al. 2009). Although we build on that

Table 5.1   Summary Comparison of Conventional Decision Case Learning and Open-Ended Learning Strategies Used in Courses in the United States and Nordic Agroecology Programs

	Conventional Decision Case Method Learning	Open-Ended Cases for Learning in Agroecology Courses
Goal	Develop solutions from a predetermined situation.	Envision potential solutions to real-world situations.
Process	Follow a series of defined steps to uncover known solution.	Follow a discovery process to envision alternatives.
Information	Provided by instructors in a logical/sequential manner.	Students seek out needed information from key clients in field/community.
End product	Rational solution that may correspond to actual situation.	Multiple possible future scenarios and their potential impacts.
Type of learning	Closed learning cycle to seek what is known to instructor.	Open colearning by students and instructor to explore unknown.
Evaluation of learning	How closely does solution relate to the "real answer?"	How creative are future scenarios and evaluations of potential impacts?
Ownership of process	Instructors know the answers and determine student success.	Students own the learning and set their own criteria for success.
Learning culture	Conventional search by students to find fixed answers.	Open-ended search to develop future options and predict impacts.
Institutional setting	Stimulus from teacher and response from students.	Multiple sources of stimulation, continuous interaction toward goals.
Role of instructor	Design the logical steps to reach the known (right) answers.	Open a learning landscape for creative discovery of alternatives.
Role of student	Active learner, engaged in a comfortable process.	Autonomous learners find discomfort in a challenging, open-ended situation.
Responsibility for learning	Starts with instructor, passed to students in case study.	Primarily rests with students, who are free to pursue different options.
Applicable mainly to	Past and present situations that are known.	Future situations that are complex and unknown.
Appropriate mostly for	Lower hierarchical system levels.	Higher order hierarchical system levels.
Most useful for	Simple, well-defined systems and situations.	Complex, ill-defined systems and situations.
Answers and solutions	Mostly fixed and predetermined by instructor.	Mostly open and dependent on multiple factors and context.
Major sources of inspiration	Hard facts and discrete systems that are well known.	Hard facts and social methods, plus human judgment and creativity.

Source: Francis, C., J. King, G. Lieblein, T.A. Breland, L. Salomonsson, N. Sriskandarajah, et al. "Open-ended cases in agroecology: Farming and food systems in the Nordic region and the U.S. Midwest." *Journal of Agricultural Education and Extension* 15 no. 4 (2009): 385–400. With permission.

foundation plus case studies in business and law schools, the approach to open cases is sufficiently different to require more explanation. It is useful to recount our instructor experiences and reflective learning that led us to this method.

Our choice of learning methods builds on experience from prior courses in Norway and elsewhere, as described earlier. The immediacy and relevance of shared experiences with clients have been apparent in the student team dialogues, oral and written reports on the farm visits, and individual student learner documents at the end of short courses and the autumn semester when students reflect on the learning process. We are told consistently that participation in farm tasks and interviews with farmers are one practical dimension that is essential to learning.

As the process developed in M.Sc. agroecology classes from 2000 to 2008, we studied the farming system in the first half of the semester and then the food system in communities in the second half. Since 2009, we have integrated the two activities so that students stay in rural communities for 1 week, twice during the semester, living in the context of the farm and community, interacting with stakeholders, and conducting semi-structured interviews along with making observations of the present situation. Every evening and in travel time to the locations, student teams engage in a continuous personal as well as team reflection on what is observed and discussed, an emergent property of the organization of these field components of learning.

Farmers are preselected for each team, and they have been briefed by the instructors on the goals of the course and what the students are likely to ask in the interviews. Students set up their own schedules for farm interviews and transect walks to absorb as much of the farm context as possible, to make their time with the farmer efficient, and to learn about the farm's resources and current operations. It is essential that they learn about the farmer's philosophy and long-term goals for the farm, if they intend to analyze the present and outline potential scenarios for the future.

To initiate study of the community food system, we provide the name of a key client who is often the agricultural officer for the county [*fylke*], and a person well acquainted with both public and private sector involvement in the food system. This key person helps to identify other players in the system who would be useful to interview. Depending on the community, the most important and involved people often include those in food procurement for schools, municipal canteens, hospitals, military posts, and care homes, as well as those in the private sector such as wholesale and retail food businesses. With information on production and consumption in a community, students diagram the system with a rich picture and begin to identify key elements, interactions, strengths, and weaknesses of the system relative to the goals of the community. This learning process leads to design of potential future food system elements and strategies that will help stakeholders meet their long-term goals. We call these the team's "future-oriented scenarios" and they are presented back to the clients for consideration.

We urge students not to be prescriptive in their observations or recommendations, nor to jump to quick conclusions about the best possible solutions to what they perceive initially as key challenges, but to continue to observe, evaluate, reflect, and consider possible options from the perspectives of the stakeholders. Important tools and skills to learn that can help students build on their observations and reflections are those embedded in the process of visioning future systems (Lieblein et al. 2011). This process to some extent requires that students, clients, and instructors suspend current reality, and then look into the future to imagine a desired and improved situation that will help the farmers or food system participants realize their goals. This project activity builds on prior student team experiences such as the transect walks across farms or through communities, semistructured interviews, mind, and concept mapping, and use of metaphors to discuss and communicate ideas. Analyses include the application of strengths, weaknesses, opportunities, threats (SWOT) and force field analyses, use of various development models, and sustainability analyses using multiple criteria. A number of these learning methods have been published recently in the *North American Colleges and Teachers of Agriculture Journal* (*NACTA Journal*) as "Teaching Tips" (www.nactateachers.org/journal.html). We have found these practical learning methods to be effective with groups of students who come from different cultures, have various prior academic study backgrounds, and represent a range of past experiences in agriculture and food systems.

Important to our continued improvement of the course and study program has been evaluation of the agroecology M.Sc. by peers, and especially by students. Peer evaluation has been intense over more than a decade as colleagues have reviewed our submitted journal articles and book chapters and posed questions at educational conferences where we have presented summaries of the program and teaching methods. Student evaluation is based on our reading and reflecting on their student team documents, oral presentations of team results and scenarios, and individual learner documents of each semester. The program continues to attract a full cohort of 25 students each autumn, and in 2014 there are more than twice the number of applicants than we can accommodate. Graduates of the program provide valuable feedback about how they are using the methods and tools from the course in their further study or job situations. The program has been formally recognized by peers and institutions with the Nordic Veterinary and Agriculture University (NOVA) award for education in 2005 and the NMBU all-university award for excellence in teaching in 2012. We continue to study the student learning process, improve the program, and ponder the elements necessary to make a conversion from conventional teaching to involved and reflective learning in other programs. And, of course, we continue to learn.

5.5 CONVERSION FROM CONVENTIONAL TEACHING TO AGROECOLOGICAL LEARNING

There is ample information in the literature related to the need for experiential learning as described by Dewey (2005) and reviewed by Moncure and Francis (2011). In fact the concepts are scarcely new, as Gentry (1990) cites quotes from Sophocles in 400 BC that "one must learn by doing the thing, for though you think you know it you have no certainty, until you try"; he also quotes George Santayana: "The great difficulty of education is to get experience out of ideas." Yet there is scant reference to what qualities are needed in educators who are compelled to teach in an experiential learning situation such as what we have established in the agroecology learning landscapes. In the following section, we explore what we consider important for instructors who want to transition to different ways of teaching, how we propose that instructors learn and practice these competencies, and what institutional factors can promote or constrain the process.

5.5.1 What Is Needed for Transition to Occur?

We have observed the challenges of conversion from one type of farming system to another. For example, conventional monoculture and chemical-intensive agriculture to an organic system that depends on renewable and internal resources or conversion from increased dependence on global food chains to greater emphasis on local foods. The first and often most difficult step is the mental conversion, or the raising of *awareness* that something needs to change based on new knowledge about resource scarcity, short-term success, but long-term futility of present conventional farming or food strategies, and/or lack of equity in food access in the current food system, among others. This awareness could be a result of new research results related to climate change, to growing concerns about broad impacts of chemical pesticides, or to more evidence of impacts of diet on personal health and health care costs to society. These represent only a small sample of the multiplicity of drivers of change and awareness that impact us every day in a world "grown smaller" by the instant communication of the Internet and other modern information technology methods. This is an information domain, an awareness of "what is happening."

Once there is an awareness of the need to explore creative alternatives, and a concern of how this could be incorporated into education, a vital step is to consider a multiplicity of opinions and determine one's own *attitude* toward needed change. If we accept the status quo and see the future as only an extrapolation of present trends, the easy route for many is to continue to immerse in current systems as they are, assume that technology will continue to solve any emerging challenges, and place confidence in corporations and governments whose vested interest in present systems compels them to assure us that things are fine, and whose message is frequently that "we only need to make small adjustments to do well in the future." There is a certain level of comfort in the status quo, and many people are reluctant to change: "If the system is working, why change it?" In contrast, if one firmly believes that there are impacts and consequences of the current farming and food system that are not sustainable, that do not have the resilience to withstand shocks to the system such as severe weather or political events, or that are not equitable in terms of food availability for all people, this reality can generate a different attitude toward the future. Many of us in agroecology have an awareness that future systems will be quite different from those of today, that change has to come from within and from the grassroots, and that we can design a positive and sustainable future food system rather than letting the *status quo* continue. We need to encourage students to explore their own attitudes toward the compelling issues of our time and then to move on to a next step of action. This arena could be called "so what?" and, if there are needed changes, are we committed to make them happen?

With an awareness of the current reality and the constraints that will confront us in future farming and food systems, and an attitude that drives us to seek positive change and to encourage students to carefully examine the issues, we move to education for promoting *responsible action* as a vital step in conversion. Perhaps, the focus on action is what distinguishes the agroecology educational program from others in the sciences—biophysical, social, or economic—that concentrate on knowledge and skills. We consider the future-oriented focus and call to action so important that this becomes integral to all learning activities through the autumn semester, in subsequent courses of semesters two and three, and in the choice of topic and conduct of research. Although we focus on phenomenology, validating, and building on previous student experience, and knowledge learned from stakeholders, it is important to emphasize that we rely on the rigors of science in choosing people and methods for interviews, in conducting analyses, and in reporting results in the M.Sc. thesis. This is a science-based education program, and not only a training ground for later activities. Hence, the instructors consider themselves scientists and educators as well as advocates and activists who promote learning for later responsible action. This may not be apparent to colleagues in our mainstream university departments who see the focus on critique of current farming systems and concern with contemporary social issues as an arena commonly populated by nonprofit, activist, or political organizations. We are concerned about action, but insist that this must be based on solid science and careful consideration of a range of potential alternatives. For this reason, we use the open-ended case approach (Francis et al. 2009), which encourages students to work closely with stakeholders and instructors to explore a number of relevant options in their analysis of farming and food systems; also it is one that will help our clients meet their long-term objectives. Next we examine the competencies needed by educators to catalyze this type of educational process and how we could promote change in educational focus.

5.5.2 How Do We Promote New Educator Competencies?

If this educational agenda has currency in the university community and if different knowledge, skills, and attitudes are crucial for educators to fill new roles as instructors, we need to first enumerate the competencies needed and then the motivations that are important to make change happen. One definition of agroecology is *the ecology of food systems* (Francis et al. 2003). In contrast to narrowly focused, discipline constrained studies of components of the farming and the food systems, agroecology promotes a broad educational approach that transcends disciplines and seeks to build understanding of the interfaces between disciplines. These could be called "educational ecotones" and represent the important interactions among elements of systems. These dynamic interactions result in emergent properties—unexpected outcomes—that often are not predictable based on understanding of individual elements and their behaviors. To understand system properties and behaviors, including both structure and functions, requires both broad knowledge and sets of skills that often go beyond those found in any single discipline. In addition, we find from experience that exploring attitudes and a willingness to validate and affirm student's prior experiences are crucial qualities needed in agroecology instructors. For an instructor, this means putting aside one's own ego and recognizing that it takes a special attitude to not always assume the role of expert, even though we need to feel fully competent and prepared to answer technical questions, as well as to guide and catalyze conversations in the classroom and in the field that continually focus on students and learning, rather than on instructors and teaching. In fact, we have found this to be a powerful way of teaching, although we become vulnerable and may feel that we are at times losing control of the agenda when students truly become responsible for their own learning. Such an approach to education could be called "a pedagogy of no mercy," as we welcome challenges and expect students to clearly articulate their opinions based on prior experience even if these do not agree with our own.

It is obvious that instructors need to have a broad understanding of systems approaches and how components fit together to make systems function well. Often agroecology instructors come from a wide range of different disciplines—in Norway our faculty team includes a soil microbiologist, an organic farming and weed management specialist, a plant breeder, a basic plant physiologist, and a cultural anthropologist. Two of the five are women, and four have extensive prior experience working and teaching in other countries, cultures, and languages. Although we come from disparate technical areas, it has become important to learn from each other, to respect and practice the methods used by people from other technical domains, and especially to elevate the importance of the social sciences, including capacity for communication, interest in the opinions and knowledge of others, and appreciation of the importance of inclusiveness in discussions and in student team activities. In fact, many of these qualities of team players and respect for others' opinions should be apparent to students as they observe the ways that our team of instructors interacts and learns from each other, and how we entertain and foster a healthy level of disagreement in discussions in class.

We have written extensively about this process since the M.Sc. degree program began in 2000 (see Lieblein et al. 2008, 2012, and prior articles), and there have been three new educational programs designed from the model in Norway and introduced in Sweden, Ethiopia, and Uganda. Instructors in these programs have benefited from participation in teaching workshops and annual meetings of instructors, where there has been candid sharing of experiences and learning from each other. It is likely that one of the best ways to learn how to be an effective instructor in an agroecology program using the open-ended case method is to observe and participate in teaching in one of the established university programs. This is the way that we as instructors have learned by doing, first, by reading about the methods used by other teachers and the history of experiential learning (Moncure and Francis 2011) and then by immersing ourselves in the educational milieu and avidly learning from students about what works best and what could be improved. A key source of feedback for us as teachers and educational guides is to listen carefully to the comments of students in the weekly review and reflection sessions, where we structure a time when people can relate about not only what they have learned but how the learning process has worked for them. The idea of learning by doing is equally important for teachers to appreciate and practice designing and implementing agroecology courses, as it is for students to learning about farming and food systems. How then can universities promote a change in the learning environment, including encouraging instructors to evaluate and adopt new methods of pedagogy?

5.5.3 How Do We Infuse New Educational Values in Institutions?

Teachers in agriculture, farming, and food systems respond to incentives in the reward systems of their educational institutions, and we are all sensitive to the opinions of colleagues in our departments. Thus it is important to consider the institutional framework and context of education where agroecology programs currently reside as well as those where they are likely to be introduced. What incentives are there for change, and what elements need to be in place for instructors to be motivated to change their own teaching styles and to encourage others in their universities? We have experienced lengthy discussions within our own group, lamented on the lack of acceptance of experiential methods by most of our friends and other instructors, and reflected on why this is so. Here, we speculate on the conditions that are necessary for change.

First, there must be a healthy respect for academic freedom and willingness on the part of administrators and teaching coordinators to allow space and even provide encouragement for new methods to be tested and adopted. Most universities have a forward-thinking mission statement in their promotional literature about the institution, but this may not be reflected in the practice of education by teachers. We recognize in ourselves many of the incentives that led us into education, including inspiring lecturers who served as role models at some point in our pursuit of academic degrees. To assume a similar role as effective lecturer was one of our own motivations to become university teachers. Once we have a responsibility for a given course and develop the teaching materials for that course, and we experience a degree of success as reflected by teaching evaluations by students and administrators, we may be reluctant to throw that preparation out and adopt a whole new method of teaching. Given the potentials of the Internet to provide ready sources of information, it is unlikely that many students will consider today's instructor as the sole or even principle authority in a given field. A student recently stated, "OK, prof, but if it is not on the Internet, for me it does not exist!" So much for late nights in the library, careful reading of lecture notes or even Power Points, or spontaneous small discussion groups of students who attempt to bring meaning to their lecture notes from class. In fact, one of the principle motivations to attend class may be to find out what will be on the exam. So at least they will come to class.

More importantly we find a need to build urgency for change by including thoughtful discussion with students about the major issues of the day such as climate change, fossil fuel, and water scarcity, land grabs by those with available financial resources, food safety and security, food sovereignty, and equity of access to food by all. These can all be infused into discussions of topics in agroecology, and they are enriched by topics that students volunteer as their major concerns. This is an open approach to learning and to structuring education. The evaluations we receive by students, the final reflection documents they prepare, and the applications of theory and methods that appear in their thesis research are all indications of success in the program. The autumn course in Norway continues to be fully subscribed each year, and most students continue for the full 2-year M.Sc. degree in agroecology.

One important dimension that has contributed in recognition of the program, and that provides an extension of the ideas to other educators, is our strong emphasis on evaluation and documentation of educational results (see Francis et al. 2011; Lieblein et al. 2012; Østergaard et al. 2010 for additional references). It is this publication record, in part, that has provided professional recognition for us as individuals and for the agroecology teaching team in our universities and allowed us to be accepted as productive members of the academy. There are many good teachers in our profession but many focus on students and improving their education and do not dedicate time to evaluation and writing up results. We consider all of this an essential part of university education and an important role of educators, and it has served us well to institutionalize courses in agroecology in our universities.

5.6 WIDER IMPLICATIONS FOR EDUCATION IN AGROECOLOGY AND OTHER FIELDS

Although our recent experience has been focused on courses in agroecology in Norway and the United States, with observations and evaluations of new programs in three other countries, we all have prior and current experiences in other courses, including organic farming, loss of farmland, agroecosystems analysis, international and internal land grabs, food systems, and food security. From this broad experience in teaching other topics, we are convinced that many of the lessons learned as instructors have a wide application in education related to practical and applied fields. Here are some of the ideas we are discussing in conversations related to future changes in agroecology courses, in research and teaching in other fields, and in shaping institutions to provide relevant education in the future.

5.6.1 Pragmatic and Radical Phenomenology

With a learning foundation in phenomenology, we aim at bringing students' learning processes as close as possible to lifeworld phenomena. In the concrete learning setting, however, phenomenology is practiced in two different manners: *pragmatic phenomenology* implies choosing phenomena to illustrate already existing knowledge. That is, the phenomena become examples of existing knowledge. In contrast to this, in *radical phenomenology* lifeworld phenomena are explored openly to create rich experiences as a basis for theoretical considerations. That is, existing knowledge and theories are examples of how lifeworld phenomena can be explained and understood. One might argue that phenomenology per definition is always radical because true understanding of theory presupposes an experiential lifeworld foundation. However, these two approaches to phenomenological learning in agroecology are complementary rather that mutually exclusive. The pragmatic phenomenology might very well be used as an entrance into radical phenomenology. In both approaches, the idea is to bridge the world of experiences and action and the theoretical, abstract world. For both students and for us as teachers, the radical phenomenology, because of its open-endedness, forms a much greater challenge compared to the pragmatic approach.

5.6.2 Creating Urgency

A review of definitions of agroecology as a science, a set of practices, and a movement (Wezel 2009) provides examples of how this term is used in different countries and by disparate groups to describe a wide range of activities related to research, application, and activism. It has not been uncommon for academic colleagues to challenge those of us in agroecology for including in our learning program an explicit goal of leading students to "responsible action"; they suggest that we are confusing or not drawing a line "between education and advocacy." If one insists that university education should consist of learning facts, knowledge, and skills then this could be a valid criticism. On the other hand, if we consider education as important in building awareness about the large challenges that face society, and essential to this awareness is a set of critical thinking skills to sort out priorities, then the approach we have chosen in agroecology is a relevant one. At the very least, we hope students will sense our feeling of urgency to solve some of the large and food-related issues of our time.

We do not see the role of instructors as authority figures who should impose their goals, ethical positions, or worldviews on students. However, it may be important to be transparent about such views and provide positive role models that reflect concern and compassion for others, strong advocacy for education, and for people to make up their own minds about how to become involved and knowledge and direction in addressing critical challenges. The drive toward developing critical thinking is embodied in the liberal arts tradition of academia, whereas action education is an essential component in the extension division of land grant universities in the United States and elsewhere when there is a mandate for outreach. We see agroecology as encompassing these two traditions using biophysical, economic, and social science methods to assess situations, and to consider multiple perspectives to make rational decisions on actions to improve farming and food systems.

5.6.3 Identifying Barriers to Change

Based on more than a decade of experience in our own universities and on numerous discussions with colleagues, we recognize a number of barriers to change in teaching methods and in the acceptance of agroecology as a component of the core course requirements for other fields of study. The general model of semester or quarter courses where students take several different subjects concurrently means that their time must be allocated efficiently among several courses. The types of field trips and weeklong projects for working and interviewing visits to farmers and communities are extremely difficult to schedule if there are other courses and lectures to attend. Our students are dedicated full time to agroecology during the autumn semester, making this scheduling much easier. Another curriculum management challenge relates to students from other major fields of study who would like to take agroecology. The requirements for specialized majors are already directed toward a list of specific courses that provide in-depth knowledge and skills in each area, and there is no room in the program for additional courses with a systems focus, which can be perceived to "dilute" the specialized training. If there is currency in holistic studies, then this must be appreciated as an essential dimension of study in other areas to have advisors recommend agroecology to their students.

These logistical and scheduling challenges may be minor compared to the philosophical changes that are needed to promote holistic and integrated systems approaches in more study programs in agriculture and food systems. We recognize the need for educational strategies that provide students with the methods and tools to deal with an increasingly complex world, one filled with "messy constraints" that defy solution when approached from a narrow perspective available in any single discipline. It is through the use and blending of methods from biophysical science with those from social and economic sciences that we can begin to make sense of the whole, to see challenges through the eyes of stakeholders, and to seek solutions that will be appropriate to the multidimensional and multigoal situations faced in the real world. It is this change in thinking that is essential to promote if we are to gain more of a foothold in mainstream academia.

5.7 CONCLUSIONS

Agroecology in our universities has been developed as an experiential education program that uses observation, participation, dialoging, and visioning as well as thoughtful and structured reflection at each step of the learning process. In addition to introducing a holistic systems approach to education, we distinguish our approach to agroecology from many discipline-specific courses and programs of study by an explicit goal of learning for responsible action. Over more than a decade of our own teaching and learning experiences, we have observed that action education, particularly in the field, has proven a successful strategy for students learning about farming and food systems, especially while embedded with stakeholders who are seeking solutions to their most pressing challenges. It is this broad and systemic focus that leads us to define agroecology as *the ecology of food systems* (Francis et al. 2003).

As described in detail earlier, we consider phenomenology as a key element in designing our methods of learning. We nurture open-minded students who are able to approach either the farming or the larger food system without preconceived solutions to major challenges and will "let the system speak for itself" and reveal elements and interactions that could be invisible to the learner who seeks quick answers and simple cause–effect relationships. This is one reason why we have used the open-ended case approach, where teams of students–instructors–clients explore the complexity of systems on the farm and in the community and seek a range of potential future scenarios that could be followed to improve those systems. The answers are not predetermined or known to the instructor, but the quest for solutions is a cooperative effort that more closely resembles the types of challenges that students will face later on the job.

Our own conversion from conventional lectures and field trip activities to one of systemic and holistic approaches to understand farming and food systems helps inform us about the process that will be needed for others to follow this path. A number of conditions appear to be important, including awareness that there are other methods to be tested, an attitude of openness to those possibilities, and specific action steps to introduce these into a course or curriculum. It is also useful if there is both administrative and collegial support for introducing and evaluating new approaches to teaching. These may reflect a change in attitude toward greater emphasis on values and ethical issues connected to learning to enrich the already robust educational focus on facts and skills. We identify a number of competencies we consider essential for instructors to use in a program of experiential learning: technical capabilities, openness to student opinions, willingness to explore innovative learning approaches, and empowering students to take responsibility for their education.

Finally, we use our own visioning skills to look into the future of education for responsible action, including an examination of radical phenomenology, where the actual situation in the field is given primacy over theories and knowledge in the classroom. The question of urgency is paramount in the design of agroecology courses, as we hope to instill in graduates the spirit and energy to make positive changes in the world. This includes an explicit search for both the barriers to change as well as those forces that will help to bring positive improvements in food systems. We take seriously the admonition in the education of "So what?" Will the world be a better place, will there be greater food sovereignty and less hunger, and will people's well-being improve as a result of our educational programs in agroecology? We hope that the answer is an unequivocal "yes."

REFERENCES

ASA. *Case Studies Published in Journal of Natural Resources and Life Sciences Education, 1992–2005.* Madison, WI: American Society of Agronomy, 2006.

Bennett, J., C. Gräsel, I. Parchmann, and D. Waddington. "Context-based and conventional approaches to teaching chemistry: Comparing teachers' views." *International Journal of Science Education* 27 (2005): 1521–1547.

Bruner, J. *Actual Minds, Possible Worlds.* Cambridge, MA: Harvard University Press, 1986.

Bruner, J. *Making Stories.* New York, NY: Farrar, Strauss, and Giroux, 2002.

Cobern, W.W. and G.S. Aikenhead. "Cultural aspects of learning science." In *International Handbook of Science Education.* B. Fraser and K.G. Tobin (Eds.). Dordrecht, The Netherlands: Kluwer Academic Publishers, 1998, 39–52.

Dahlin, B. "The primacy of cognition—Or of perception? A phenomenological critique of the theoretical bases of science education." *Science and Education* 10 (2001): 453–475.

Dahlin, B., E. Østergaard, and A. Hugo. "An argument for reversing the bases of science education—A phenomenological alternative to cognitionism." *Nordina* 5 (2009): 201–215.

Dewey, J. *Art as Experience.* London: Penguin Books, 2005.

Donnelly, J.F. "Schooling Heidegger: On being in teaching." *Teaching and Teacher Education* 15 (1999): 933–949.

Francis, C., G. Lieblein, S. Gliessman, T.A. Breland, N. Creamer, R. Harwood, et al. "Agroecology: The ecology of food systems." *Journal of Sustainable Agriculture* 22 (2003): 99–118.

Francis, C., J. King, G. Lieblein, T.A. Breland, L. Salomonsson, N. Sriskandarajah, et al. "Open-ended cases in agroecology: Farming and food systems in the Nordic region and the U.S. Midwest." *Journal of Agricultural Education and Extension* 15 no. 4 (2009): 385–400.

Francis, C., L. Salomonsson, G. Lieblein, and J. Helenius. "Serving multiple needs with rural landscapes and agricultural systems." In *Agroecosystems Analysis.* D. Rickerl and C. Francis (Eds.). Madison, WI: American Society of Agronomy, 2004.

Francis, C., S. Morse, T.A. Breland, and G. Lieblein. "Transect walks across farms and landscapes." *North American Colleges and Teachers of Agriculture Journal* 56 no. 1 (2012): 92–93.

Francis, C.A., N. Jordan, P. Porter, T.A. Breland, G. Lieblein, L. Salomonsson, et al. "Innovative education in agroecology: Experiential learning for a sustainable agriculture." *CRC Critical Reviews in Plant Science* 30 no. 1 and 2 (2011): 226–237.

Gentry, J.W. "What is experiential learning?" In *Guide to Business Gaming and Experiential Learning*. J.W. Gentry (Ed.). Dubuque, IA: Nicholls Publ. Co., 1990.

Harvey, C.W. *Husserl's Phenomenology and the Foundations of Natural Science*. Athens, GA: Ohio University Press, 1989.

Husserl, E. *The Crisis of the European Sciences and Transcendental Phenomenology*. Evanston, IL: Northwestern UP, 1970b (First published in 1936).

Husserl, E. *Logical Investigations*, vol. 1. London: Routledge, 1970a (First published in 1900/1901).

Kolb, D. *Experiential Learning: Experience as the Source of Learning and Development*. Englewood Cliffs, NJ: Prentice-Hall Publishing, 1984.

Lieblein, G., C. Francis, W. Barth-Eide, H. Torjusen, S. Solberg, L. Salomonsson, et al. "Future education in ecological agriculture and food systems: A student-faculty evaluation and planning process." *Journal of Sustainable Agriculture* 16 no. 4 (2000): 49–69.

Lieblein, G., C.A. Francis, and H. Torjusen. "Future interconnections among ecological farmers, processors, marketers, and consumers in Hedmark County, Norway: Creating shared vision." *Human Ecology Review* 8 no. 1 (2001): 60–71.

Lieblein, G., E. Østergaard, and C. Francis. "Becoming an agroecologist through action education." *International Journal of Agricultural Sustainability* 2 no. 3 (2004): 1–7.

Lieblein, G., T.A. Breland, C. Francis, and E. Østergaard. "Agroecology education: Action-oriented learning and research." *Journal of Agricultural Education and Extension* 18 no. 1 (2012): 27–40.

Lieblein, G., T.A. Breland, E. Østergaard, L. Salomonsson, and C. Francis. "Educational perspectives in agroecology: Steps on a dual learning ladder toward responsible action." *North American Colleges and Teachers of Agriculture Journal* 51 no. 1 (2007): 37–44.

Lieblein, G., T.A. Breland, L. Salomonsson, N. Sriskandarajah, and C. Francis. "Educating tomorrow's agents of change for sustainable food systems: Nordic agroecology MSc program." *Journal of Hunger and Environmental Nutrition* 3 no. 2 (2008): 309–327.

Lieblein, G., T.A. Breland, S. Morse, and C. Francis. "Visioning future scenarios." *North American Colleges and Teachers of Agriculture Journal* 55 no. 4 (2011):109–110.

Merleau-Ponty, M. *The Phenomenology of Perception*. London: Routledge, 1992.

Moncure, S. and C. Francis. "Foundations of experiential education for agroecology." *North American Colleges and Teachers of Agriculture Journal* 55 no. 3 (2011): 75–91.

Moran, D. *Introduction to Phenomenology*. London: Routledge, 2000.

Neuhauser, P.C. *Corporate Legends and Lore: The Power of Storytelling as a Management Tool*. New York, NY: McGraw-Hill, 1993.

Østergaard, E. "How can science education foster students' rooting?" *Cultural Studies of Science Education* (2015):10:515-525).

Østergaard, E., A. Hugo, and B. Dahlin. "From phenomenon to concept: Designing phenomenological science education." In V. Lamanauskas and G. Vaidogas (Eds.). In *proceedings from the 6th IOESTE Symposium for Central and Eastern Europe*, Siauliai, Lithuania, 2007, pp. 123–129.

Østergaard, E. and B. Dahlin. "Sound and sensibility. Science teacher students bridging phenomena and concepts" In *Proceedings from the 2009 NARST Annual International Conference*, April 17–21, 2009, Garden Grove, CA, p. 328.

Østergaard, E., C. Francis, and G. Lieblein. "Practicing and preparing for stakeholder interviews." *North American Colleges and Teachers of Agriculture Journal* 57 no. 1 (2013): 97–99.

Østergaard, E., G. Lieblein, T.A. Breland, and C. Francis. "Students learning agroecology: Phenomenon-based education for responsible action." *Journal of Agricultural Education and Extension (Wageningen)* 16 no. 1 (2010): 23–37.

Rickerl D. and C. Francis. "Multidimensional thinking: A prerequisite to agroecology." In *Agroecosystems Analysis*. D. Rickerl and C. Francis (Eds.). Madison, WI: American Society of Agronomy, 2004.

Rossiter, M. "Narrative and stories in adult teaching and learning." ERIC Digest, No. 241, EDO-CE-02-241. Columbus, OH: Educational Resources Information Center, 2002. http://calpro-online.org/eric/docs/dig241.pdf (accessed June 7, 2014).

Scharmer, C.O. and K. Kaufer. *Leading from the Emerging Future: From Ego-System to Eco-System Economics*. San Francisco, CA: Berret-Koehler Publishers, 2013.

Simmons, S.R., R.K. Crookston, and M.J. Stanford. "A case for case study." *Journal of Natural Resources and Life Sciences Education* 21 no. 2 (1992): 2–3.

Vaarst, M., D.K. Byarugaba, J. Nakavuma, and C. Laker. "Participatory livestock farmer training for improvement of animal health in rural and peri-urban smallholder dairy herds in Jinja, Uganda." *Tropical Animal Health and Production* 39 (2007a): 1–11.

Vaarst, M., T.B. Nissen, S. Østergaard, I.C. Klaas, T.W. Bennedsgaard, and J. Christensen. "Danish stable schools for experiential common learning in groups of organic dairy farmers." *Journal of Dairy Science* 90 (2007b): 2543–2554.

Wagenschein, M. (Ed.). *Kinder auf dem Wege zur Physik [Children on their Way Towards Physics]*. Weinheim, Germany: Beltz Verlag, 1990.

Wezel, A., S. Bellon, T. Dore, C. Francis, D. Vallod, and C. David. "Agroecology as a science, a movement or a practice." *Agronomy for Sustainable Development* 29 no. 4 (2009): 503–516.

CHAPTER 6

Complexity in Tradition and Science
Intersecting Theoretical Frameworks in Agroecological Research

John Vandermeer and Ivette Perfecto

CONTENTS

6.1 INTRODUCTION

Our proposition is dual: (1) traditional small-scale farmers have a knowledge base that is fundamentally sound and (2) that knowledge base is structurally similar to the growing scientific understanding of ecological complexity. It is a proposition that we suspect will stir two distinct responses, at least initially. On the one hand, will be those who say, that is obvious, and it is simply not news for anyone vaguely familiar with anthropological or rural sociological work, whether or not they are aware of or appreciate work done over the past decade in ecology. On the other hand, will be those who say that excessive reliance on traditional knowledge is nothing more than romantic drivel, that managing agriculture requires hard-nosed chemistry and engineering, not a return to preenlightenment mysticism, whether or not they are aware of or appreciate work done over the past decade in ecology.

First, if our proposition seems too obvious to bother with, we argue that it has been only recently that advances in the field of ecology have changed the way we look at ecosystems (Green et al. 2005; Vandermeer et al. 2010). Rather than the ordered equilibrium-like processes formerly thought to underlay assemblages of species, new analytical techniques have been brought to bear on ecosystem dynamics. We now understand that issues such as complex network structures, spatial dynamics, nonlinearities, stochasticity, and time lags, all create unexpected outcomes and challenge older notions of stability and sustainability. Furthermore, new molecular tools have provided a new lens on processes as they happen in nature, complimenting the experimental approach that ecology had adopted in the decades of the 1970s and 1980s (Burton 1999). Putting these two approaches together, complex theoretical methods and new examination tools, we have a new ecology, one based as much on the insights of complex systems as on natural history. And the epistemological framework of this new ecology begins to look very much like the framework engaged by more traditional systems of knowledge applied to natural ecosystems.

Second, if our proposition itself seems too romantic, we argue that the transformation of world agriculture at the end of World War II (Russell 2001) ignited a passion of irrational exuberance that has led to meltdown after meltdown, from massive pesticide resistance to hypoxic ocean zones, such that the wisdom of the traditionalists, even on the surface, is worth reconsidering. Furthermore, a discerning historical lens reveals a structure that has long been with us. None other than Robert Boyle noted that insights from the "trades," when coupled with systematic scientific structures provide a nutritious recipe for fundamental scientific discoveries (Conner 2005).[*] That is, the "wisdom of the ages" is wiser than we think. It just uses different words to describe phenomena. Richard Levins has noted what we refer to here as the Levins paradox—traditional agricultural knowledge is profound but local, while scientific knowledge is general but superficial (Lewontin and Levins 2007). The idea that advanced scientific knowledge can be seen as in accord with some of the principles long held by traditionalists should not really be a surprise to anyone not religiously committed to the modernist myth.

6.2 FOOD SOVEREIGNTY AS A UNIFYING CONCEPT

Both the idea of an ecological focus of research in agroecosystems and the generation of knowledge directly by farmer scientists and their interactions are given political life in the movement for food sovereignty. As stated in four of the six principles of food sovereignty from *La Via Campesina* and the Nyéléni 2007—Forum for Food Sovereignty (http://www.nyeleni.org/spip.php?article334), Food sovereignty:

- *Focuses on Food for People:* Food sovereignty ... rejects the proposition that food is just another commodity or component for international agribusiness.
- *Puts Control Locally:* Food sovereignty places control over territory, land, grazing, water, seeds, livestock, and fish populations on local food providers and respects their rights... it promotes positive interaction between food providers in different regions and territories and from different sectors that helps resolve internal conflicts or conflicts with local and national authorities; and rejects the privatization of natural resources through laws, commercial contracts, and intellectual property rights regimes.
- *Builds Knowledge and Skills:* Food sovereignty builds on the skills and local knowledge of food providers and their local organizations that conserve, develop, and manage localized food production and harvesting systems, developing appropriate research systems to support this and passing on this wisdom to future generations.
- *Works with Nature:* Food sovereignty uses the contributions of nature in diverse, low external input agroecological production and harvesting methods that maximize the contribution of ecosystems and improve resilience and adaptation, especially in the face of climate change; it seeks to heal the planet so that the planet may heal us; and, rejects methods that harm beneficial ecosystem functions.

---

[*] "Trade" here refers to a skill or craft. Robert Boyle, seventeenth-century philosopher, chemist, and physicist, and namesake of Boyle's law of gases, was keen on understanding the way in which common tradesmen and women accumulated knowledge that was systematic, organized, and insightful, much as the modern scientific method. This point is discussed in detail by Conner (2005).

The first principle, that food is different, is critical. Contemporary economic orthodoxy assumes as a basic principle, that all goods and services may, should, be viewed as tradable commodities. As part of the underlying historical transition from feudalism to capitalism, this idea is central. When the product–money–product (PMP) model was replaced by the money–product–money (MPM) model, viewing a product for its utility, its use value, was replaced with viewing a product for its potential to be bought and sold for money, its exchange value. Markets then are viewed not as places where products are bought and sold so as to increase the utility of those products in the society, in general, but rather as places where profit is to be generated. How to "monetize" products has become a buzzword in modern economic circles for everything from "economic service instruments" to water. Food sovereignty takes as a fundamental principle the assumption that access to food should be a human right, not a function of either the accident of birth or vicissitudes of economic fortunes. Like the air we breathe, societies should be organized such that no human being goes hungry. The UN declaration in 1994 that food security should be a basic human right is identical to this aspect of food sovereignty.

But food sovereignty goes further. Its second key element is the collective right to decide how to provide that security. This element, while simple in concept, generates important complications and contradictions. It means, for example, that a community must be self-defined and collectively see the food production system as part of community norms. Communities must reject dumping, for example. This implies a community consciousness that transcends the simple assumptions of *Homo economicus*, that all individuals seek to maximize utility. It effectively views the past 500 years the way Dutch merchants might have viewed the seemingly intractable feudal system at the dawn of the Bourgeois Revolution, as a transitional period that eventually will give way to a more rational mode of organizing society. Although it was rational to replace the PMP system with the MPM system during the time of feudal domination, food sovereignty suggests that times have changed, that the modern world no longer needs to concern itself with overthrowing the assumption of royal authority, that the new system that needs to be overthrown is the assumption that corporate structure must decide how food is produced and distributed. The challenge of food sovereignty is to take control of that production and distribution process and organize it for the sake of producing nutrition for people rather than profits for corporations.

The third element emphasizes the role of peasant farmers in the production of knowledge of agroecosystems. The famous farmer-to-farmer methodology of traditional technology transfer has been wildly successful, especially in Latin America, and emphasizes the fact that local farmers have very deep knowledge of their local systems and have a lot to teach one another. Interchanges among farmers from different regions have a cross-fertilization effect of expanding the generalization of knowledge through sharing experiences.

The fourth element recognizes the importance of the natural world in providing guidance and inspiration for the development of agricultural technology, a central component to our vision of the intersection of the complicated vision of small-scale farming operations and the science of ecology.

6.3 HISTORICAL CONTINGENCY DROVE THE INDUSTRIAL AGRICULTURAL SYSTEM

It is not difficult to see the outlines of the problems we must address, even though they are enormous: one in three children is unhealthy because of food promoted by advertisers whose main concern is people's wallets, not their health (Nestle 2007); pesticide residues linger at levels that are deemed safe only through corporate lobbying, causing an unknown number of health problems (Pimentel et al. 1992); ocean dead zones result from massive artificial fertilizer applications (Diaz and Rosenberg 2008; Nassauer et al. 2007); global warming is exacerbated from many elements of the industrial agricultural model (Lin et al. 2011). In short, our problem is the production of food that is unhealthy for people with methods that are unhealthy for the environment. How did we get into this situation?

For at least 90% of our existence as a species we were hunters and gatherers (Lee and Daly 1999). The energy we required to do what needs to be done came from the substances acquired through hunting and gathering directly from nature—large vertebrate herbivores, fruits, tubers, grubs, and similar natural items. The adoption of agriculture enabled a far more efficient way of obtaining that energy and promoted a dramatic increase in our numbers and leisure (Cowan et al. 2006). We began a grand manipulation of nature, but were necessarily constrained by ecological laws and could only produce within the constraints of those laws. We engaged in what might be referred to as "natural systems agriculture," as it has been referred to more recently (Jackson 2002).

But then something rather drastic happened. Beginning during the early decades of the last century, and culminating in earth-shattering changes in the postwar years, our species forced into the agricultural enterprise the tools of the recent, spectacularly successful, Industrial Revolution (Hendrickson and James 2005). We automated, regularized, commoditized, monetized, and chemicalized the process of generating food. What had been done in industry was now done in agriculture, human labor, and ecological processes were replaced with fossil fuels. We applied, in myriad ways, industrial energy to the process of producing food. In the end and largely as an unintended consequence of the giddiness of the Industrial Revolution's successes, we transformed the system that made our acquisition of energy more efficient into a system that effectively used more energy than it produced—from an "energy producing" system to an "energy consuming" system (Martinez-Alier 2011; Pimentel and Pimentel 1979; Pimentel et al. 1973, 2005).

Furthermore, as a consequence of industrializing food production, it seemed quite natural to industrialize food consumption as well. A key problem was the ability to produce more food than people generally wanted to eat, or at least, more food than people needed to eat to stay healthy (i.e., when considering food as a commodity, it is inelastic). A tomato must be eaten within a few days after harvest, or at best a week or two under refrigeration, or it is basically lost to nature's recycling ways. But people would not cooperate with the new agricultural economics—they insisted on eating only the number of tomatoes that made them full. Two strategies evolved to deal with this problem. First, food preservation technology, long a traditional activity, especially in the North, became industrialized. The tomatoes were converted to tomato sauce that could be canned and stored virtually in perpetuity and second, people were encouraged to eat more and more. Food scientists not only invented creative ways of extending shelf life, they also came to understand the basic human responses to taste and texture and thus how to manipulate those responses to encourage people to want more and more. Consequently, we had a revolution that resulted in food being processed into what Michael Pollen calls "food-like substances" and people converted into consumption machines that ever increased the limits of their intake capacities (Pollan 2007). Indeed, in the modern food system people are referred to as simply "consumers."

Today the environmental crisis created by the industrial agricultural system is beginning to receive the same scholarly attention as climate change. Direct emissions of greenhouse gasses from industrial agriculture are now appreciated and juxtaposed with the troubling fact that there is little hard evidence that intensification has led to a global increase in food security, no matter how defined (Patel 2010). These observations, along with other environmental insults coming from the industrial system, have generated a number of critical reports. The most notable was the release of the UN and World Bank–sponsored report, the International Assessment of Agricultural Knowledge, Science and Technology for Development (IAASTD) in 2009 (IAASTD 2009). That report, similar to the early IPCC reports, noted a human and environmental health disaster on the horizon if the industrial agricultural system continued its trajectory. In a press release comment on that report, Robert Watson, one of the IAASTD cochairs (and former chair of the IPCC) noted "Business as usual is not an option" referring to the industrial agricultural system.

Similar to the IPCC report and the many tobacco reports previously, powerful political actors immediately sprang to action with the same questionable tactics used by the tobacco industry and energy industries previously (Orestes and Conway 2010). And when Michelle Obama planted an organic garden on the White House lawn, Crop Life, one of Industrial Agriculture's main lobbying groups, immediately objected, claiming that such an act sent a "bad signal" about pesticides (Burros 2009).

Throughout this history, the enterprise we call science has been central. Unfortunately, a reactionary tendency has emerged in some quarters in which science itself is portrayed as the culprit in the dystopia that is the modern agricultural system, and sometimes occult and mystical forms of understanding are championed as a better alternative for development of a rational agriculture. We certainly disagree with such a tendency. Rather, we argue, it is that the science that has been applied has been the wrong science. Is it really true that the plant nutrition is a subject of inorganic chemistry? Is food consumption really a subject of physiology? Is food distribution a subject of economics? Is pest control a subject of organic chemistry? Our position is that due to historical accident the drastic transformation of human society unleashed by the Industrial Revolution happened at a time when certain sciences were themselves undergoing dramatic transformations. It seemed quite natural that agricultural subjects should be divided up into the sciences that had proved so successful and were continuing to contribute to new and seemingly more efficient technologies. Of particular, mechanical engineering (and the physics on which it was based), genetics, and chemistry were important. Ecology did not really even exist as a formal science. Indeed the word itself was not common currency until the later years of the nineteenth century. In these early years of agricultural transformation, the only true ecologists seem to have been the farmers themselves. And this is where a remarkable store of ecological knowledge was nurtured.

6.4 REFLECTIONS ON THE TRADITIONAL

In the 1990s, a Guatemalan entomologist, Helda Morales, began research for her doctoral dissertation among traditional Mayan maize producers in the Guatemalan mountains. In seeking to understand and study traditional methods of pest control, she began by asking the question, "What are your pest problems?" She was surprised to find almost unanimity in the responses of most of the farmers she interviewed—"We have no pest problems." Taken aback, she reformulated her questionnaire and asked, "What kind of insects do you have in your *milpa*?" to which she received many answers, including all the main characteristic pests of maize and beans in the region. She then asked why these insects, known to be pests by professional entomologists, were not pests according to the Mayan farmers. Again she received all sorts of answers, mostly in the form of how the agroecosystem was managed. The farmers were certainly aware that these insects could be problems, but they also had ways of managing the agroecosystem such that the insects remained below levels that would categorize them as pests. Morales' initial plan was probably influenced by her early training in agronomy and classical entomology, but her interactions with the Mayan farmers caused her to change her approach. Rather than study how Mayan farmers solve their problems, she focused on why the Mayan farmers do not have problems in the first place (Morales and Perfecto 2000).

The lessons from the Morales studies are many. And most point to the inadequacy of standard agricultural research. The classical agronomist's approach is fixed by the idea that farmers always face "problems" that need solutions (or in the more decorative rhetoric of the post–World War II chemical industry, farmers have enemies that must be vanquished). Finding solutions to those problems can hardly be questioned on moral or ethical grounds, even as those solutions always seem to benefit their funders. Their focus is on solving problems, real or imagined. In contradiction, part of a rational agroecological research agenda should take a clue from the Morales studies. We need to understand why problems do NOT exist. That is, we need to understand the ecological complexity inherent in most traditional agricultural production systems, not place our focus on extant problems only.

Along with the recognition that functioning farms indeed do function within ecological principles, and part of the job of the researcher is to understand those principles, is the fact that farmers themselves have long been scientists and their knowledge is, although perhaps narrow in scope, quite deep in regard to their particular farm and farming system, frequently having benefitted from the accumulated knowledge of generations of their ancestors (Grossman 2003; Richards 1985; Toledo and Barrera-Bassols 2008; Wilken 1987). But more important is the fact that farmers act as scientists in another, perhaps more important, way. Much as science is accomplished through associations and scientific societies, which is to say science in the end is a social activity, farmer scientists have always engaged in interchanges (Leitbeg et al. 2011). The farmer in valley X who tries a particular mode of planting cassava and finds that it works efficiently in resisting the onslaught of a particular pest, invariably shares that knowledge with a farmer in valley Y when they meet in their common marketplace. Based on this obvious idea, some action-oriented researchers have promoted the idea of farmer-to-farmer interchange as a vehicle for development (Bentley et al. 2003; Holt-Giménez 2006; Röling and van de Fliert 1994) and an important social tool for the generation of new scientific knowledge (Stuiver et al. 2004), an idea that it is difficult to question. Indeed, formal evaluations, such as the IAASTD, specifically acknowledge this mode of deriving new knowledge.

6.5 ECOLOGICAL COMPLEXITY

It is not new to acknowledge that ecosystems are extremely complicated. But in the past 30 years, we have come to appreciate this complexity in a far more sophisticated way than before. Naturalists, with a much more direct connection to the natural world, rightly criticized early theoretical ecologists for their seemingly uncritical acceptance of the mechanical universe made so popular by seventeenth- and eighteenth-century physicists, what one might call "vulgar Newtonism." Jokes about the naiveté of mathematical economists frequently could be levied almost word-for-word against the early attempts at constructing a mechanical version of the ecosystem, whether through simple equations describing changes in population growth or energy flowing among organisms as if in a series of tubes. Nature was clearly not easily amenable to such naïve approaches.

Yet allowing for a dialectical interpenetration of several sciences, ecologists started examining a variety of epistemological approaches, to say nothing of developing some of their own sophisticated breakthroughs (some of which became cornerstones of other sciences[*]). The consequence is that we today stand on a completely different platform to view the natural world, a platform that is to some extent enabled by the modern technology of computers, but also one that acknowledges that very interpenetration of different sciences. Furthermore, as we discuss further below, this new platform begins to look like the same platforms that many agroecologists had been advocating for years, from Albert Howard to Wes Jackson to Miguel Altieri to Steve Gliessman and others, promoting the idea that we all should take seriously what both nature and traditional farmers say about ecology. More on that later, but for now, it is worth a quick review of some of the elements characteristic of that new approach to complexity.

One of the major ideas that is still today transforming ecology was a brilliant insight of the British mathematician Alan Turing. Even casual observers of nature understand that the

---

[*] Robert May, a theoretical ecologist at Oxford, noted that one of the ecologist's favorite conceptual tools, the logistic map, is perhaps the simplest equation that can generate chaos. That illustration is now included in most introductions to chaos, regardless of the subfield of science in which they are embedded.

distribution of animals and plants is not a random pattern—certain species of plants tend to grow together, some corals tend to be grouped in species-specific clusters, pests of crops frequently attack some farms but not others in a given landscape. It has been the assumption of both early naturalists and modern ecologists that such nonrandom patterns reflected something about the background habitat. But then Turing suggested a very general mechanism whereby a spatial pattern could be formed by agents of that pattern themselves. Although he was thinking of chemicals associated with the ontogeny of morphology and trying to model the emergence of spots and stripes on animals, his approach potentially applies to populations of organisms also. Two components are required, an "activator" (in the case of chemicals a chemical reaction that produces a local concentration of some chemical) and a "repressor" (a chemical that is produced when the first chemical reaches a critical concentration), which cancels the activity of the activator. Then, when the two chemicals are put into a spatial context, when the diffusion of the activator in the space is smaller by a critical amount than the diffusion of the repressor, a distinct pattern (such as the spots on a jaguar's coat or the stripes on a zebra) will form spontaneously. Therefore, if we have populations that behave as if they are activating and another agent (a natural enemy, for example) behaving as if it were a repressor, spontaneous patterns of both populations could form in nature. A direct application of Turing's equations to a predator–prey situation has demonstrated how prey and predator populations can form spatial patterns from their interactions alone, even in a perfectly uniform environment (Alonso et al. 2002), and we suggested that the predatory ant *Azteca sericeasur* (formally *instabilis*) along with its natural enemies could do the same, in an organic coffee farm (Figure 6.1). It is worth noting that this particular ant is involved with a complex ecological community that is arguably responsible for the autonomous control of three of the most important coffee pests in the region (Vandermeer et al. 2010), an example of autonomous biological control (Lewis et al. 1997).

An equally important result emerges from relaxing the naïve Newtonian assumption that interactions are generally linear and occur effectively in reduced dimensionality (i.e., one or two species involved, for example, a crop and its pest, a pest and its natural enemy, a plant disease and its disperser, etc.). If we examine the classic ideas of these two-dimensional systems in the context of nonlinear effects, some surprising consequences rapidly emerge (for a brilliant exposition of this idea, see Levins 1979). For example, the simple predator–prey models introduced in elementary ecology classes have a form in which predator and prey oscillate regularly. The oscillations either eventually damp down to a single equilibrium point or form a permanent oscillation (a limit cycle). This basic framework has for years been the major way in which we conceptualize predator–prey equations whether in the context of harvesting resources or controlling pests. Yet, what predator system occurs in isolation? If we add a top predator to the system, some rather interesting things happen. What we should expect is not obvious at first glance. The original predator and prey oscillate—but so do the top predator and prey. What will be the pattern of the overall three-dimensional system? The details are relatively complicated (Hastings and Powell 1991), but formally the system may be chaotic, which means that precise prediction of any of the three population densities is impossible no matter how well the parameter and variable values may be estimated. The formal declaration that precise predictions are, in principle, impossible may imply that there is no use even trying to understand the system, that if a system is chaotic we must throw up our hands and admit defeat. Indeed, if a system is chaotic it does not matter how precise one measures variables or parameters nor how closely one monitors the natural system, precise prediction of what will happen in the long term is not possible. Yet in that chaotic behavior there is a sensible pattern that we can describe, suggesting that our appropriate response is not to give up in despair, but to change our sense of what "understanding" the system means in the first place. For example, in Figure 6.2 we see the population densities of all three populations over time, when in the chaotic mode (the herbivore is a caterpillar, the predator is a spider, and the top predator is a bird). For this system, it can be rigorously shown that, even

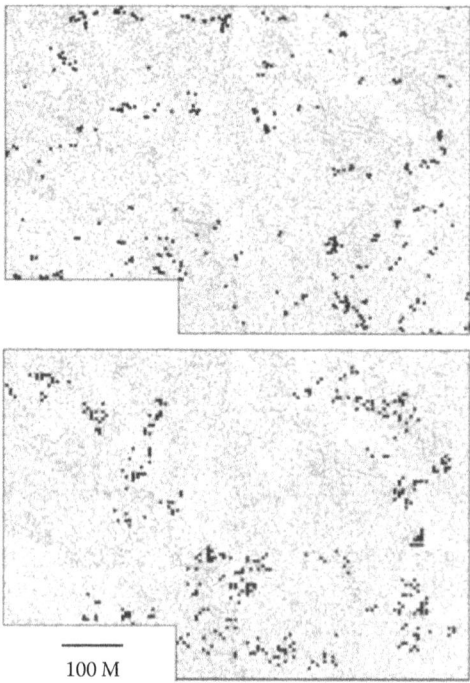

Figure 6.1   Distribution of *Azteca sericeasur* in a coffee plantation in southern Mexico. Gray symbols are shade trees, any one of which could harbor an ant nest, and dark shaded symbols are shade trees that do harbor an ant nest. (a) The distribution of nests on the farm in 2004. (b) One snapshot of the distribution of nests based on a cellular automata model with Turing-like dynamics (From Vandermeer et al., *Nature*, 451, 2008). Note the qualitative similarity between the computer manifestation and the actual distribution.

knowing the populations today and knowing everything there is to know about those populations down to the finest detail, it is impossible to predict where any of those populations will be 10 years from now. But what indeed can be known, and what we argue is evident from a simple glance at the data, is that the traces of the original population behavior can be discerned within the chaos. The birds and spiders oscillate (as they should since the birds eat the spiders) and the spiders and the caterpillars oscillate (as they should since the spiders eat the caterpillars), and one can see "episodes" of the dominance of either one or the other oscillatory modes. We abandon the naïve Newtonian assumption of precise prediction, but seek a new level of understanding by asking qualitative questions about the system. This is the essence of what the chaos revolution has brought—qualitative understanding of complex systems is possible but the goal of precise predictions characteristic of the vulgar Newtonians should be placed in the trash bin of history.

Closely related to the chaos revolution in ecology are a series of new approaches to ecological networks. The simplest ecological networks are food webs, long appreciated for their complexity in the real world. More complicated ideas of ecological networks include indirect interactions (e.g., it is not sufficient to understand that the parasitic wasp attacks the caterpillar pest, but is important to note that the parasitic wasp responds to the chemical volatiles the plant emits when the caterpillar attacks). Given the current ubiquity of high-speed computers, there has been an explosive growth in attempts to understand the consequences of various kinds of network structure (Hsieh et al. 2012; Golubski and Abrams 2011). As with chaos itself, this new research strongly suggests that the notion of precisely predicting the course of events should not be a particular research goal, but rather that a qualitative understanding of overall structure and change

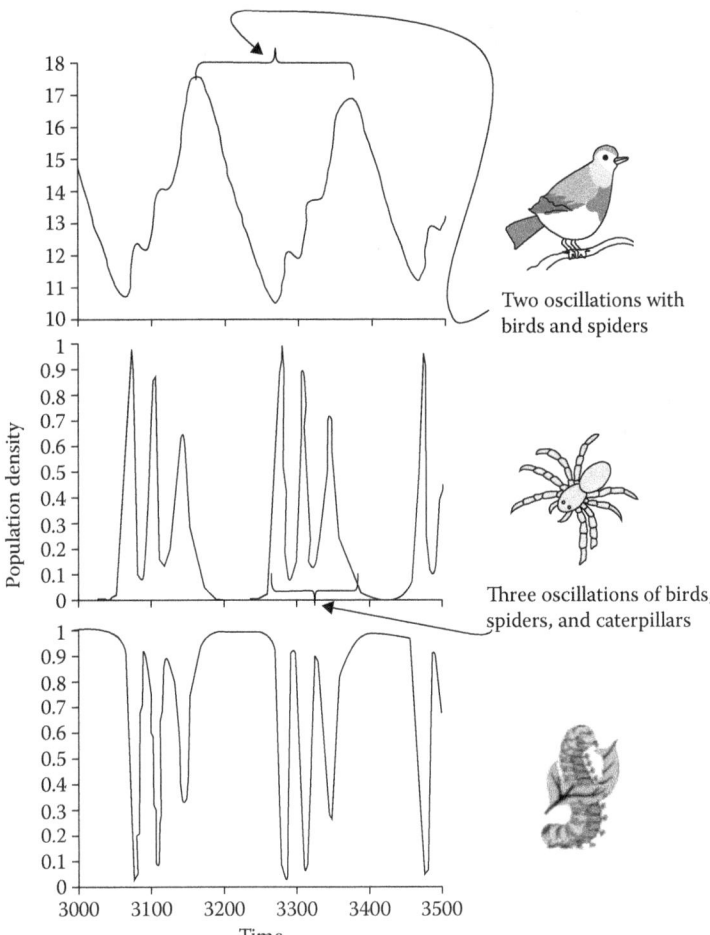

Population density

Two oscillations with birds and spiders

Three oscillations of birds, spiders, and caterpillars

Time

Figure 6.2 The chaotic attractor of Hastings and Powell (so-called "teacup" attractor). Note that although formal prediction of the system is completely impossible, one can see the shadow of the two original oscillatory processes embedded in this attractor.

can emerge from particular network structures. Yet, it remains a frustrating limitation on understanding that our ability to say anything either quantitative or qualitative about a system remains captive of a science that is in its infancy (i.e., complexity in ecology). Our optimism derives from more recent work that seeks a more sophisticated understanding than is commonly attained by traditional agronomic epistemology. We quote from the conclusion of one of our recent articles (Vandermeer et al. 2010):

> In the end, this model system suggests that the vision of the natural world as harmonious and balanced is wrong if we naively accept an unreconstructed Newtonian world view of balance—ecosystems are not like a marble coming to rest at the bottom of an inverted cone. However, through the spatially explicit complexity of myriad interactions, many of which are multiply nonlinear, a higher notion of balance emerges—not the balance of Newton, but rather the balance of a shifting sand dune whose detailed structure changes minute to minute, but whose fundamental nature as a 'sand dune' is never in doubt. Our understanding becomes not the crude, positivist logic that must identify a singular enemy to conquer, and a magic bullet with which to do so, but rather the holistic vision of a new kind of 'balance' emerging from the very complexity that traditional farmers intuitively understood from the beginning.

6.6 ECOLOGICAL COMPLEXITY INTERSECTING WITH TRADITIONAL KNOWLEDGE

At the Land Institute in Salinas Kansas, Wes Jackson (2002) has been promoting "natural systems agriculture." The idea is that the local natural ecosystem provides us with the vision of how an agroecosystem ought to be designed. Jackson's idea gains considerable force from tradition. More or less the same idea was elaborated in a more simplified form by Sir Albert Howard when he was dispatched to India by Queen Victoria to teach Indian farmers how to do agriculture. He discovered deep traditions, mainly based in knowledge of local ecology that he judged did a better job than the "modern" agriculture the Victorian scientists were promoting. Other examples could also be cited (e.g., Altieri 1990, 2004; Berkers et al. 2000; Denevan 1995; Ewel 1986; Funes et al. 2002; Gliessman et al. 1981; Sevilla Guzmán 1991; Toledo 1990; Toledo and Barrera-Bassols 2008; Wilken 1987). But Jackson brings to the table an explicit search for the dialectical relationship between the modern science of genetics and ecology and the structure of natural ecosystems. He notes that grain belt farming in North America seeks to impose an annual monoculture in an environment that has, at least since the Pleistocene, been characterized by a perennial polyculture. The problem, he notes, is that perennial grasses have not had the sort of genetic modifications that traditional farmers imposed on the annual grasses that make up the idea of annual monocultures, and set upon a program of genetic modification to create higher yielding perennial grasses (or, the perennialization of the classic grains).

It could be argued that "natural systems agriculture" has been practiced by more traditional farmers the world over for as long as agriculture has been in existence. The new twist in Jackson's version is his vision for the use of modern science in pursuit of the underlying goals of natural systems agriculture. The modern science available in the nineteenth century was relatively primitive and was thus applied in a naïve brute force fashion, effectively attempting to cancel the laws of ecology with substitutions such as pesticides for the predators that normally control pests, herbicides for what had been a balance between various plant competition effects, or industrial chemicals for soil management with legumes and various forms of compost. Jackson sees the use of modern concepts of genetics and ecology as recapturing the base for agriculture. His Land Institute focuses on the genetics and ecology of perennial polycultures, an immensely complicated process, but one that the local natural system has already designed.

In most tropical regions of the world one can see the influence of natural systems agriculture, practiced as a matter of course. For example, when coffee was brought to Latin America (at least in the northern part of the region), farmers began cultivating it beneath a canopy of shade trees (and frequently even under a natural forest canopy), knowing that its natural state is as a forest understory plant. Further development in the region lead to the development of what are now called coffee forests, well known to be a major refuge for biodiversity (Moguel and Toledo 1999; Perfecto et al. 1996). A similar evolution characterized cacao production in Brazil (Faria et al. 2006) and elsewhere and rubber in West Africa and Indonesia (Suyanto et al. 2001). We have been involved in the study of traditional forms of coffee production and our conclusion, as noted in the quote above, is that small farmers the world over

… have a universal and evident sense that the natural world offers ecosystem services that contribute to the stability, productivity, and sustainability of their farms. (Vandermeer et al. 2010)

6.7 DISCUSSION

The knowledge contained in the theory and practice of traditional farmers the world over is encyclopedic to be sure. As farmers continue learning from the experiments and understanding of each other and previous generations it is certain that more rational systems of agriculture will develop, even as the industrial system pushes its unrelenting propaganda on them. And it is likely that the industrial model will continue with that unrelenting propaganda. The ecological alternative that we favor and that combines current ecological theory and traditional knowledge to date has had limited, albeit growing, influence. A problem that seems to be universally recognized is the dramatic level of uncertainty involved in our understanding of the ecological systems involved. The folly of following old research techniques is evident to all except those whose career depends on them (recalling Upton Sinclair's admonition, "It is difficult to get someone to understand something when his or her salary depends on misunderstanding it"). Yet, we must acknowledge that since World War II there have been hundreds, if not thousands, of researchers leveraging billions of dollars of research in support of the furtherance of the industrial system. They are exceedingly good at making that industrial system perform as best it can. In contrast, the ecological study of agroecosystems remains in its infancy. When that same billions of dollars are spent in trying to untangle the enormous complexity of ecosystems, when thousands of researchers have the same level of support, and when that cutting-edge ecological research joins force with the traditional knowledge of farmers that have benefited from thousands of years of trial and error and experimentation, we can envision the day where we will be far better able to muster the ecological principles of agroecosystems in support of agroecological planning.

Thus, we envision a future where the science of ecology, especially as applied to agroecology, will become ever more enlightening. At the same time, we envision a future in which small-scale farmers will have control of their own production systems, which is to say will have a full plate of food sovereignty, and will continue their own development of science. A major challenge, as we see it, is to creatively engage the Levins paradox. This will involve creative engagement on all sides of the issue.

We argue that the modern science of ecology has a great deal to offer the growing agroecosystem revolution. Indeed, we argue that, as the science of chemistry is the basis of chemical engineering, the science of ecology is (or should be) the basis of agroecology. Yet, it is also the case that the accumulated knowledge of the world's millions of small-scale farmers has a great deal to offer the modern science of agroecology. Indeed a common definition of agroecology incorporates traditional knowledge as one of the bases of agroecology. As Conner (2005) elaborated in his "A People's History of Science," the practical necessities of actually producing things (i.e., not ethereal trickery such as financial "instruments," but real goods and services that get used by people) has, through the ages, motivated people to understand how the world works. And science at its core is about that understanding. Indeed, we agree with Robert Boyle that

> as the naturalist may ... derive much knowledge from an inspection into the trades, so by virtue of the knowledge thus acquired ... he may be as able to contribute to the improvement of the trades,

a principle that is probably more important than his famous law about gases. Indeed, it is perhaps the most important scientific principle of all—the Levins paradox. Traditional knowledge is deep but local, while modern ecological knowledge is general but shallow. Is it too much to promote a research agenda that seeks to combine those two? To have at least as the ultimate goal (dream), the generation of knowledge that is simultaneously deep and general?

REFERENCES

Alonso, D., F. Bartumeus, and J. Catalan. "Mutual interference between predators can give rise to turing spatial patterns." *Ecology* 83 no. 1 (2002): 28–34.

Altieri, M.A. "Linking ecologist and traditional farmers in the search for sustainable agriculture." *Frontiers in Ecology and the Environment* 2 (2004): 35–42.

Altieri, M.A. "Why study traditional agriculture?" In *Agroecology*. C.R. Carrol, J.H. Vandermeer, and P.M. Rosset (Eds.). New York: McGraw Hill, 1990, 551–564.

Bentley, J.W., E. Boa, P. Van Mele, J. Almanza, D. Vazquez, and S. Eguino. "Going public: A new extension method." *International Journal of Agricultural Sustainability* 1 (2003): 108–123.

Berkers, F., J. Colding, and C. Folke. "Rediscovery of traditional ecological management as adaptive management." *Ecological Applications* 10 (2000): 1251–1262.

Burros, M. "Industry is critical of Michelle Obama's organic garden." Politico June 17, 2009. http://www.Politico.com/news/stories/0609/23838.html.

Burton, R.S. "Molecular Tools in Marine Ecology." *Journal of Experimental Marine Biology and Ecology* 200 (1999): 85–101.

Conner, C.D. *A People's History of Science: Miners, Midwives and "Low Mechanicks."* New York: Nation Books, 2005.

Cowan, C.W., P.J. Watson. *The origins of Agriculture: An International Perspective.* Tuscaloosa: University of Alabama Press, 2006.

Denevan, W.M. "Prehistoric agricultural methods as models for sustainability." *Advances in Plant Pathology* 11 (1995): 21–43.

Diaz, R.J. and R. Rosenberg. "Spreading dead zones and consequences for marine ecosystems." *Science* 321 (2008): 926–929.

Ewel, J.J. "Designing agroecosystems for the humid tropics." *Annual Review of Ecology and Systematics* 17 (1986): 245–271.

Faria, D., R.R. Laps, J. Baubarten, and M. Cetra. "Bat and bird assemblages from forests and shade cacao plantations in two contrasting landscapes in the atlantic forest of Southern Bahia, Brazil." *Biodiversity and Conservation* 15 (2006): 587–612.

Funes, F., L. Garcia, M. Bourque, N. Perez, and P. Rosset. *Sustainable Agriculture and Resistance: Transforming Food Production in Cuba.* Oakland, CA: Food First, 2002.

Gliessman, S.R., R.E. Garcia, and M. Amador. "The ecological basis for the application of traditional agricultural technology in the management of tropical agro-ecosystems." *Agro-Ecosystems* 7 (1981): 173–185.

Golubski, A.J. and P.A. Abrams. "Modifying modifiers: What happens when interspecific interactions interact?" *Journal of Animal Ecology* 80 no. 5 (2011): 1097–1108.

Green, J.L., A. Hastings, P. Arzberger, et al. "Complexity in ecology and conservation: Mathematical, statistical, and computational challenges." *BioScience* 55 (2005): 501–510.

Grossman, J.M. "Exploring farmer knowledge in organic coffee systems in Chiapas, Mexico." *Geoderma* 111 (2003): 267–287.

Hastings, A. and T. Powell. "Chaos in a three-species food chain." *Ecology* 72 no. 3 (1991): 896–903.

Hendrickson, M.K. and H. S. James. "The ethics of constrained choice: How the industrialization of agriculture impacts farming and farmer behavior." *Journal of Agricultural and Environmental Ethics* 18 (2005): 269–291.

Holt-Giménez, E. *Campesino a Campesino: Voices from Latin America's Farmer to Farmer Movement for Sustainable Agriculture.* Oakland, CA: Food First, 2006.

Hsieh, H.Y., H. Liere, E. J. Soto, and I. Perfecto. "Cascading trait-mediated interactions induced by ant pheromones." *Ecology and Evolution* 2 no. 9 (2012): 2181–2191.

IAASTD (International Assessment on Agriculture, Science and Technology for Development). *Agriculture at a Crossroads.* Washington, DC: Island Press, 2009.

Jackson, W. "Natural systems agriculture: A truly radical alternative." *Agriculture, Ecosystems and Environment* 88 (2002): 111–117.

Lee, R.B. and R.H. Daly. *The Cambridge Encyclopedia of Hunters and Gatherers.* Cambridge: Cambridge University Press, 1999.

Leitbeg, F., F.R. Funes-Monzote, S. Kummer, and C.R. Vogl. "Contribution of farmer's experiments and innovation to Cuba's agricultural innovation system." *Renewable Agriculture and Food Systems* 26 (2011): 354–367.

Levins, R. "Coexistence in a variable environment." *American Naturalist* 114 no. 6 (1979): 765–783.

Lewis, W.J., J.C. Van Lenteren, S. Phatak, and J. Tumlinson. "A total system approach to sustainable pest management." *Proceedings of the National Academy of Sciences* 94 no. 23 (1997): 12243–12248.

Lewontin, R. and R. Levins. "The maturing of capitalist agriculture: Farmer as proletarian." In *Biology under the Influence: Dialectical Essays on Ecology, Agriculture, and Health.* New York: Monthly Review Press, 2007.

Lin, B.B., J.M. Chappell, J. Vandermeer, et al. "Effects of industrial agriculture on climate change and the mitigation potential of small-scale agroecological farms." *CAB Reviews* 6 (2011): 1–18.

Martinez-Alier, J. "The EROI of agriculture and its use by Via Campesina." *Journal of Peasant Studies* 38 (2011): 145–160.

Moguel P. and V.M. Toledo. "Biodiversity conservation in traditional coffee systems of Mexico." *Conservation Biology* 13 (1999): 11–21.

Morales, H. and I. Perfecto. "Traditional knowledge and pest management in the Guatemalan highlands." *Agriculture and Human Values* 17 (2000):49–63.

Nassauer, J.I., M.V. Santelmann, and D. Scavia (Eds). *From the Corn Belt to the Gulf: Societal and Environmental Implications of Alternative Agricultural Futures.* Washington, DC: Resources for the Future Press, 2007.

Nestle, M. *Food Politics*. Berkeley, CA: University of California Press, 2007.

Orestes, N. and E. Conway. *Merchants of Doubt: How a Handful of Scientists Obscured the Truth on Issues from Tobacco Smoke to Global Warming.* New York: Bloomsbury Press, 2010.

Patel, R. *Stuffed and Starved: Markets, Power and the Hidden Battle for the World's Food System.* Canada: Harper Collins, 2010.

Perfecto, I., R.A. Rice, R. Greenberg, and M. Van der Voort. "Shade coffee: A disappearing refuge for biodiversity." *BioScience* 46 (1996): 598–608.

Pimentel, D., P. Hupperly, J. Hanson, D. Douds, and R. Seidel. "Environmental, energetic, and economic comparisons of organic and conventional farming systems." *BioScience* 55 (2005): 573–582.

Pimentel, D., H. Aquay, M. Biltoen, et al. "Environmental and economic costs of pesticide use." *BioScience* 42 (1992): 750–760.

Pimentel, D., L.E. Hurd, A.C. Bellotti, et al. "Food production and the energy crisis." *Science* 182 (1973): 443–449.

Pimentel, D. and M. Pimentel. *Food, Energy and Society*. London: Arnold, 1979.

Pollan, M. *The Omnivore's Dilemma: A Natural History of Four Meals.* New York: The Penguin Press, 2007.

Richards, P. *Indigenous Agrarian Revolution*. Boulder, CO: Westview Press, 1985.

Röling, N. and E. van de Fliert. "Transforming extension for sustainable agriculture: The case of integrated pest management in rice in Indonesia." *Agriculture and Human Values* 11 (1994): 2–3.

Russell, E. *War and Nature: Fighting Humans and Insects with Chemicals from WWI to Silent Spring.* Cambridge: Cambridge University Press, 2001.

Sevilla Guzmán, E. "Hacia un Desarrollo Agroecológico Delsde el Campesinado." *Política y Sociedad* 9 (1991): 57–72.

Stuiver, M., C. Leeuwis, and J.D. van der Ploeg. "The power of experience: Farmer's knowledge and sustainable innovations in agriculture." In *Seeds of Transition: Essays on Novelty Production, Niches and Regimes in Agriculture.* J.S.C. Wiskerke, J.D. Ploeg. Assen: Royal Van Gorcum, 2004, 93–117.

Suyanto S., T.P. Domich, and K. Otsuka. "Land tenure and farm management efficiency: The case of smallholder rubber production in customary land areas of Sumatra." *Agroforestry Systems* 50 (2001): 145–160.

Toledo, V.M. "The ecological rationality of peasant production." In *Agroecology and Small Farm Development.* M. Altieri, S. Hecht (Eds.). Boca Raton, FL: CRC Press, 1990.

Toledo, V.M. and N. Barrera-Bassols. *La Memoria Biocultural: La Importancia Agroecológica de las Sabidurias Tradicionales.* Barcelona: Icaria Editorial, 2008.

Vandermeer, J., I. Perfecto, and S. M. Philpott. "Clusters of ant colonies and robust criticality in a tropical agroecosystem." *Nature* 451 (2008): 457–459.

Vandermeer, J., I. Perfecto, and S.M. Philpott. "Ecological complexity and pest control in organic coffee production: Uncovering an autonomous ecosystem service." *BioScience* 60 (2010): 527–537.

Wilken, G.C. *Good Farmers: Traditional Agricultural Resource Management in Mexico and Guatemala.* Berkeley, CA: University of California Press, 1987.

CHAPTER 7

Agroecology, Food Sovereignty, and the New Green Revolution

Eric Holt-Giménez and Miguel A. Altieri

CONTENTS

7.1 HUNGER, THE CORPORATE FOOD REGIME, AND THE RETURN OF THE GREEN REVOLUTION

The global food crisis of 2008 returned in 2010 with devastating impacts on the world's poor—most of whom are peasant farmers (Collier 2008; FAO 2011). Hunger resulted not from a lack of global food stocks but from food price inflation (Bailey 2011; Brown 2011). Volatility and high food prices have led institutions in the *corporate food regime* to call for a 70% increase in food production by 2050 (Conforti 2010; FAO 2011).

A food regime is a "rule governed structure of production and consumption of food on a world scale" (Friedmann 1993, 30–31, in McMichael 2009, 142). The present corporate food regime (McMichael 2009, 2013) is made up of the global food system's government ministries, global institutions, agrifood monopolies, land grant universities, think tanks, and big philanthropy that generate the technologies, the discourse, and enforce the regime's "rules" (e.g., the free trade agreements [FTAs], the US Farm Bill, and the European Common Agricultural Policy—CAP).*

With the food crisis, international institutions produced a steady stream of assessments calling for investment in biotechnology and a new Green Revolution (Baulcombe et al. 2009; Beddington 2011;

---

* The construction of the corporate food regime began in the 1960s with the Green Revolution that spread the high-external-input, industrial model of agricultural production to the Global South. The World Bank and International Monetary Fund's (IMF) structural adjustment policies followed in the 1980s, privatizing state agencies, removing barriers to northern capital flows, and dumping subsidized grain into the Global South. The FTAs of the 1990s and the World Trade Organization (WTO) enshrined SAPs within international treaties. The cumulative result was massive peasant displacement, the consolidation of the global agrifood oligopolies, and a shift in the global flow of food; while developing countries produced a billion dollar yearly surplus in the 1970s, by 2004 they were importing US$11 billion a year (Holt-Giménez et al. 2009: 23–59).

Bertini and Glickman 2008; McIntire et al. 2009; Von Braun 2007; World Bank 2007). With the notable exception of the International Assessment of Agricultural Knowledge, Science and Technology for Development (IAASTD), these and subsequent reports rest on the following several problematic suppositions: that grain-fed meat consumption will expand in emerging economies; that arable land will be diverted to agrofuels; that financial speculation and price volatility in food commodities will continue unchecked; that production increases depend on transgenic, proprietary technologies and external inputs; that liberalized, global trade is essential to food security; and that the inevitable urbanization of the planet will require cheap, industrial food to feed the growing communities of poor people in the world's cities.

These suppositions buttress the political–economic assertion behind the 70% by 2050 call: it is not proprietary, Green Revolution agriculture and liberalized global markets that have caused the food crises *per se*, but their inefficient or inadequate *application*. Therefore, the solution is to do more of the same, over a greater area, more efficiently.

As Amin (2011) indicates, this neoliberal strategy is

> [supported] by the "absolute and superior rationale" of economic management based on the private and exclusive ownership of the means of production.... According to this principle, land and labor become merchandise like any other commodity, and are transferable at the market price in order to guarantee their best use for their owners and for society as a whole. This is nothing but a mere tautology, yet it is the one upon which all critical economic discourse is based.

Faced with stagnant global economic growth, this paradigm views the peasantry as a site for "accumulation by dispossession" (Harvey 2005: 137) and as a sector for potential market expansion. Because their numbers are growing at 8% a year, market access to the 2.5 billion farmers at the "Base of the Pyramid" has become attractive for global capital (World Economic Forum and Boston Consulting Group 2009).

As in the 1960–1980s, capital's key to the peasantry's land, factor, and commodity markets is once again the Green Revolution. Similar to the role once played by the Ford and Rockefeller foundations, the Bill and Melinda Gates Foundation is the Green Revolution's new philanthropic flagship, tasked with resurrecting the Consultative Group on International Agricultural Research (CGIAR) and obtaining broad social, financial, and government agreement (Holt-Giménez 2008; Patel et al. 2009). The new "Doubly Green Revolution" (Conway 1997, 2012) retains the same proprietary genetic foundations as the original Green Revolution, but has added transgenic technologies, global markets, environmental concerns, and a leading role for the private sector (Holt-Giménez 2013). The U.S. Agency for International Development's "Feed the Future," the Gates Foundation's "Alliance for a Green Revolution in Africa," and industry's "New Vision for African Agriculture" initiatives, for example, feature value chains, public–private partnerships, microfinance, village "agro-dealers," and smallholder contract farming (Gates Foundation 2008; World Economic Forum and Boston Consulting Group 2009).

Widespread social, environmental, and agricultural critiques of the Green Revolution notwithstanding (see Bello 2009; Freebairn 1995; Holt-Giménezet al. 2009; Magdoff and Tokar 2010; Patel 2013; Soil Association 2010; Toulmin et al. 2011; Winders 2009; Wittman et al. 2010), food regime institutions have steadily converged around the new Green Revolution agenda.

7.2 PEASANT AGRICULTURE AND AGROECOLOGY: A MEANS AND A BARRIER FOR THE GREEN REVOLUTION

The planet's smallholders and the practice of agroecology both constitute *a means and a barrier* to the expansion of capitalist agriculture. Smallholders subsidize capitalist agriculture with cheap labor and supply a vast, low-end factor market. This "functional dualism" between peasant and capital-intensive agriculture accelerates industrial expansion, resulting in the differentiation and displacement of the peasantry and the subsumption of peasant agriculture to capitalist agriculture (De Janvry 1981). At the same time, family labor, small farm size, diversified farming, and knowledge systems and smallholder's pluriactive livelihood strategies preserve peasant farming systems, presenting barriers and competition for capitalist agriculture (Netting 1993; Wilken 1988), resulting in the "persistence of the peasantry" (Edelman 2000: 14; Van Der Ploeg 2010, 2013).

Traditional agriculture was the cultural and ecological basis for the development of agroecology as a science (Altieri 1995; Gliessman 2007). Because it is rooted in smallholder systems and relies on agroecosystem management rather than external inputs, agroecology is also a barrier to Green Revolution technologies. Agroecology is knowledge intensive (rather than capital intensive) and tends toward small, highly diversified farms and emphasizes the ability of local communities to generate and scale-up innovations through farmer-to-farmer research and extension approaches (Holt-Giménez 2006).

The first Green Revolution drew in millions of smallholders, many of whom were forced out of farming by larger, better capitalized farmers, or went bankrupt after their soils became sterile and subsidized credit disappeared (see Hewitt de Alcántara 1976; Shiva 1991). Over 70% of the world's agrobiodiversity—largely held *in situ* in smallholder agroecosystems—was lost from farming (FAO 2009). When smallholder farms began crashing under Green Revolution methods in the 1970s, many farmers turned to agroecology in an effort to restore soil organic matter, conserve water, restore agrobiodiversity, and manage pests (Altieri 2004). Since the early 1980s, hundreds of nongovernmental organizations (NGOs) in Africa, Latin America, and Asia have promoted thousands of agroecology projects that incorporate elements of traditional knowledge and modern agroecological science (Altieri et al. 1998; Pretty 1995; Uphoff 2002). With the growing food, fuel, and climate crises, the importance of the ecological and social services provided by agroecological peasant agriculture is becoming widely recognized (de Schutter 2010; Holt-Giménez 2002).

In Latin America, the expansion of agroecology has produced cognitive, technological, and sociopolitical innovations, intimately linked to new political scenarios such as the emergence of progressive governments in Ecuador, Bolivia, and Brazil, and peasants/indigenous resistance movements (Ruiz-Rosado 2006; Toledo 1995). Thus, agroecology's "epistemological, technical, and social revolution" are mutually constitutive with social movements and political processes from below (Altieri and Toledo 2011: 587).

While the Green Revolution has been "greening" itself since its highly publicized renewal (CGIAR 1997), its champions have criticized agroecology's alleged low productivity and for not "scaling up." These criticisms ignore the evidence demonstrating the high productivity and resilience of agroecologically managed peasant agriculture (Badgley et al. 2009; Holt-Giménez 2002; Pretty 1995; Pretty and Hine 2000), and forget that scaling up the first Green Revolution required the massive structural mobilization of state and private-sector resources (Jennings 1988).

In spite of the fact that agroecology has spread widely through the efforts of NGOs, farmers' movements, and university projects, it remains marginal to official agricultural development plans, is still not a part of agricultural policy (see González de Molina 2013), and is dwarfed by the resources provided to the Green Revolution. In contrast, the remarkable scaling up of agroecology in Cuba stems in large part from the government's strong structural support (Rosset et.al. 2011). Asking "why can't agroecology scale up?" begs the question, "what is holding agroecology back?"

7.3 THE GREEN REVOLUTION AND AGROECOLOGY: MARRIAGE OR FUNCTIONAL DUALISM?

Given its popularity and its potential, some governments, universities, and even big philanthropy are selectively incorporating technical aspects of agroecology that do not challenge the power structures of the Green Revolution. The senior spokesperson for this position is Gordon Conway, whose

1997 book *The Doubly Green Revolution: Food for All in the 21st Century*, called for a less environmentally damaging Green Revolution. More recently, in *One Billion Hungry: Can We Feed the World?*, Conway, a pioneer in the field of integrated pest management and ecological resilience, again argues for feeding the world, this time through an apolitical mix of sustainable intensification, high-yielding varieties, biotechnology, and agroecological techniques. Similar positions are found among some organic farmers (Roland and Adamchak 2009) and ecologists (Foley 2011) who advocate a marriage between agroecology, organic farming, and biotechnology. Others seek to close yield gaps while reducing industrial agriculture's environmental footprint by increasing efficiency of inputs and/or deploying climate-smart genetic varieties (Baulcombe et al. 2009). The Gates Foundation is adding-on integrated soil fertility management to its projects (Gates Foundation 2008). Advocates for these approaches suggest that because of the severity of the food crisis, "we need all solutions," for example, ostensibly productive genetically modified crops (GMOs) and supposedly unproductive (but greener) agroecological practices (Gates 2009). Invariably, agroecology receives a fraction of the funding provided to Green Revolution technologies (GM Freeze 2011).

Agroecology is further subordinated to conventional agriculture by revisionist academic projects that erase its history, stripping it of its political content (e.g., Tomich et al. 2011). By coopting agroecology, relegating it to the margins of science and niche markets of the corporate food regime, these strategies advance a form of "functional dualism" (De Janvry 1981: 174). The political intent behind this technocratic reductionism (see González de Molina [2013], Chapter 4 this volume) is to mitigate the Green Revolution's socioenvironmental contradictions without altering the balance of power within the corporate food regime (McMichael 2013).

7.4 AGROECOLOGY AND FOOD SOVEREIGNTY MOVEMENTS

A new Green Revolution could conceivably concentrate food production on some 50,000 industrial farms worldwide (Amin 2011). Given the best land, subsidized inputs and favorable market access, these farms could produce the world's food (though not very sustainably). But how would 2.5 billion displaced smallholders buy this food? The alternative—smallholder-driven agroecological agriculture—was recognized by the authors of the IAASTD as the best strategy for rebuilding agriculture and ending rural poverty and hunger:

> [The wealth] of agricultural knowledge, science, and technology the world has built up ... should be targeted toward agroecology strategies that combine productivity with protecting natural resources like soils, water, forests, and biodiversity. In particular, the research and development efforts must now target and include in a participatory manner small scale and family farmers, since they make up the major part of the poor and hungry, while they also represent the major part of the stewards for the environment. Agricultural practices like organic, biodynamic, conservation, and agroecological are ... options that address the main constraints to food and nutrition security as well as food sovereignty issues (Herran and Hilmi 2011).

To be an effective strategy, major changes must be made in policies, institutions, and research priorities to create an enabling environment for peasant-based, agroecological development (Gonzáles de Molina, ibid). This transformation will likely require a combination of extensive on-the-ground agroecological practice *and* strong political will to overcome opposition and co-optation from the Green Revolution.

What could bring about the political will?

Smallholders working with movements like *Campesino a Campesino* (farmer-to-farmer) of Latin America and NGO networks for farmer-led sustainable agriculture like participatory land use management of Africa have restored degraded farmland using highly effective agroecological practices on hundreds of thousands of acres of land (Holt-Giménez 2006; Wilson 2011).

At the same time, peasant organizations fighting for agrarian reform have confronted commodity dumping, market-based land reform and more recently, extensive land grabs (Borras and Franco 2012; Rossetet al. 2006). The international peasant federation, Via Campesina, has called for *food sovereignty*, "The right of people's to healthy and culturally appropriate food, produced through ecologically sound and sustainable methods, and their right to define their own food and agricultural systems" (Patel 2009: 666). The cross-border globalization of these movements (Keck and Sikkink 1998) responds in part to the intensification of capital's enclosures and is partly a strategic decision to engage in global advocacy (Borras 2004).

The need for structural support for smallholders in locally based agroecology networks, and the globalized agrarian demands of the food sovereignty movement are complementary areas of strategic synergy (Holt-Giménez 2010). The food crisis is drawing them toward convergence.

When NGO federation PELUM brought over 300 farmer extensionists to Johannesburg to speak on agroecology at the World Summit on Sustainable Development, farmers formed the Eastern and Southern Africa Farmers Forum to address agrarian issues (Wilson 2011). Following the Rome food crisis meeting in 2008, Via Campesina met in Mozambique, where they signed a declaration for a smallholder's agroecological solution to the food crisis. Developments like this (and many others) suggest that the international call for food sovereignty is beginning to take root in smallholders' agroecology networks. Similarly, Via Campesina is steadily spreading agroecological approaches throughout its own farmer organizations (Martinez-Torres and Rosset 2010; Via Campesina 2010).

As local networks for agroecological practice merge with the transnational agrarian movements for food sovereignty they generate massive social pressure—pressure that is needed to tip the scales of political will in favor of food sovereignty and agroecology. This pressure can take the form of constitutional reform, for example, Ecuador's food sovereignty law (Patel 2009), grassroots campaigns, and civil society declarations linking agroecological practice to political practice (Via Campesina 2012), or the adoption of agroecology as a development strategy, for example, Brazilian Landless Worker's Movement's (Movimento dos Trabalhadores Sem Terra [MST]) schools and training programs.

However, this convergence faces historical divisions between agrarian-based farmer organizations and the NGO-based agroecology networks. The latter are more easily co-opted into technical and political approaches to agricultural development. This has led long-time agroecology practitioners to call for a shift in NGO behavior and priorities, from technology-led agendas, to strategies that support farmer-led political organizations (Batta et al. 2011).

The call for strategic alliances also comes from peasant leaders. Alberto Gómez of Mexico's National Union of Peasant Organizations affirms, "We have to form alliances with technicians or with NGOs that complement our activities … our struggle is not only in the political arena, in movement building, it is also about building local alternatives. It is about creating a different context for agriculture and peasant life. In this sense, there are complementarities" (Gómez in Holt-Giménez 2010: 228).

7.5 DISCUSSION

Like the capitalist economic system, the corporate food regime goes through periods of *liberalization* characterized by unregulated markets and massive capital concentration, followed by devastating busts and social upheaval. These are followed by *reformist* periods in which markets are regulated in an effort to restabilize the regime. Although these phases appear politically distinct, they are actually two sides of the same system. As Polanyi (1944) observed, if unregulated capitalist markets ran rampant indefinitely, they would eventually destroy the social and natural resource base of capitalist production. However, necessary reforms do not result from the good intentions of reformists. As liberal markets undermine society and environment, social conditions deteriorate, giving rise to strong *counter-movements* that force governments to reform their markets and institutions.

Holt-Giménez and Shattuck (2011) identify Neoliberal and Reformist trends within the corporate food regime. Both share a power base rooted in G-8 governments (France, Germany, Italy, Japan, United Kingdom, and the United States), multilateral institutions, monopoly corporations, and big philanthropy. The Neoliberal trend is hegemonic, grounded in economic liberalism, driven by corporate agrifood monopolies and managed by institutions such as the USDA (under Secretary of Agriculture Tom Vilsack), the CAP, the WTO, the private sector financing arm of the World Bank (International Finance Corporation), and the IMF. The Reformist trend is much weaker and managed by subordinate branches of the same institutions (e.g., Deputy Secretary of Agriculture Kathleen Merrigan and the public sector financing arm of the World Bank).

While the mission of Reform is to mitigate the excesses of the market, its "job" is identical to that of the Neoliberal trend: the reproduction of the corporate food regime. Reformists call for mild reforms like social safety nets, fair trade/organic niche markets, and apolitical, technology-focused renderings of agroecology.

Global food movements are characterized by two major trends: progressive and radical. Many actors within the progressive trend advance practical alternatives to industrial agrifoods, such as sustainable, agroecological, and organic agriculture. The radical trend also calls for practical alternatives, but focuses more on structural reforms to markets and property regimes, and class-based, redistributive demands for land, water, and resources, that is, food sovereignty (Holt-Giménez and Shattuck 2011).

Partly due to its academic and NGO-based history, agroecology has largely resided within the progressive trend. As such, agroecology is exposed to financial and political co-optation from the food regime's reformist projects. Nonetheless, many agroecologists work with radical peasant organizations and identify with food sovereignty, agrarian sustainability, and political agroecology. Radical, movement-based agroecology is shunned by the food regime in favor of depoliticized and project-based agroecology that is easily subsumed under Green Revolution agendas. Given the political and financial power of the corporate food regime, many academic programs and NGOs "follow the money" in difficult economic times, depoliticizing their work and accommodating to Green Revolution/global market objectives. However, the unchecked neoliberal expansion of industrial agriculture also radicalizes agriculture (and agroecology) on the ground, as smallholders fight for survival.

7.6 CONCLUSION: PREVENTING CO-OPTATION, STRENGTHENING AGROECOLOGY

The functional dualism of capitalist agriculture utilizes the new Green Revolution to convert smallholders and agroecology into means (rather than barriers) for the expansion of industrial agriculture. The resulting Neoliberal enclosure of seeds, land, and markets will likely destroy the livelihoods of most of the planet's 2.5 billion smallholders, further reduce agrobiodiversity, and severely weaken global agroecosystem resilience. These developments will increase global hunger and limit our ability to mitigate and cope with climate change.

Agroecology has a pivotal role to play in the future of our food systems. If agroecology is co-opted by reformist trends in the Green Revolution, the corporate food regime will likely be strengthened, the counter-movement weakened, and the substantive reforms to the corporate food regime unlikely. However, if agroecologists build strategic alliances with radical food sovereignty struggles, the counter-movement to the corporate food regime could be strengthened. A strong counter-movement could generate considerable political will for the transformative reform of our food systems. The livelihoods of smallholders, the elimination of hunger, the restoration of the planet's agrobiodiversity, and agroecosystem resilience would all be better served under this scenario.

REFERENCES

Altieri, M. *Agroecology: The Science of Sustainable Agriculture*. Boulder, CO: Westview Press, 2004.

Altieri, M. "Linking ecologists and traditional farmers in search for sustainable agriculture." *Frontiers in Ecology and the Environment* 2 (1995): 35–42.

Altieri, M., P. Rosset, and L.A. Thrupp. *The Potential of Agroecology to Combat Hunger in the Developing World. 2020*. Washington, DC: IFPRI/International Food Policy Research Institute, 1998.

Altieri, M., and V. Toledo. "The agroecological revolution in Latin America: Rescuing nature, ensuring food sovereignty and empowering peasants." *Journal of Peasant Studies* 38 no. 3 (2011): 587–612.

Amin, S. "Food sovereignty: a struggle for convergence in diversity." In *Food Movements Unite! Strategies to Transform our Food Systems*. E. Holt-Giménez (Ed.) (Oakland, CA: Food First Books, 2011), xi–xviii.

Badgley, C., J. Moghtader, E. Quintero, E. Zakem, M.J. Chappell, K. Avilés-Vázquez, et al. "Organic agriculture and the global food supply." *Renewable Agriculture and Food Systems* 22, no. 2 (2009): 86–108.

Bailey, R. *Growing a Better Future: Food Justice in a Resource-Constrained World*. London: OXFAM, 2011.

Batta, F., S. Brescia, P. Gubbels, B. Guri, J.B. Cantave, and S. Sherwood. "Transforming NGO roles to help make food sovereignty a reality." In *Food Movements Unite! Strategies to Transform our Food Systems*. E. Holt-Giménez (Ed.) (Oakland, CA: Food First Books, 2011), 93–114.

Baulcombe, D., I. Crute, B. Davies, J. Dunwell, M. Gale, J. Jones, et al. *Reaping the Benefits: Science and the Sustainable Intensification of Global Agriculture*. London: The Royal Society, 2009.

Beddington, J. *Foresight. The Future of Food and Farming. Final Project Report*. London: The Government Office for Science, 2011.

Bello, W.F. *The Food Wars*. London: Verso, 2009.

Bertini, C., and D. Glickman. *Renewing American Leadership in the Fight Against Global Hunger and Poverty*. The Chicago Initiative on Global Agricultural Development. Chicago Council on Global Affairs, 2008. http://www.thechicagocouncil.org/UserFiles/File/GlobalAgDevelopment/Report/gadp_final_report.pdf (accessed March 28, 2012).

Borras Jr., S.M. *La ViaCampesina: An Evolving Transnational Social Movement*. Amsterdam: Transnational Institute, 2004.

Borras, Jr., S.M., and J.C. Franco. "Global land grabbing and trajectories of agrarian change: A preliminary analysis." *Journal of Agrarian Change* 12 no. 1 (2012): 34–59.

Brown, L.R. "The new geopolitics of food." *Foreign Policy*, 2011. http://www.foreignpolicy.com/articles/2011/04/25/the_new_geopolitics_of_food (accessed July 10, 2012).

CGIAR Secretariat. *CGIAR Annual Report: CGIAR 25 Years, 1971–1996*. Consultative, Group on International Agricultural Research. Washington, DC, 1997.

Collier, P. "The politics of hunger: How illusion and greed fan the food crisis." *Foreign Affairs* 187 (2008): 67–79.

Conforti, P. *Looking Ahead in World Food and Agriculture*. Food and Agriculture Organizationof the United Nations, 2010. http://www.fao.org/docrep/014/i2280e/i2280e00.htm (accessed March 28, 2012).

Conway, G. *The Doubly Green Revolution*. Oxford: Penguin Books, 1997.

Conway, G. *One Billion Hungry: Can We Feed the World?* Ithaca, NY: Cornell University Press, 439 pp., ISBN 978-0-8014-7802-4, 2012.

De Janvry, A. *The Agrarian Question and Reform in Latin America*. Baltimore and London: John Hopkins University, 1981.

de Schutter, O. *Report Submitted by the Special Rapporteur on the Right to Food. Human Rights Council 16th Session*. United Nations General Assembly, New York, NY: United Nations, 2010.

Edelman, M. "The Persistence of the Peasantry." *North American Congress on Latin America Report on the Americas* 33 no. 5 (2000).

FAO. *First Fruits of Plant Gene Pact*. Food and Agriculture Organization of the United Nations, 2009. http://www.fao.org/news/story/0/item/20162/icode/en/ (accessed April 7, 2012).

FAO. *The State of Food Insecurity in the World: How Does International Price Volatility Affect Domestic Economies and Food Security?* Food and Agriculture Organization of the United Nations, 2011. http://www.fao.org/docrep/014/i2330e/i2330e.pdf (accessed April 7, 2012).

Foley, J.A. "Can we feed the world, sustain the planet?" *Scientific American* 305 no. 5 (2011): 60–65.

Freebairn, D. "Did the Green Revolution concentrate incomes? A quantitative study of research reports." *World Development* 23 no. 2 (1995): 265–279.

Freeze, G.M. *The Bill and Melinda Gates Foundation, Biotechnology and Intensive Farming. GM Freeze*, 2011. http://www.gmfreeze.org/site_media/uploads/publications/Gates_brief_final.pdf (accessed March 10, 2012).

Friedmann, H. The political economy of food: A global crisis. *New Left Review* 197 (1993): 29–57.

Gates Foundation. *Agricultural Development Strategy, 2008–2011*. Seattle: Bill and Melinda Gates Foundation, 2008.

Gates, B. World Food Prize Symposium. Des Moines, IA, 2009. http://www.gatesfoundation.org/speeches-commentary/Pages/bill-gates-2009-world-food-prize-speech.aspx (accessed April 4, 2012).

Gliessman, S.R. *Agroecology: The Ecology of Sustainable Food Systems*. New York, NY: Taylor and Francis Group, 2007.

González de Molina, M. "Agroecology and politics: How to get sustainability? About the necessity for a political agroecology." *Agroecology and Sustainable Food Systems* 37 no. 1 (2013): 45–59.

Harvey, D. *A Brief History of Neoliberalism*. Oxford: Oxford University Press, 2005.

Herren, H., and A. Hilmi. "Agriculture at a crossroads." In *Food Movements Unite! Strategies to Transform our Food Systems*. E. Holt-Giménez (Ed.) (Oakland: Food First Books, 2011), 243–256.

Hewitt de Alcántara, C. *Modernizing Mexican Agriculture: Socioeconomic Implications of Technological Change, 1940–1970*. Geneva: United Nations Research Institute for Social Development, 1976.

Holt-Giménez, E. "Measuring farmers' agroecological resistance after Hurricane Mitch in Nicaragua: A case study in participatory, sustainable land management impact monitoring." *Agriculture, Ecosystems & Environment* 93 (2002): 87–105.

Holt-Giménez, E. "Out of AGRA: The Green Revolution returns to Africa." *Development* 51 no. 4 (2008): 464–471.

Holt-Giménez, E. *Campesino a Campesino: Voices from Latin America's Farmer to Farmer Movement for Sustainable Agriculture*. Oakland, CA: Food First Books, 2006.

Holt-Giménez, E. "Grassroots voices." *Journal of Peasant Studies* 37 (2010): 226–229.

Holt-Giménez, E. "Review of *One billion hungry: Can we feed the world?* by Gordon Conway." *Agroecology and Sustainable Food Systems* 37 no. 8 (2013): 968–971.

Holt-Giménez, E., R. Patel, and A. Shattuck. *Food Rebellions! Crisis and the Hunger for Justice*. Oakland, CA: Food First Books, 2009.

Holt-Giménez, E., and A. Shattuck. "Food crises, food regimes and food movements: Rumblings of reform or tides of transformation?" *Journal of Peasant Studies* 38 no. 1 (2011): 109–144.

Jennings, B. *Foundations of International Agricultural Research: Sciences and Politics in Mexican Agriculture*. Boulder, CO: Westview Press, 1988.

Keck, M.E., and K. Sikkink. *Activists Beyond Borders: Advocacy Networks in International Politics*. Ithaca, NY: Cornell University Press, 1998.

Magdoff, F., and B. Tokar. (Eds.) *Agriculture and Food in Crisis: Conflict, Resistance, and Renewal*. New York, NY: Monthly Review Press, 2010.

Martinez-Torres, M.E., and P. Rosset. "La ViaCampesina: The birth and evolution of a transnational peasant movement." *Journal of Peasant Studies* 37, no. 1 (2010): 149–176.

McIntire, B., H. Herren, J. Wakhungu, and R. Watson. *Agriculture at a Crossroads: International Assessment of Agricultural Knowledge, Science and Technology for Development*. Washington, DC: Island Press, 2009. http://www.agassessment.org/.

McMichael, P. "A food regime genealogy." *Journal of Peasant Studies* 36 no. 1 (2009): 139–169.

McMichael, P. *Food Regimes and Agrarian Questions*. Halifax: Fernwood Publishing, 2013.

Netting, R.M. *Smallholders, Householders: Farm Families and the Ecology of Intensive Sustainable Agriculture*. Stanford, CA: Stanford University Press, 1993.

Patel, R., E. Holt-Giménez, and A. Shattuck. "Ending Africa's hunger." *The Nation* 21 (2009): 17–22. http://www.foodfirst.org/en/node/2556 (accessed April 7, 2012).

Patel, R. "Grassroots voices: What does food sovereignty look like?" *Journal of Peasant Studies* 36 no. 3 (2009): 663–706.

Patel, R. "The long Green Revolution." *Journal of Peasant Studies* 40 no. 1 (2013): 1–63.

Polanyi, K. *The Great Transformation*. Boston, MA: Beacon Press, 1944.

Pretty, J. *Regenerating Agriculture; Policies and Practice for Sustainability and Self-Reliance.* London: Earthscan Publications, 1995.

Pretty, J., and R. Hine. *Feeding the World with Sustainable Agriculture: A Summary of New Evidence. Final Report from SAFE-World Research Project.* Colchester: University of Essex, 2000.

Roland, P.C., and R.W. Adamchak. *Tomorrow's Table: Organic Farming, Genetics and the Future of Food.* Oxford: Oxford University Press, 2009.

Rosset, P., R. Patel, and M. Courville. *Promised Land: Competing Visions of Agrarian Reform.* Oakland, CA: Food First Books, 2006.

Rosset, P., B. Sosa, A. Jaime, and D. Lozano. "The Campesino-to-campesino agroecology movement of ANAP in Cuba: Social process methodology in the construction of sustainable peasant agriculture and food sovereignty." *Journal of Peasant Studies* 38 no. 1 (2011): 29–30.

Ruiz-Rosado, O. "Agroecologia; Una Disciplina que tiende a la Transdisciplina." *Interciencia* 31 no. 2 (2006): 140–145.

Shiva, V. *The Violence of the Green Revolution: Third World Agriculture, Ecology and Politics.* London: Zed Books, 1991.

Soil Association. *Telling Porkies: The Big Fat Lie about Doubling Food Production.* Bristol: Soil Association, 2010.

Toledo, V. *Peasantry, Agroindustriality, Sustainability: The Ecological and Historical Basis of Rural Development.* Working Paper. Morelia, Mexico: Interamerican Council for Sustainable Agriculture, 1995.

Tomich, T., S. Brodt, F. Ferris, R. Galt, W. Horwath, E. Kebreab, et al. "Agroecology: A review from a global-change perspective." *Annual Review of Environment and Resources* 36, no. 15 (2011): 1–30.

Toulmin, C., P. Bindraban, S. Borras Jr., E. Mwangi, and S. Sauer. *Land Tenure and International Investments in Agriculture.* Rome: Committee on World Food Security, 2011.

Uphoff, N. *Agroecological Innovations: Increasing Food Production with Participatory Development.* London: Earthscan, 2002.

Van Der Ploeg, J.D. "The peasantries of the twenty-first century: The commoditization debate revisited." *Journal of Peasant Studies* 37 (2010): 1–30.

Van Der Ploeg, J.D. *Peasants and the Art of Farming: A Chayanovian Manifesto.* Halifax: Fernwood Publishing, 2013.

Via Campesina. "*Sustainable Peasant's Agriculture.*" Via Campesina, 2012. http://viacampesina.org/en/index.php?option=com_content&view=category&layout=blog&id=17&Itemid=42 (accessed July 24, 2012).

Via Campesina. "*Sustainable Peasant and Family Farm Agriculture Can Feed the World.*" Via Campesina (2010). http://www.foodmovementsunite.org/addenda/via-campesina.pdf (accessed July 24, 2012).

Von Braun, J. *The World Food Situation: New Driving Forces and Required Actions.* Washington, DC: International Food Policy Research Institute, 2007.

Wilken, G. *Good Farmers: Traditional Agricultural Resource Management in Mexico and Central America.* Berkeley, CA: University of California Press, 1988.

Wilson, J. "Irrepressibly toward food sovereignty." In *Food Movements Unite! Strategies to Transform our Food Systems.* E. Holt-Giménez (Ed.) (Oakland, CA: Food First Books, 2011), 71–92.

Winders, B. "The vanishing free market: The formation and spread of the British and US food regimes." *Journal of Agrarian Change* 9 (2009): 315–344.

Wittman, H.K., A.A. Desmarais, and N. Wiebe. *Food Sovereignty: Reconnecting Food, Nature and Community.* Oakland, CA: Food First Books, 2010.

World Bank. *World Development Report 2008: Agriculture for Development.* Washington, DC: World Bank, 2007.

World Economic Forum and Boston Consulting Group. *The Next Billions: Business Strategies to Enhance Food Value Chains and Empower the Poor.* Geneva, Switzerland: World Economic Forum, 2009.

CHAPTER 8

The Intercultural Origin of Agroecology
Contributions from Mexico

Francisco J. Rosado-May

CONTENTS

8.1 INTRODUCTION

The number of publications on agroecology in Mexico has notably expanded during the last 20 years of the twentieth century and continues to grow in this new century. Many of the early publications present a definition of agroecology and mention related concepts, such as the *agroecosystem*. More recently, several articles also present the history of the concept to develop a theoretical framework that better categorizes where each contribution is located within the field (Wezel et al. 2009). Nevertheless, only a few papers present an epistemological analysis on the construction of agroecological thinking, and none discuss the importance of intercultural processes in the construction of both the concept and the field of agroecology. Notable contributions include Hecht (1999) who wrote "La evolución del pensamiento agroecológico (The evolution of agroecological thinking)" in the book *Agroecología: Bases Científicas Para Una Agricultura Sustentable (Agroecology, the Scientific Basis for Sustainable Agriculture)* Altieri (1999); Ruiz-Rosado (2006), who argues that agroecology is a transdisciplinary field; and Guadarrama-Zugasti (2007), who explores paradigms from biotechnology, computing sciences, and others, in the construction of sustainable agriculture, and contrasts them with the development of agroecology. In 2010, SOCLA published a book edited by Altieri (2010b), *Vertientes del Pensamiento agroecológico: Fundamentos y Aplicaciones (Approaches to Agroecological Thinking: Foundations and Applications)*, in which the use of new technologies, social equity, and the design of sustainable agroecosystems are discussed. The contributions by Wezel and Soldat (2009), Wezel et al. (2009), and Tomich et al. (2011) have been reviewed by Gliessman (2013) who argues that agroecology was also the result of a movement to resist the penetration of Green Revolution technology in the humid tropics of Mexico. Gliessman presents agroecology as a viable alternative to confront global challenges related to food production, environmental degradation, and improvement of life conditions.

Based on the above, it is possible to say that agroecology has grown significantly in importance. It has captured the attention not only of those producing food, but also of those in decision-making positions. Agroecology offers new and creative approaches and new thinking to address issues regarding food production, social equity, and environmental concerns. Thus, it is very important to understand, epistemologically, the origins and the conceptualization of agroecology, as well as the processes that led to its construction as a field. This understanding has the potential to expand the present avenues and to open new ones, especially for understanding future processes related to food production.

8.2 THE EVOLUTION OF AGROECOLOGICAL THINKING: THE ROLE OF MEXICO

The term "agroecology" is not recent (Gliessman 2013, 2015; Wezel et al. 2009). Authors such as Klages and Bensin independently used the term in 1928 and 1930. Both are considered in the English language agroecological literature as the first ones to use the term. Bensin's use of the term "agroecological" refers to research carried out in Mexico on the ecological needs for the management of local varieties of corn. In most papers, reviewing the contribution of authors to the development of agroecology, with the exception of Gliessman (2013), Mexican authors are rarely included. However, an in-depth review of the literature shows that in 1926 the "First Agroecological Congress" was carried out in Meoqui, Chihuahua, Mexico, organized by Enrique Peredo Reyes y Mario MatíasVillada (Sánchez Martínez 2012, p. 26).

The concern for ecological/environmental factors affecting food production has been strongly present in the English language literature since the early 1920s. In those years, Mexico was just ending an internal war known as the Mexican Revolution. In 1921, Rómulo Escobar, founder of the Higher Agricultural School in Ciudad Juárez, Chihuahua, made visible the need for technical studies before giving land to farmers. In 1923, the federal government opened the Office for Experimental Fields to be used for food production and training technicians in soil science through the scientific method. Very soon it was determined that, to improve food production, there had to be a better understanding of the environment, especially water management. The need for an ecological approach for food production was in the mind of the organizers of the First Agroecological Congress in Meoqui, Chihuahua, in 1926.

Subsequently, and with the word "agroecological" in mind, other processes took place many years later than this initial exploration in 1926. For the following 60 years, both the term and concept of agroecology seemed to remain dormant in Mexico and in other countries. It was "rediscovered" in the 1970s, when Mexican researchers revisited agricultural ecosystems in an effort to better understand their complexities. In doing so, they found that the complexity of all food systems was beyond ecological or environmental factors. The need for a broader understanding was very apparent and much needed, which led to the development of different avenues and approaches for the study of food production systems.

Among the most used definitions of agroecology in Mexico, the following best reflect key factors related to agroecological epistemological thinking: Altieri (1995) considers agroecology as the "science of sustainable agriculture;" Francis et al. (2003) indicate that agroecology is "the integral study of the ecology of the food system, including ecological as well as economical and social factors;" Gliessman (2007) considers that agroecology is "the application of concepts and ecological concepts and principles to the design and management of sustainable food systems."

In exploring the epistemological bases of agroecological thinking, it is important to mention other authors. Hecht (1999), for example, acknowledged the importance of empirical, nonscientific, knowledge (e.g., pre-Hispanic), and identified three processes that have contributed to the denial of the importance of traditional knowledge: (1) destruction of the means of transmission of knowledge form one generation to the next, (2) the dramatic changes in aboriginal societies, and (3) the emergence and positioning of positivism. According to Hecht, these were the basic conditions that led to the development of agroecology. Altieri (2010b), on the other hand, considered that agroecology provides a much broader approach and allows for a better understanding of the complexities of modern food production systems, from a foundational technical approach to including social, economic, political, cultural, and environmental dimensions. Today's concern is for the sustainability of the entire food system.

The literature reviewed, so far, may lead one to conclude that the different approaches, positions, or epistemological considerations of agroecology are the result of a process that was developed almost entirely in developed countries, mainly the United States and Europe. A reader with a little background in the subject may also think that less developed countries, such as Mexico, had very little or nothing to do with the evolution of agroecology. Contrary to these conclusions, this chapter presents the following hypotheses:

1. Mexican scientists played very important roles in the creation and evolution of agroecology. Their role has not been fully acknowledged.
2. The process that led to the construction of agroecology in Mexico was a highly sophisticated one.
3. Agroecology is the result of an intercultural process, which was affected by the unique conditions that took place in Tabasco, Mexico. These included the following: (a) the opening of the *Colegio Superior de Agricultural Tropical* (College of Tropical Agriculture [CSAT]); (b) the study of traditional agriculture at CSAT; (c) the contributions and leadership of Efrain Hernández Xolocotzi in the study of traditional agriculture; (d) the movement to resist the introduction of the Green Revolution; and (e) a very strong intercultural component, which integrated agroecology science with indigenous agricultural knowledge, all led to the creation of modern agroecology.

All these factors were unique in space and time and did not occur or coexist anywhere else. I believe that by understanding this congruence and explicitly embracing intercultural processes in the construction, application, and transmission of knowledge, agroecology could have a stronger impact and presence worldwide.

To test this hypothesis, the following elements must be considered:

1. To know the historical development of agroecology in Mexico, more research on literature that has been previously unknown should be reviewed; personal files and interviews with actors who have been involved in the process through significant research, teaching, and publishing need to be considered.
2. Interconnections between authors should be developed and focus on looking to influence academia, creating a network that could show something similar to hubs of knowledge.
3. Identification and epistemological analysis of key concepts used by scientists and farmers.
4. Comparison and integration of different forms of knowledge construction, including both local and scientific methods.

8.3 IMPORTANT CONTRIBUTIONS BY MEXICAN AUTHORS

In Mexico, different authors interested in understanding how food systems work have made contributions from different disciplines, including anthropology (e.g., González Jácome 2004), using methodologies proposed by Palerm (1967), combining anthropology and biology (González Jácome and Del Amo Rodríguez 1998), combining anthropology with agronomy and ecology (Mariaca Méndez et al. 2010), ethnobotany (Martínez Alfaro 1988), chemical ecology (Anaya 2007;

Espinosa-García 2001), political ecology (Trujillo-Ortega 2009), community development (Jiménez-Osornio et al. 2008), and phytopathology (García Espinosa 2010; Rosado-May et al. 1985). There is no doubt that there are many more authors and papers than those listed here. The idea of presenting this list was to show that in Mexico there is an important and diverse body of relevant literature for the field of agroecology. An interesting idea would be to carry out an exercise similar to the one presented by Wezel and Soldat (2009), using only the Mexican literature. The results might not only be similar to the ones reported by those authors, but it can be hypothesized that epistemological approaches might be found for understanding the complexity of food systems in Mexico.

The greatest pillars that made possible the development of agroecology in Mexico, and eventually other countries, while constructing a broad understanding of food production systems are Faustino Miranda González, Enrique Beltrán, Alfredo Barrera Marín, Efraím Hernández Xolocotzi, Jerzy Rzedowsky, Arturo Gómez-Pompa, Ángel Palerm, José Sarukhán Kermez, Teresa Rojas Rabiela, and Alba González Jácome, among others. The pre-Hispanic dimension in this process is reflected, for instance, in the works by Gómez-Pompa et al. (1990), who reported cacao production systems in Yucatan, Mexico; and Gómez-Pompa and Kaus (1999) proposing that the future of conservation of natural resources rests in understanding and applying pre-Hispanic knowledge. In 1976, Sarukhán proposed an agroecosystem paradigm to guide the development of food production in Mexico. Later, Caballero and Sarukhán (1987) proposed again the importance of ecological factors in food production. Mass et al. (1988) demonstrated the importance of good management for the conservation of tropical soils.

The work of Efraim Hernández Xolocotzi deserves special attention, as the one who called attention to how traditional agriculture was in fact a very sophisticated technology. With the exception of Gliessman (2013), the role played by Hernández X, as he is known in Mexico, in the development of both the modern concept and methodology of agroecology has been largely ignored in the international agroecological literature. Hernández X had been the recipient of teachings from Faustino Miranda (botanist), Alfredo Barrera (anthropologist), and Enrique Beltrán (biologist). In turn, Hernández X had great influence over Arturo Gómez-Pompa, José Sarukhán, Stephen R. Gliessman, and Alba González Jácome, among other imminent scientists. Figure 8.1 shows the connections among authors, based on the academic influence over different generations. The nodes that form around Hernández X and Gliessman are evident, and both of these scholars had great influence in the development of agroecology in Mexico and beyond.

A prolific author, Hernández X published his first work in English in 1946, describing traditional agricultural systems. In 1951, he coauthored with Wellhausen the description of corn varieties and their geographical distribution. In 1954, he affirmed that agriculture represents a farmer-based natural resource management system. In 1960, he demonstrated that agricultural systems must be studied not only by the natural sciences but also other fields, especially sociology and economics, because food systems represent a core factor in the social development of Mexico. In 1961, Hernández X demonstrated the importance of applying biology in agricultural systems. In 1967, he did the same for ecology. In 1972, Hernández X described how ethnobotany could articulate elements from different fields of science and apply them to agricultural systems. From 1973 to 1978, the emphasis of his research was on the technology used by traditional farmers. In this period of time, under the influence of the agroecosystem concept presented by Janzen (1973), Hernández X and Ramos Rodríguez (1977) advanced the concept of the agroecosystem within the context of the technology applied in traditional agricultural systems and their role in modern society.

In 1977, Hernández X and Ramos Rodríguez published a seminal paper entitled "Methodology for the Study of Agroecosystems with Emphasis on the Technology Used in Traditional Agriculture." In this paper, the authors present a very well-organized and elegant system with which to understand the structure and function of agroecosystems, allowing a highly reliable means to organize, articulate, and apply information in the management and decision making of food production systems.

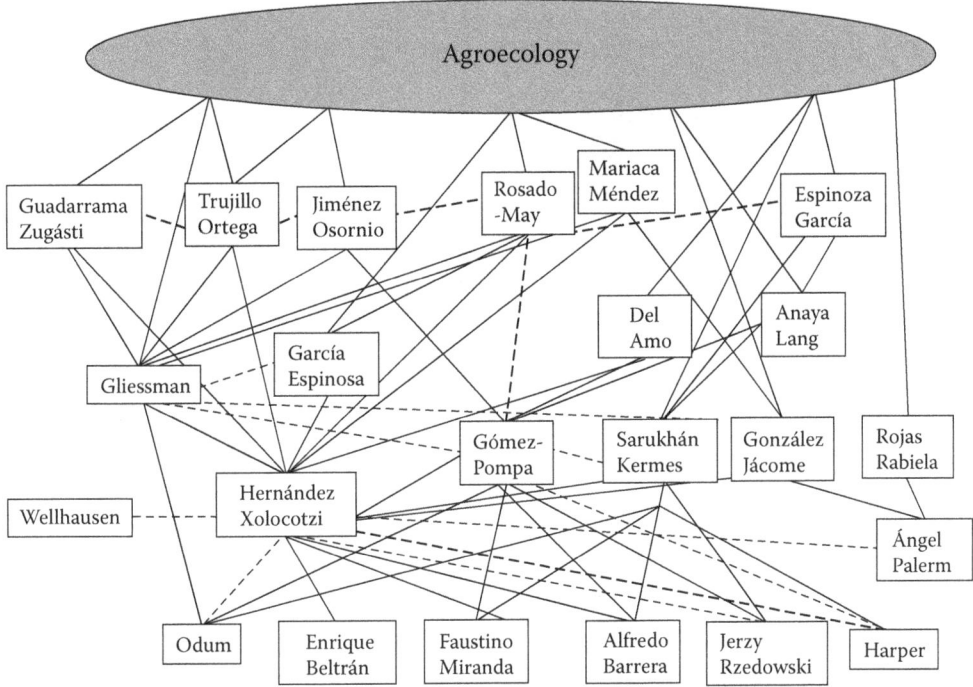

Figure 8.1 Sources of influence between authors that made possible the development of agroecology in Mexico.

The following year, 1978, was critical for the development of concepts around agroecology. In Cárdenas, Tabasco, Mexico, both the Colegio de Posgraduados and the CSAT, two academic institutions dedicated to the study of agriculture in the humid tropics of Mexico, held a meeting on Traditional Agricultural Technology (TAT) (Gliessman 1978). In this meeting, Hernández X was highly conspicuous for two reasons. First, he presented for the first time the concept that food production systems should be seen as a process, meaning that agricultural systems have a dynamic equilibrium through time, such as ecosystems. Second, along with other authors (most of them his own graduate students), Hernández X presented a conceptual paper considered the most important of all of his contributions to the development of agroecology, in which they posed the hypothesis that the structure and function of agroecosystems can be explained by understanding the interaction of three factors: the ecological, the technological, and the socioeconomic (Figure 8.2). In other words, to understand agroecosystems, a holistic, integrated vision is required, and it must be placed in the context of a dynamic process in space and time. This vision and process are mediated by how humans decide to manage the agroecosystem. The same figure is presented by Gliessman (2013) to sustain the argument that agroecology was also the result of a movement of resistance to the penetration of the Green Revolution in rural Mexico. Although the Green Revolution emphasized primarily the economic and technological factors of agroecosystems, agroecology integrates ecological, social, and cultural factors as well, providing a broader understanding of both the system and the production process.

In 1978, there was a boom of research on agroecology in Mexico, from different disciplines (e.g., Guadarrama-Zugasti 2007; Jimenez-Osornio et al. 2008; Rosado-May and García Espinosa 1986; Rosado-May et al. 1986; Ruiz-Rosado 2006; Trujillo-Ortega 2009), all of which confirm the concepts and hypothesis presented by Efraim Hernández Xolocotzi. All of this work was clearly

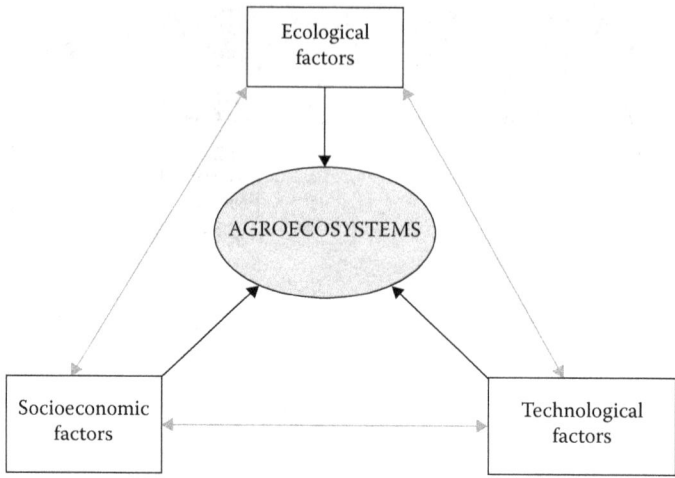

Figure 8.2   Model proposed by Hernández X (1978) to understand agroecosystems. The structure and function of an agroecosystem are explained by the dynamic interaction of the ecological, the socioeconomical, and the technological factors.

influenced by "Xolocotzian" thinking. Hernández X took from Odum (1971) the idea that the ecosystem, with the elements of structure and function, is the fundamental unit of study in ecology. In the same way, the agroecosystem is the unit of study for agroecology, only that any agroecosystem's structure and function is determined by the interaction of the three factors presented in Figure 8.2. Most definitions of agroecology include the three factors: the ecological, technological, and socioeconomic. But most importantly, as parts of the socioeconomic realm, Hernández X considered factors such as the political, cultural, and even the ethical.

8.4 THE TERM AGROECOLOGY

Based on the above discussion, it is possible to conclude that there was a significant conceptual and methodological development, in Mexico, focused on better understanding agroecosystems. It is also important to mention that the study of traditional agricultural systems, both in the present and the pre-Hispanic era, has led to an understanding that today's agricultural systems are the result of complex processes that began generations ago. This fact contributed significantly to the development of concepts and methodologies around agroecology in Mexico, and implies a process of an intercultural construction of new knowledge.

Based on the literature reviewed, it is possible to observe that the first term used in Mexico was not agroecology as such, but rather it was *agroecológico* (Peredo and Matías 1926). Bensin used the same term in 1928, in English as *agroecological*. It was not until 1973 when Janzen presented the term *agroecosystem* that the ecosystem concept began to be applied more broadly. In Mexico, the evolution of agroecological thinking and the progressive development of terms can be traced by the use of terminology, which began with "traditional agriculture," followed by "traditional agriculture technology," "agricultural production system," which changed to "agrosystem," then to "agroecosystem," all of them presented by Hernández X. This progression showed an exceptional process of evolution in thinking, both conceptual and methodological (Figure 8.3), preparing, slowly but surely, the foundation for the term agroecology, which would represent a new holistic understanding of the structure, function, and management of agroecosystems.

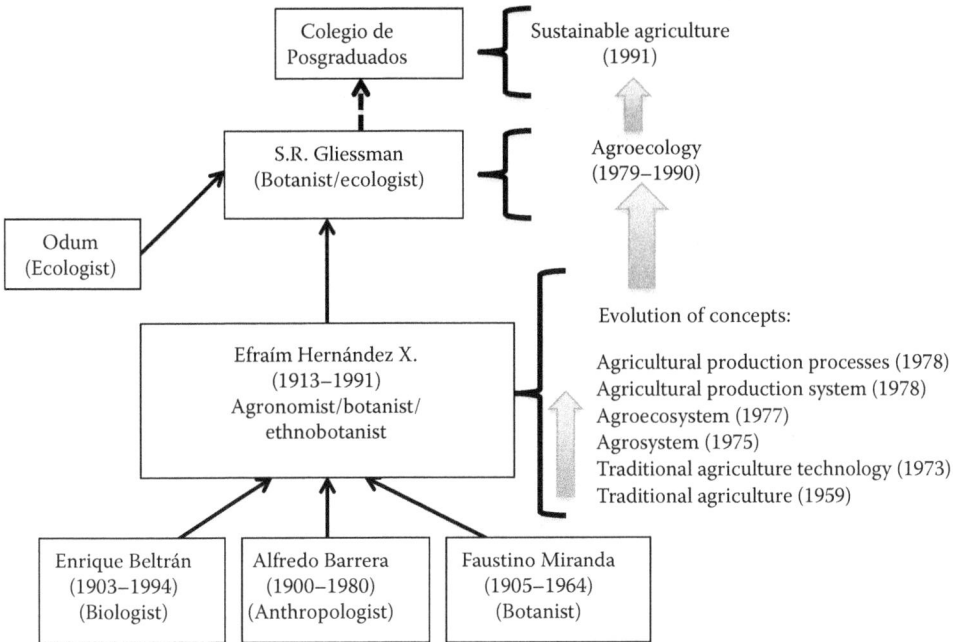

Figure 8.3 Evolution of agroecological thinking in Mexico.

It might seem obvious today that the best term to reflect the philosophy, the complexity, and the holistic structure and function of agroecosystems is agroecology. It was not obvious back in the 1960s and early 1970s in Mexico; terms such as agrobiology and ethnoecology, among others, were tried unsuccessfully (Gliessman 2013). In the international literature, Papadakis (1938) introduced "Crop Ecology," Janzen (1973) introduced the term "agroecosystem," and Cox and Atkins (1979) introduced "Agricultural Ecology." In Mexico, between 1979 and 1981, Gliessman organized and coordinated "International Courses on Tropical Ecology with an Agroecological Approach." These courses took place at the CSAT based in Tabasco, Mexico, and were financed by the Mexican Council of Science and Technology. Gliessman (2013) (Chapter 2 in this volume) describes well the work in agroecology carried out in CSAT. The students of those courses, who were mostly Mexican, interacted with scientists from Mexico and the United States, including Efraim Hernandez Xolocotzi, Alba Gonzalez Jácome, Miguel Martinez Alfaro, Ricardo Almeida Martínez, Francisco Maldonado, Roberto García Espinosa, Steve Gliessman, Miguel Altieri, John Vandermeer, Deborah Letourneau, Stephen Risch, and Douglas Boucher, among others. Several of the materials presented in the proceedings of those international courses contained the term "agroecology." Agroecology was freely used to reflect the idea of understanding agroecosystems beyond the technological and economic factors promoted by the Green Revolution. It was also used as a way to imply that other factors must also be incorporated to understand the complexity of food systems. It provided a framework that was used to look for viable alternatives to deal with field production problems and to offer a real alternative to the Green Revolution.

It is important to note that Gliessman was hired in 1976 as an ecologist to teach at CSAT. This situation was very exceptional. CSAT was designed to strengthen the advancement of the Green Revolution in the humid tropics of Mexico, but among its faculty there were ecologists, teaching the application of ecology to tropical agriculture. In 1981, Gliessman et al. published what has become an important foundational work in agroecology: *The Ecological Basis for the Application of Traditional Agricultural Technology in the Management of Tropical Agroecosystems*. The paper's main focus was to ecologically validate traditional agriculture, not to directly confront the Green Revolution. However, this step was critical for the future development of agroecology as an alternative movement.

When it was demonstrated that TAT had strong ecological foundations, it was interpreted as also meaning that pre-Hispanic agricultural systems reflected in local traditional systems had a strong rationale similar to modern ecology. In 1984, Gliessman published *Agroecology and Agroforestry* and in 1990, the edited book *Agroecology: Researching the Ecological Basis for Sustainable Agriculture*. Thus, the term agroecology began to become visible in major publications in English. This offered a new vision, solid framework, and methodology to back up the research that looks for alternatives to agricultural systems that have negative impacts, not only on the environment, but also on the social and economic fabric of society. At this point, knowing the important influence that CSAT had in the process, it is safe to consider that, in Mexico, Gliessman is known as the "Father of Agroecology." Gliessman's key achievements were that he was able to analyze the great body of Mexican research, was able to articulate it with knowledge from elsewhere in the world, and had the cultural sensibility to understand the greater picture of a process that involved several components, as follows: (1) local knowledge; (2) an institutional setting; (3) ecological training; (4) the ability to sense the movement of resistance to the Green Revolution; and (5) the need to create a viable alternative, reflected in the term agroecology. As Hernández X did in 1978, Gliessman inspired a significant generation of research and publications related to agroecology, which were also highly visible internationally, starting with the publication of his book in 1990.

The term agroecology also represents an important and inspiring cohesive factor among different research disciplines and people from different cultural backgrounds. Its evolution points toward ecology as the root of the term. Almost all authors reviewed in this chapter started their journey into agroecology from an ecological perspective, and with a specific concern for the environment, and Gliessman was not the exception. More recently, many authors, guided by the framework of agroecology, have embraced a broader conceptulization that includes social, economic, cultural, political, ethical, and technological factors, much like Hernández X did in the 1970s. In almost all the cases, most authors acknowledge the importance of understanding traditional agriculture in the process of creating agroecology, thus acknowledging, albeit not explicitly, the intercultural nature of agroecology.

8.5 THE INTERCULTURAL NATURE OF AGROECOLOGY

Interculturality is the result of a process in which different systems of constructing knowledge coexist (e.g., indigenous and scientific), in a safe environment, creating new knowledge that articulates the views and understanding of the cultures involved, to understand the same phenomena/issue. Figure 8.3 shows the evolution that agroecology had in Mexico, both conceptually and methodologically. Embedded in this evolution is the interaction of cultures. Agroecology is the result of an intercultural process involving the pre-Hispanic culture that was still alive in Tabasco during which CSAT was created and Gliessman was hired. This was reflected in the traditional agricultural systems that were being researched, the influence of authors from other countries, especially the United States, the role of Hernández X, the role of the faculty and the director of CSAT, the role of students at CSAT, and the role of the Mexican Council for Science and Technology. This unique set of actors and conditions at CSAT were not present anywhere else. From the positive interaction of people from different cultures, interested in the same purpose, a new academic discipline emerged. All of those conditions make possible the argument that the cradle of agroecology was in Tabasco, Mexico, where a vision developed, along with the ability to articulate it.

By mentioning the interaction of different cultures, it is necessary to establish at least two dimensions. The first one is the academic and the scientific. This is the visible dimension. Efraim Hernández Xolocotzi was trained as an agronomist/botanist, but was influenced by anthropology

(Alfredo Barrera), biology (Alfredo Barrera), and botany (Faustino Miranda). Equipped with the insights of these disciplines, Hernández X studied traditional agroecosystems. He also was influenced by Odum (1971), Harper (1974), and Janzen (1973), but nevertheless Hernández X's contributions to agroecology were basically of Mexican origin. His research most likely was influenced too by his culture, the Nahuatl. Gliessman, on the other hand, represented "Western" thinking in the process, applying Odum's ecology to the study of agricultural systems.

From an agroecological point of view, Gliessman and Amador's important publications began in 1980. He was a professor at CSAT, hired in 1976, and after 4 years of research on traditional agricultural systems Gliessman started a prolific and important set of agroecological publications. The papers he published, with García Espinosa and Amador Alarcon in 1981 and with Chacón in 1982, became agroecological classics. Although in his article published in 2013, Gliessman refers to the content of those papers, he does not address how those papers reflect what I interpret as an elegant intercultural process of constructing knowledge. In my opinion, prior to Gliessman's work, this intercultural process had not been visible in the development of agroecology.

The paper published by Gliessman et al. (1981) acknowledges that traditional agricultural systems have been working sustainably for a longer periods of time, starting in pre-Hispanic times, and were based on ecological principles. This Western academic reasoning is reinforced in the article by Gliessman and Chacón (1982) describing the concept of *monte*, or plants that colonize agroecosystems, from the traditional farmers in Tabasco. To those farmers the concept of "weed" does not exist. There is good or bad *monte*, depending on the effect it has on a crop, and this effect depends on the time of the year, the weather, and the growth stage of the crop. For local farmers, good monte "cools" the soil, whereas a bad monte "heats" the soil. Based on this knowledge, the farmers have developed agroecosystem management strategies that do not include the eradication of the *montes*, but instead use them to deal with other crop production limiting factors, such as nematodes or soilborne pathogens (Rosado-May et al. 1986; Rosado-May and García Espinosa 1986). With his ecological background, Gliessman had the vision and sensibility to correctly interpret the farmers' concept of *monte* and the ability to explain it in western terms. He found that the *monte* that heats the soil had allelopathic properties and that the farmers in Tabasco knew the effect, but did not know the mechanism behind it. This is how the positive interactions between the "local" and the "Western" understanding of a system, in Tabasco, Mexico, made possible the development of the science of agroecology. This is a very elegant process, fitting clearly the definition of interculturality that I have presented previously (Rosado-May 2013). The stakeholders of this process, from different cultural background, had the vision, the opening, the sensibility, and the flexibility to understand the same phenomena from different perspectives, without the exclusion of one for the other. An intercultural synergy gave birth to a new field of science.

The visible dimension, described above, is complemented by the intangible one, which also plays an important role in the process. The roles of Efraim Hernández Xolocotzi and Stephen R. Gliessman are very clear. Not so visible is the role of the CSAT. Almost against the nature of an institution of agriculture, this College for Tropical Agriculture "allowed" for the intervention of ecologists in the training of future agronomists, and "facilitated" the development of research projects on traditional agroecosystems, contrary to the nature of the Green Revolution. CSAT hosted seminars, group discussions, congresses, workshops, and courses that not only were not aligned with the Green Revolution, but some that were also highly critical of that approach to food production. Allowing, facilitating, and hosting activities and people not related to the Green Revolution can be explained without the role of the founding director of CSAT, Ángel Ramos Sánchez. Ramos Sánchez was most likely subjected to great pressure from interest groups supporting the Green Revolution, but he managed to encourage activities as diverse as the creation of new rice varieties and the use of pesticides and fertilizers, the development of technological packages for banana, rice, and sugarcane, while at the same time giving space to alternatives to the Green Revolution. Today, CSAT is recognized for its role in the creation of agroecology and almost not at all for any other of its contributions to tropical agriculture. It can be said that CSAT made possible the birth of agroecology. This resonates to the writings of Bacon (1605), when he presented in his work *The Advancement of Learning*, that the institutionalization of knowledge is critical for its development.

What other nontangible factors can explain the institutionalization of agroecology at CSAT? Some of these factors are the following:

1. Efraim Hernández Xolocotzi was very much appreciated as a scientist, with strong influence and authority that was supported by his research. He was able to share, in a very efficient way, his vision for agriculture. His great ability for analysis and synthesis created a "school of thought" very useful to agroecology. He was also very proud of his Náhuatl origin. His indigenous pride still inspires many generations of scientists researching the hidden, but still alive, traditional knowledge of agriculture and natural resources management.
2. The founding director of CSAT, Ángel Ramos Sánchez, had a vision for the college that encouraged not only research for the benefit of the Green Revolution but also encouraged alternatives, based on his understanding that tropical ecosystems are fragile. It was highly unusual for a conventional school of agriculture to allow both perspectives to be pursued. Ramos Sánchez was an agronomist with a master's degree in botany, specialized in tropical grasses, a former student of Hernández X, and of indigenous origin as well. He was proudly a Mixteca from Oaxaca, Mexico.
3. This chapter assumes that the cultural origin of a person has influence on his/her work and decision-making processes and his/her perception and explanation of the social and economic processes around him/her. Thus, it is important to make apparent the cultural backgrounds of Hernández X and Ramos Sánchez.
4. Gliessman's contribution to this process can be explained by his sensibility in accepting other ways of understanding and explaining the structure and function of agroecosystems. Consciously or not, Gliessman facilitated the encounter of traditional knowledge with modern science through ecology. He has continued to integrate the participation of other scientific disciplines in this process.

How can we understand the indigenous visions of Hernández X or Ramos Sánchez? How can we understand, or at least start to understand, the vision that the Chontal Maya in Tabasco have of their agricultural systems? It is important to remember that the traditional agriculture studied by Gliessman in Tabasco, as well as in Chiapas, Quintana Roo, and Campeche, is of Maya origin. It is important to also understand, linguistically, the meaning of agriculture to the Maya. To Western culture, "agriculture" is a word that has Latin roots: *agri* = field, *culture* = nursing. Agroecology has a Greek etymology: *agro* = field, *oikos* = house, *logos* = knowledge. Neither agriculture nor agroecology has a direct translation in Maya (*) or Nahuatl (**).*† In Maya, the closest expression is *MeyajbilK'aax*, which means working with nature. In Nahuatl, the word *Millakayotl* also means cultivating with nature. In both cases, the interpretation is much closer to agroecology than to the Green Revolution agriculture. It is highly possible, then, that the indigenous understanding of agriculture (working with nature) had an important influence on the work of Hernández X and the decisions of Ramos Sánchez. To understand the intercultural origin of agroecology, it is important to know and understand the above discussion.

8.6 THE INTERCULTURAL APPROACH IN AGROECOLOGY

If we accept that agroecology has an intercultural nature, it is important to advance both conceptually and methodologically. For example, Gliessman (2015) points out five levels of transition and the integration of components of agroecology needed for the transformation to a sustainable

---

* http://www.mayas.uady.mx/diccionario/k_glotal_maya.html
† http://aulex.org/nah-es

world food system. Using this framework, it is possible to present the importance of an intercultural approach (Table 8.1).

One of the most commonly used approaches in agroecological research is participatory action research (PAR). By incorporating intercultural elements, PAR has the potential to be much more effective than it currently is. PAR has changed through time, beginning most notably, in 1994 when Chambers introduced participatory rural appraisals (PRA). In 2008, this same author considered that PAR is not a monolithic body of ideas and methods, but a dual orientation for the construction of knowledge and social change. More recently, Chevalier and Buckles (2013) indicate that the part of participation in PAR refers to life in society, action to experience, and research to new knowledge. These authors refer to an intercultural process without acknowledging it. This lack of visibility could be because the concept of interculturality is rather more recent or because a methodology has not been developed to incorporate it.

PAR can be carried out without intercultural elements. A researcher could arrive in a rural community, accepted by the people, conduct research, and publish the results. In this scenario, the publication could reflect results and interpretation biased by the researcher's own culture. This is not an example of an intercultural research process. There is a difference between a participatory intercultural interaction and just participatory interaction. Agroecology could not have been developed with the application of PAR without the intercultural dimension.

The intercultural dimension of PAR is not new, but it has mostly been implicit. In 1995, FalsBorda, a Colombian sociologist, gave a conference in Atlanta to the Southern Society of Sociology, addressing PAR. In his talk, FalsBorda insisted that social scientists neither monopolize nor impose their knowledge on others, especially in the communities where they did their research; he asked that in the research process both the scientist and the community members be considered as coauthors of the research and of the knowledge that emerges in this process. FalsBorda is acknowledged as a pioneer of PAR (Hall 2005; Ortiz and Borjas 2008).

Did the Maya in the Yucatan Peninsula, Mexico, have a kind of PAR in their process of constructing knowledge? The Maya coined the term *tsikbal*, still in active use today. This term refers to the process of transmitting and updating knowledge through time in a very active and participatory manner. The vehicle was both words in the act of conversation, and action through trial and error, practice, and more practice (Rosado-May et al. 2015).

Table 8.1 Gliessman's Levels of Conversion Toward Sustainable Food Systems (Gliessman 2015), and Their Relation to Explicit, Society-wide "Intercultural Moments"

Level	Intercultural Moment
I. Increase Efficiency of Industrial Practices	Society does not acknowledge its cultural diversity nor does it promote conditions to achieve it. It only, or mostly, acknowledges conventional technology (genetically modified organisms, pesticides, etc.) applied to food production.
II. Substitute Alternative Practices and Inputs	Society begins to acknowledge its cultural diversity and promotes conditions to explore it and to potentiate it. It acknowledges alternative food production systems, not only the conventional ones. By doing so, more options for finding efficient substitution for conventional elements are available.
III. Redesign the Whole Agroecosystems	Beginning the construction of intercultural knowledge, combining Western with local.
IV. Reestablish Connection Between Growers and Eaters, Develop Alternative Food Systems	Society develops new paradigms that lead to new knowledge, alternative technology, viability, and sustainability for food systems.
V. Rebuild the Global Food System so that it is Sustainable and Equitable for all	Advanced society, intercultural. New paradigms in socioeconomic relationships are in place. Not necessarily both producer and consumer are from the same cultural group. Level IV is facilitated.

The intercultural dimension should be an intrinsic part of PAR, but is not visible in either the English or the Spanish literature. This might be because both the term and the concept are new. Nevertheless, being aware of the intercultural concept allows the visibility of different systems for constructing knowledge, all valid when it comes to understand the structure and function of an agroecosystem. The intercultural dimension is especially important when it comes to communicating knowledge that reflects processes between different cultures interested in understanding and studying the same phenomena.

8.7 CONCLUSIONS

On the basis of the evidence and arguments presented in this chapter, I conclude with the following points:

1. Mexican scientists played a major role in the creation and development of agroecology.
2. The process of construction of agroecology in Mexico was a highly elegant and sophisticated one.
3. Tabasco, Mexico, was the cradle of agroecology in the world. This was due to the unique set of conditions, which were present, including (a) the seminal contribution by Efraim Hernández Xolocotzi; (b) the research on the ecological basis of traditional sustainable agroecosystems carried out by S.R. Gliessman; and (c) the safe space offered by both the CSAT and its founding director A. Ramos Sánchez. The interaction that took place at CSAT through research and teaching between Mexican and US scientists, both with the attitude of learning from and interacting with traditional farmers, allowed an intercultural synergy. These unique conditions have not happened elsewhere in the world.
4. In Mexico, Stephen R. Gliessman can be considered as the "Father of Agroecology," not only for his extensive research contributions into the field but because of his ability to integrate and articulate concepts and methods, analysis and synthesis, throughout extensive research, on the structure and function of traditional agroecosystems, also incorporating in this process, farmers, students, and fellow scientists from different cultures.

REFERENCES

Altieri, M.A. *Agroecology, the Science of Sustainable Agriculture*, 2nd ed. Boulder, CO: Westview Press, 1995.

Altieri, M.A. *Agroecología, Bases Científicas Para Una Agricultura Sustentable*. Montevideo, MN: Editorial Nordan-Comunidad, 1999.

Altieri, M.A. (Ed.). *Vertientes del Pensamiento Agroecológico: Fundamentos y Aplicaciones*. Bogotá, Colombia: Sociedad Científica Latinoamericana de Agroecología, 2010a.

Altieri, M.A. "El Estado del Arte de la Agroecología." In *Vertientes del Pensamiento Agroecológico: Fundamentos y Aplicaciones*. M.A. Altieri (Ed.). Bogotá, Colombia: SOCLA, 2010b, 77–104.

Anaya, A.L. *Ecología Química*. México: Instituto de Ecología, UNAM and Plaza y Valdés Editores, 2007.

Bacon, F. *Of the Proficience and Advancement of Learning, Divine and Human*. Renascence Editions (http://darkwing.uoregon.edu/~rbear/adv1.htm), 1605 (Accessed on july 18th, 2015).

Bensin, B.M. *Agroecological Characteristics Descriptions and Classification of the Local Corn Varieties Chorotypes*. Book (Publisher unknown), 1928.

Caballero, J. and J. Sarukhán. "Opciones para la Alimentación Futura en México: Inestabilidad en la Especialización o Estabilidad en la Diversificación." In *La Alimentación del Futuro 2*. R. Carvajal and J.M. Vergara (Eds.). Mexico: Universidad Nacional Autónoma, 1987, 223–235.

Chambers, R. "Participatory rural appraisal (PRA): Analysis of experience." *World Development* 22 no. 9 (1994): 1253–1268.

Chambers, R. "PRA, PLA and pluralism: Practice and theory." In P. Reason and H. Bradbury (Eds.). *The Sage Handbook of Action Research: Participative Inquiry and Practice*, 2nd ed. (London: Sage, 2008), 297–318.

Chevalier, J.M. and D.J. Buckles. *Participatory Action Research: Theory and Methods for Engaged Inquiry.* New York: Routledge, 2013.

Cox, G.W. and M.D. Atkins. *Agricultural Ecology.* San Francisco, CA: Freeman, 1979.

Espinosa-García, F.J. "La Diversidad de los Metabolitos Secundarios y la Teoría de la Defensa Vegetal." In *Relaciones Químicas entre Organismos. Aspectos Básicos y Perspectivas de su Aplicación* A.L. Anaya, F.J. Espinosa-García and R. Cruz-Ortega (Eds.). México: Universidad Nacional Autónoma de México y Plaza y Valdés, 2001, 231–249.

FalsBorda, O. *Research for Social Justice: Some North-South Convergences*, Plenary Address at the Southern Sociological Society Meeting, Atlanta, April 8, 1995.

Francis, C., G. Lieblein, S. Gliessman, T.A. Breland, N. Creamer, R. Harwood, L. Salomonsson, J. Helenius, D. Rickerl, R. Salvador, M. Wiendehoeft, S. Simmons, P. Allen, M. Altieri, J. Porter, C. Flora, and R. Poincelot. 2003. Agroecology: The ecology of food systems. *Journal of Sustainable Agriculture,* 22:99–118

Gliessman, S.R. "Agroecología y Agroforestería." *Investigación de Técnicas Agroforestales Tradicionales.* In J. Beer and E. Somarriba (Eds.). Boletín Técnico No. 12, Costa Rica: CATIE, 1984.

Gliessman, S.R. *Agroecology: Researching the Ecological Basis for Sustainable Agriculture.* Series in Ecological Studies No. 78. New York: Springer-Verlag, 1990.

Gliessman, S.R. *Agroecology: The Ecology of Sustainable Food Systems*, 2nd ed. Boca Raton, FL: CRC Press, 2007.

Gliessman, S.R. "Agroecology: Growing the roots of resistance." *Agroecology and Sustainable Food Systems* 37 (2013): 19–31.

Gliessman, S.R. *Agroecology: The Ecology of Sustainable Food Systems*, 3rd ed. Boca Raton, FL: CRC Press/ Taylor & Francis, 2015.

Gliessman, S.R. *Memorias del Curso Intensivo de Ecología Tropical.* Tabasco, México: Colegio Superior de Agricultura Tropical, Cárdenas, 1979–1981.

Gliessman, S.R. (Ed.). *Seminarios Regionales sobre Agroecosistemas con Énfasis en el Estudio de Tecnología Agrícola Tradicional.* Tabasco, México: Colegio Superior de Agricultura Tropical, Cárdenas, 1978, 29–33.

Gliessman, S.R. and J.E. Chacón. "The use of the 'non-weed' concept in traditional agroecosystems of tropical Mexico." *Agro-Ecosystems* 8 (1982): 1–11.

Gliessman, S.R. and M. Amador. "Ecological aspects of production in traditional agroecosystems in the humid lowland tropics of Mexico." *Tropical Ecology and Development.* In J.I. Furtado (Ed.). Kuala Lumpur: ISTE, 1980, 1383.

Gliessman, S.R., R. García Espinosa, and M. Amador. "The ecological basis for the application of traditional agricultural technology in the management of tropical agroecosystems." *Agro-Ecosystems* 7 (1981): 173–185.

Gómez-Pompa, A. and A. Kaus. "From prehispanic to future conservation alternatives: Lessons from Mexico." *Proceedings of the National Academy of Sciences* 96 (1999): 5982–5986.

Gómez-Pompa A, J. S. Flores-Guido, and M. Aliphat. "The sacred Cacao groves of the Maya." *Latin American Antiquity* 1 (1990): 247–257.

González Jácome, A. *Transformaciones en el Agro Mexicano.* México: Universidad Iberoamericana, 2004.

González Jácome A. and S. Del Amo Rodríguez. *Agricultura y Sociedad en México: Diversidad, Enfoques, Estudios de Caso.* México: Plaza y Valdéz, PROAFT and Universidad Iberoamericana, 1998.

Guadarrama-Zugasti, C. "Agroecología en el siglo XXI: Confrontando nuevos y viejos paradigmas de producción agrícola." *Revista Brasileira de Agroecología* 2 no. 1 (2007): 204–207.

Hall, B.L. "In from the cold: Reflections on participatory action research from 1970–2005." *Convergence* 38 no. 1 (2005): 5–24.

Harper, J.L. "The need for a focus on agro-ecosystems." *Agroecosystems* 1 (1974):1–12.

Hecht, S.B. "La Evolución del Pensamiento Agroecológico." *Agroecología, Bases Científicas para una Agricultura Sustentable.* In M.A. Altieri (Ed.). Montevideo: Editorial Nordan-Comunidad, 1999, 15–30.

Hernández Xolocotzi, E. *Agricultural Areas Along the México-Oaxaca Highway From Matamoros to Oaxaca City.* México: Mimeografiado, 1946.

Hernández Xolocotzi, E. *Las Zonas Agrícolas de México.* México: Nueva Agronomía (Estudios del Campo Mexicano). Ateneo Nacional Agronómico (Serie Técnica), 1954.

Hernández Xolocotzi, E. "Las ciencias naturales y el desarrollo social de México." *Revista de la Sociedad Mexicana de Historia Natural* 21 no. 1 (1960): 19.

Hernández Xolocotzi, E. *La Ecología y las Investigaciones Agropecuarias*. Chapingo, México: ENA, 1967.

Hernández Xolocotzi, E. "Exploración etnobotánica en maíz." *Fitotecnia Latinoamericana* 8 no. 2 (1972): 46–51.

Hernández Xolocotzi, E. "El papel de la tecnología agrícola tradicional en el desarrollo agropecuario." *Narxhí-Nandhá* 6/7 (1978): 14–27.

Hernández Xolocotzi, E. and A. Ramos Rodríguez. "Metodología para el Estudio de Agroecosistemas con Persistencia de Tecnología Agrícola Tradicional." *Agroecosistemas de México*. In E. Hernández Xolocotzi (Ed.). Chapingo, Mexico: Colegio de Postgraduados, 1977, 321–333.

Janzen, D.H. "Tropical agroecosystems." *Science* 182 (1973): 1212–1219.

Jiménez-Osornio, J.J., H. Estrada-Medina, W. Aguilar-Cordero, P. Montañez-Escalante, R. Ruenes-Morales, and R. Ortíz-Pech. "Ecological Land Use Planning for Sustainable Landscape in Yucatan." In *Applying Ecological Knowledge to Land Use Decisions*. H. Tiessen and J.W.B. Stewart (Eds.). Sao Paulo, Brazil: SCOPE, the Scientific Committee on Problems of the Environment IAI, the Inter-American Institute for Global Change Research, and IICA, the Inter-American Institute for Cooperation on Agriculture, 2008, 42–48.

Klages, K.H.W. "Crop ecology and ecological crop geography in the agronomic curriculum." *Journal of the American Society of Agronomy* 20 (1928): 336–353.

Mariaca-Méndez R., A. González-Jácome, and L.M. Arias-Reyes. *El Huerto Maya Yucateco en el Siglo XVI*. México: El Colegio de la Frontera Sur, Cinvestav-Mérida, Universidad Intercultural Maya de Quintana Roo, Consejo de Ciencia y Tecnología del estado de Yucatán y Fondo Mixto Conacyt de Yucatán, 2010.

Martínez Alfaro, M.A. *Contribuciones Iberoamericanas al Mundo: Botánica, Medicina, Agricultura*. Madrid: Anaya, 1988, 126.

Mass, M., C. Jordan, and J.Sarukhán. "Soil erosion and nutrient losses in seasonal tropical agroecosystems under various management techniques." *Journal of Applied Ecology* 25 (1988): 595–607.

Odum, E.P. *Fundamentals of Ecology*. Philadelphia, PA: W.B. Saunders, 1971.

Ortiz, M.and B. Borjas. "La investigación acción participativa: aporte de fals borda a la educación popular." *Espacio Abierto* 17 no. 4 (2008): 615–627.

Palerm, A. *Introducción a la Teoría Etnológica*. México: Universidad Iberoamericana, 1967.

Papadakis, J.S. *Ecologie Agricole*. In *Gembloux*. J. Duculot (Ed.), Paris: Librarie Agricole de la Maison Rustique, 1938.

Peredo Reyes, E. and M. Matías Villada. "Primer Congreso Agroecológico." In *2012, El Centro Internacional de Mejoramiento de Maíz y Trigo—México, y su Participación en la Ciencia Mexicana en el Período 1966-2010, Visto a Través de la Literatura Científica Publicada. Tesis de Licenciado en Biblioteconomía. Escuela Nacional de Biblioteconomía y Archivonomía*. U. Sánchez Martínez (Ed.). *Secretaría de Educación Pública, Dirección General de Educación Superior Universitaria, México*, Meoqui, Chihuahua, México, 1926.

Rosado-May, F.J. "Experiencias y Visiones de Futuro de la Universidad Intercultural Maya de Quintana Roo, México. Aportaciones del Modelo Intercultural a la Sociedad." In A. Wind (Ed.). *Las Universidades Indígenas: Experiencias y Visiones Para el Futuro* (La Paz, Bolivia: Instituto Internacional de Integración, 2013), 157–172.

Rosado-May, F.J. and R. García Espinosa. "Estrategias empíricas para el control de la mustia hilachoza (*Tanathephorus cucumeris Frank Donk*) de frijol común en la Chontalpa, Tabasco." *Revista Mexicana de Fitopatología* 4 (1986):109–113.

Rosado-May, F.J., R. García Espinosa, and S.R. Gliessman. "Impacto de los fitopatógenos del suelo al cultivo del frijol, bajo diferente manejos en la Chontalpa, Tabasco." *Revista Mexicana de Fitopatologia* 3 (1985): 80–91.

Rosado-May, F.J., S.R. Gliesman, and M. Alejos Peraza. "Potencial alelopático del cadillo (*Bidens pilosa* L.) y su relación con el ataque de algunos fitopatógenos del suelo al maiz." *Revista Mexicana de Fitopatología* 4 (1986): 124–132.

Rosado-May, F.J., Y. C. Olvera León, and M.C. Osorio Vázquez. "El Papel del Maya en el Aprendizaje de Inglés, ¿O Viceversa? Experiencias de la Universidad Intercultural Maya de Quintana Roo." In *Memorias del Coloquio sobre Educación Superior Indígena México-Canadá, en el marco del Día Internacional de las Lenguas Maternas*. M.C. Osorio Vázquez (Ed.). José Ma Morelos, Q. Roo, México: Universidad Intercultural Maya de Quintana Roo, 2015, 3–15.

Ruiz-Rosado, O. "Agroecología: Una disciplina que tiende a la transdisciplina." *Interciencia* 31 no. 2 (2006):140–145.

Sánchez Martínez, U. *El Centro Internacional de Mejoramiento de Maíz y Trigo-México y su Participación en la Ciencia Mexicana en el Período 1966-2010, Visto a Través de la Literatura Científica Publicada.* México: Tesis de Licenciatura, Escuela Nacional de Biblioteconomía y Archivonomía, 2012.

Sarukhán, J. "Bases Agroecosistémicas para una Filosofía de Ecodesarrollo." In *Memorias del I Simposio sobre Ecodesarrollo.* E. Leff (Ed), Primer Simposium sobre ecodesarrollo. Asociación Mexicana de Epistemología. 1977, 34–51.

E. Leff (Ed), Primer Simposium sobre ecodesarrollo. Asociación Mexicana de Epistemología. 1977, 34-51

Tomich, T.P., S. Brodt, H. Ferris, W.R. Horwath, E. Kebreab, J. Leveau, et al. "Agroecology: A review from a global change perspective." *Annual Review of Environment and Resources* 36 (2011): 193–222.

Trujillo-Ortega, L.E. "Ecología Política del Desarrollo Sostenible." In *Agroecología e os Desafíos da Transicao Agroecológica.* Sauer e Balestr (Ed.). Sao Paulo, Brazil: Editora Expresao Popular, 2009, 71–99.

Wellhausen, E.J., L.M. Roberts, and E. Hernández Xolocotzi, in collaboration with P.C. Mangelsdorf. *Races of Maize in Mexico. Origin, Characteristics and Distribution.* México: Secretaría de Agricultura y Ganadería, Oficina de Estudios Especiales, 1951.

Wezel, A., S. Bellon, T. Doré, C. Francis, D. Vallod, and C. David. "Agroecology as a science, a movement and a practice. A review." *Agronomy for Sustainable Development* 29 no. 4 (2009): 503–515.

Wezel, A. and V. Soldat. "A quantitative and qualitative historical analysis of the scientific discipline of agro-ecology." *International Journal of Agricultural Sustainability* 7 no. 1 (2009): 3–18.

CHAPTER 9

Participatory Action Research for an Agroecological Transition in Spain
Building Local Organic Food Networks

Gloria I. Guzmán, Daniel López, Lara Román, and Antonio M. Alonso

CONTENTS

9.1 INTRODUCTION

The environmental and socioeconomic crisis of industrialized agriculture worldwide has led to the emergence of agroecology as a theoretical and methodological approach that aims to increase agricultural sustainability from an ecological, social, and economic perspective (Francis et al. 2003). In Spain, organic farming (OF) is the most consistent implementation of this strategy, exceeding all other areas in Europe, with 1.62 million hectares in 2011 (Willer et al. 2013). Although in recent years, OF has undergone an increasing process of conventionalization, limiting its positive effect on agricultural sustainability (Buck et al. 1997; Darnhofer et al. 2010; de Wit and Verhoog 2007), it is also true that numerous markedly agroecological experiences have been developed under this umbrella. Common features of these experiences are productive diversification, appreciation of local resources (organic matter, farmers' knowledge, old livestock breeds and crop varieties, landscape), strengthening of community organizations, and developing short food supply chains (SFSCs) that enable farmers and consumers to establish direct relationships that benefit both (Best 2008; Goldberger 2011; Lobley et al. 2009; Milestad et al. 2010). In this article, we have given the name "agroecological transition" to the conversion process of industrial agroecosystems to level 3 (redesign of the agroecosystem so that it functions on the basis of a new set of ecological processes and relationships) and level 4 (reestablish a more direct connection between those who grow food and those who consume it, with a goal of reestablishing a culture of sustainability that takes into account the interactions between all components of the food system), as defined by Gliessman (2010).

However, the transition from "industrialized" to "agroecological" models is not easy (Lobley et al. 2009; Milestad et al. 2010). Farmers have identified a number of difficulties with an agroecological transition, which represents a complex process that links different levels (farm, local community, and wider society) and that is affected by social, economic, technological, cultural, and ecological factors (Guzmán and Alons, 2010).

Sevilla Guzmán (2006) establishes three dimensions of agroecology. These dimensions are complementary and should be smoothly articulated to dynamize the integrated processes of an agroecological transition:

The *ecological–agronomical dimension* develops an integrated, systemic vision of the productive process. In this dimension, the agroecological transition has two objectives:

1. The redesign of the agroecosystem. It aims to modify the structure and function in industrialized agroecosystems to maximize the internalization of the flows of energy, materials (mainly nutrients and water), and information (genetic, knowledge). The study of the traditional agroecosystems of the region (Guzmán and González de Molina 2009) offers relevant information to this end.
2. The design of appropriate technologies to successfully replace the techniques of the Green Revolution. Research and technological innovation are fundamental to this task. Traditional agricultural knowledge can also offer valuable information that, nonetheless, must be adapted to the twenty-first century.

The *socioeconomic and cultural dimension* are centered on the conditions of the social reproduction of rural communities. These conditions must guarantee that the communities continue to undertake agricultural activity and improve the state of natural resources. To this end, the focus is on the revitalization of local resources (organic matter, old livestock breeds and crop varieties, farmers' knowledge, etc.), the articulation and organization between local farmers, and between farming sector and other economic activities (agrotourism, environmental education, etc.), and the development of SFSCs, which allow farmers to retain more of the added value of the product.

The *sociopolitical dimension* aims to impact decision making in the agrofood system and to reconstruct the sector's capacity for dialogue and for making proposals. Therefore, it is a movement from the local to the global scale, questioning the policies that might hinder local sustainability projects and promoting other policies, which could favor them. This dimension contemplates alliances with other social groups related to agrofood. Proposals from social movements, such as food sovereignty, enrich the debate on the agroecological transition in this dimension.

Table 9.1 summarizes the difficulties most frequently encountered in Spain for achieving the agroecological transition. These difficulties are grouped by scale (farm, local community, and wider society) and dimension (agronomical–ecological, socioeconomic and cultural, and sociopolitical). In the agronomic–ecological dimension, the degradation of natural resources and the delocalization of energy and material flows stand out on all of the scales. We underline Spain's high dependency on imported animal feed, owing to its magnitude and its impact on the transition of livestock farming systems. The problems of the socioeconomic and cultural dimension can be found in the organizational and economic weakness of the agricultural sector. This weakness limits its role as an interlocutor and social and economic player and also limits to an extraordinary extent its capacity to remedy the situation. The progressive urbanization and industrialization of society have made agriculture increasingly invisible. On the sociopolitical scale, this is translated into an inability to influence local, national, and supranational (European Union) public policies. Today, these policies are clearly hostile to an agroecological transition, since they openly favor the intensification and globalization of both agricultural inputs and outputs. Numerous regulations prevent relocalization of energy, material, and information flows necessary for sustainable agricultural production. Due to their relevance, we highlight the regulations that prevent or limit the trade in reproductive material of local varieties and those that hinder the access of livestock to public pastureland. Meanwhile, these regulations also hinder the development of small-scale agroindustry and local markets while financing and promoting the opposite.

Table 9.1 shows that the three dimensions can be found on all scales and they mutually condition each other, both with regard to the integrated analysis of existing problems and with regard to the development of solutions to those problems. Farmers and agroecologists frequently face transition problems at the farm level. However, the solutions are often found at a higher scale (Dalgaard et al. 2003). This is why it is important to design farm transition strategies at the local and the wider society scale. Specifically, the articulation of the farm and local community scales is unavoidable. Furthermore, for most of Spain, the Mediterranean agroclimatic conditions favor the integration of energy, material, and information flows at the local scale, not at the farm scale (Guzmán and González de Molina 2009).

In summary, the agroecological transition of Spanish agriculture requires that a weakened, invisible sector make an enormous effort for the restoration of natural resources and the relocalization of inputs and outputs, in a hostile economic and political context, with the aim of increasing agricultural sustainability. This is a daunting task that requires not only technological but also social innovation.

The complexity of agroecological transition calls for support for farmers in the form of methodologies that bring about the necessary changes both on-farm and in local communities. Participatory action research (PAR) is a methodological approach that provides a set of action-research techniques useful for agroecological transition (Kindon et al. 2007), which has been applied especially in Latin America. PAR can be used to design and implement, in conjunction with farmers and local people, management, and social organization proposals that increase agricultural sustainability. This chapter describes the application of the PAR methodology in four case studies in Spain to show the utility of this approach to facilitate agroecological transition in the European context.

9.2 RESEARCH APPROACH AND METHODOLOGY

PAR is a methodological approach that emerged from the social sciences halfway through the twentieth century. In the agrarian field, it began with the questioning of extension and training systems that are used to modernize the farming world (Freire 1969). PAR considers that any development process undertaken will be biased if it does not incorporate the beneficiaries of this process as

Table 9.1   The Major Difficulties Facing the Agroecological Transition in Spain, by Scale and Dimension

		SCALE		
Dimension		Farm	Local Community	Society
	Ecological–agricultural	Significant degradation of natural resources (soil, water, and biodiversity) Lack of knowledge of appropriate techniques Highly specialized production, with a prevalence of products for export	Energy, material and information flows decoupled from the territory Risk of diffuse contamination (pesticides, GMOs, urban–industrial contaminants) Prevalence of monoculture Scarcity of areas of natural vegetation Degradation of agricultural infrastructure (irrigation channels, and paths)	Climate change High dependence on flows of imported energy and material High dependence on imported livestock feed
	Socioeconomic and cultural	Low investment capacity Need for investment (machinery, facilities, etc.) during the transition High cost of third-party certification for small-scale producers Erosion of traditional agricultural knowledge Low self-esteem of farmers Low level of generational renewal	Weak local food networks Disconnect between agricultural production and other local economic agents (tourism, catering, local food retailers, etc.) Organizational weakness of the cultural sector Marginalization of agricultural activity (invisibility) Significant erosion of traditional agricultural knowledge	No payment for environmental services provided by organic agriculture No penalization of environmental damage caused by industrialized agriculture Difficulty to access loans and insurance for the agroecological transition Little social commitment to generate sustainable food networks
	Sociopolitical	Isolation; low participation in agricultural organizations	Low capacity of the agricultural sector to influence local public policies Low institutional support for the consumption of organic foods and short food supply chains (SFSCs) Regulations which hinder the establishment of composting plants Regulations which hinder access to publicly owned pastureland for livestock grazing Territorial planning which threatens, rather than protects, agricultural activity	Low capacity of the agricultural sector to influence national and European Union public policies Inadequate regulations for the development of small-scale agroindustry Regulations which hinder or limit the use of and trade in reproductive material of local varieties Regulations which permit the cultivation of genetically modified organisms (GMOs) Risk of GMO contamination Little support for institutional consumption of organic foods and for SFSCs Nonexistence of official recognition of participatory guarantee systems In general, lack of public policies which support agroecological transition

protagonists of it. In general, PAR approaches seek to generate liberating knowledge that is based on popular knowledge and that explains the global situation (systemic approach), with the aim of starting or consolidating a strategy of change (transition processes), alongside an increase in political power, aimed at obtaining positive transformations for the community on a local level; and at higher levels in as far as it is able to connect with similar experiences (networks) (Fals Borda 1991).

When applied to an agroecological transition, PAR promotes technological change and, at the same time, improves ecological sustainability of farming systems, drawing from different approaches such as participatory rural appraisal (PRA) or farmer participatory research (FPR) (Chambers 1989, 1992; Farrington and Martin 1987; Rhoades and Booth 1982). These participatory methodologies try to promote the acquisition of skills and strengthening of organizing capacity by the groups involved, so that they can continue the process by themselves. PAR starts from a participatory and holistic diagnosis of the initial situation of the farm and local community, and defines an objective, realistic situation using sustainability criteria. It encourages the group to reach the proposed goals and to establish relationships with other groups, making up networks and associations that facilitate change at different levels and establish solid foundations for sustainable rural development.

PAR has developed different tools that try to incorporate the complexity of social relations, especially related to inequity and marginality in social systems, to generate collective solutions to everyday life problems. Cuéllar and Calle (2011) discuss the particular trend of applying a community-based participatory approach to action-oriented research, presenting research from all over the world, such as Tandon (2000) in India, Park et al. (1993) in North America, and Villasante et al. (2000) in Latin America. These community-oriented approaches focus on relations between social actors, as far as "it is easier to change relations between subjects than subjects themselves" (Villasante 2006, p. 315). This shift in the focus of participatory methodologies allows one to confront situations of conflict mediated by power relations in the rural scene. Technological change and natural resources management are deeply conditioned by power relations (Scoones and Thompson 1994). Farmers are marginalized social actors, especially in the postindustrial era, as far as their weak power to shape politics or food systems (Bell et al. 2010; Reed 2008). The community-based approach allows us to connect different research scales to support the agroecological transition. That is, connect the farm scale, where research is usually conducted, to the local community and wider society scales, where solutions are usually developed. This gap between different research scales has been pointed out as one of the main weaknesses within agroecological research (Dalgaard et al. 2003).

9.2.1 Mixing Methodologies in a Participatory Research Framework to Articulate Dimensions and Scales of Agroecological Transition

The community-based approach offers a phase-sequence pattern for action-research projects (Cuéllar and Calle 2011; Villasante 2006), in which it is possible to insert research techniques from different approaches: from the social to agronomic sciences, both participatory and nonparticipatory. This mixed methodology, set up over the participatory approach, allows us to develop the interdisciplinarity inherent to an agroecological approach (Francis et al. 2003) and to articulate and integrate the diverse dimensions of agroecology in its different meanings—as a scientific discipline, as a set of agricultural practices, and as a social movement (Wezel et al. 2009). Figure 9.1 shows a schematic view of our methodological proposal broken down by agroecological dimensions. The integrated methodologies have four main complementary functions: (1) to generate the data necessary to inform the process; (2) to facilitate the participation and mobilization of social actors to progress toward the agroecological transition (mainly farmers, but also other social actors with potential to bring about necessary transformations); (3) to promote subjective and symbolic transformations in local society; and (4) to monitor the process and evaluate the progress achieved toward sustainability. We describe briefly these methodologies below, specifying their main functions.

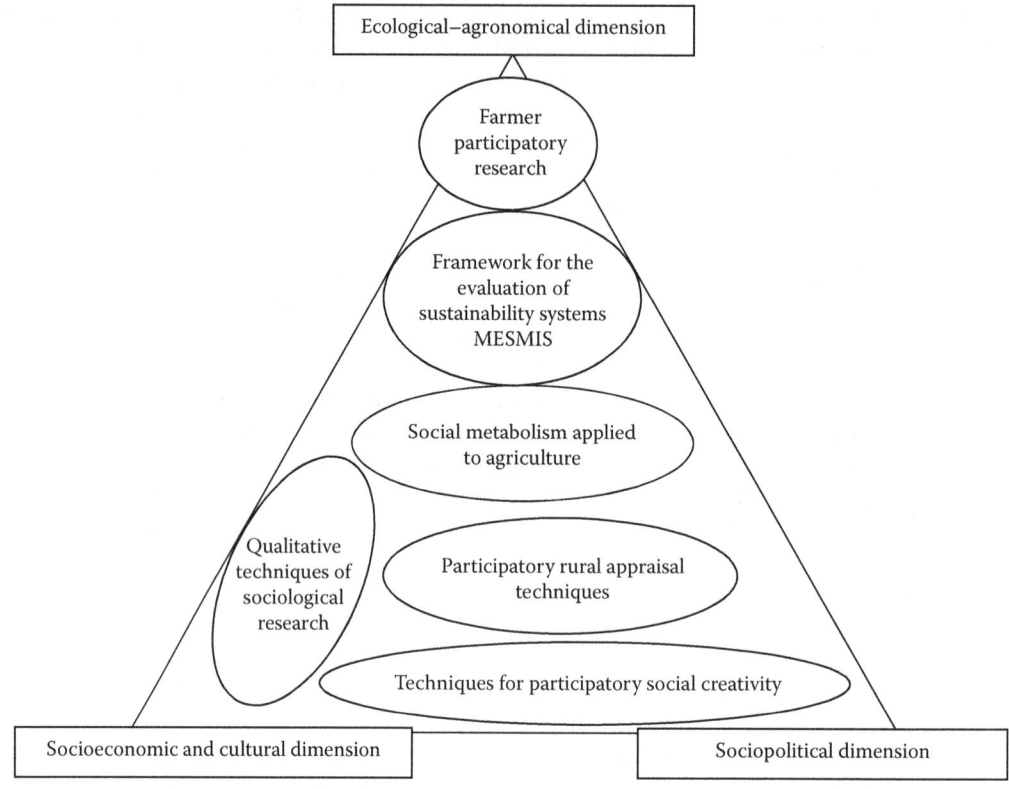

Figure 9.1   Methodologies for agroecological transition according to the dimensions of agroecology.

9.2.2 Farmer Participatory Research

FPR emerged as a response to the limitations of earlier agricultural research and extension approaches (Farrington and Martin 1987). In these earlier approaches, farmers were often considered as research subjects, components of the system under investigation, or passive recipients of extension messages. FPR advocates farmers' involvement as decision makers at all stages of the research process. Selener (1997) classifies research conducted on farms according to the level of control and management exercised by farmers and researchers. Only two categories (3 and 4) are considered FPR. In the third category, farmers and researchers work together in this approach on problem definition, design, management, and implementation of trials, and evaluation research will focus on farmers' needs. In the fourth category, farmers are the main actors and decision makers. The scientist's role is to supply the information and resources necessary for farmers to undertake the research. The proposal for the dissemination of knowledge between farmers, called "from peasant to peasant" (Holt Giménez 2008), is considered under the FPR umbrella.

This methodology is of enormous interest to farmers providing a strong incentive to participation. Furthermore, it generates very useful information for the design of appropriate technologies.

9.2.3 Framework for the Evaluation of Natural Resource Management Systems Incorporating Sustainability Indicators (MESMIS in Spanish)

This methodology attempts to evaluate the sustainability of agroecosystems using indicators, which reflect the status of sustainability attributes (productivity, stability, resilience, reliability, adaptability, equitability, and self-dependence) (López-Ridaura et al. 2002). MESMIS does not explain the functioning of agroecosystems, nor does it have the capacity to link the different scales of analysis. These two limitations reduce its usefulness for the in-depth evaluation of agricultural sustainability. Nevertheless, it is very useful when applied in a participatory manner at the farm scale. MESMIS encourages farmers to examine their farms from different perspectives (economical, social, and environmental), to define the processes that could undermine the viability of the farm, to propose an improvement plan, and to select monitoring indicators that will allow progress to be verified. It, therefore, encourages the participation of farmers and generates information, which is useful to them. It also allows for the evaluation of the agroecological transition process at the farm scale.

9.2.4 Social Metabolism Applied to Agriculture

Social metabolism refers to the exchange of energy, materials, and informations established by all societies with their physical environment, with the aim of producing and reproducing its material conditions of existence. Recently, González de Molina and Toledo (2014) proposed social metabolism as a powerful analysis methodology to study the processes of agroecological transition. The application of a metabolic focus to agriculture means that "agrarian metabolism" should be understood to mean that part of the social metabolism that specializes in the generation of biomass and environmental services for human consumption. It uses techniques such as material and energy flow accounting and energy return on investment. These techniques allow us to investigate the capacity of agroecosystems to generate the flow of both energies (from sunlight) and materials (nutrients, water) necessary for its functioning, and also to meet the needs of the society that depends on them, taking into account the agrofood system as a whole. It therefore links together the different scales of analysis. Some examples can be found in Guzmán and González de Molina (2009), Guzmán et al. (2011), García-Ruiz et al. (2012), and Tello et al. (2014). The application of these tools requires relatively complex mathematical calculations. Therefore, the involvement of the farmers in the first phase is simply in the capacity of informants. However, once the calculations have been made, and the information obtained has been returned, it becomes a powerful tool to stimulate debate, guide the redesign of agroecosystems, and evaluate in depth the progress achieved toward sustainability, both on the farm scale and on the local community scale.

9.2.5 Participatory Rural Appraisal Techniques

PRA has generated many visualization techniques (community planning map, transect walk and diagramming, and timeline) to working with groups of people with varying academic backgrounds and types of education and ensuring their effective participation (Chambers 1992; Geilfus 1997). As well as making the participation of farmers and the local population central to the process, these techniques allow information to be contributed and shared (objective data and subjective views), explaining the current situation of the agroecosystems from the viewpoint of the local community by means of simple, inclusive graphic codes. This shared knowledge is the basis for the participatory construction of the diagnosis and the proposals for agroecological transition.

9.2.6 Qualitative Sociological Research Techniques

Structured or semistructured interviews, focus groups, participant observation, deliberative polling, and so forth, allow the researcher to obtain qualitative information on specific matters (positioning on the social map, opinions, reactions, complicity, and discourse). They can also be used to monitor subjective and symbolic transformations in local society generated during the transition process.

9.2.7 Participatory Social Creativity Techniques, such as the Sociogram and the Situational Flowchart

A sociogram is applied as a qualitative technique that allows us to reflect collectively about the nature and structure of relations between social networks existing in a certain local environment. A situational flowchart is used for collective analysis of cause–effect relations of self-defined problems in a certain group or institution and a certain social situation (Villasante 2006; Villasante et al. 2000).

Another interesting contribution of participatory social creativity techniques (PSCs) is the use of analyzers and analyzer–mobilizers. The analyzers (Lapassade 1971) are present or past events, such as conflicts, shifts in certain relations, social mobilizations, and so forth that question and analyze by themselves the local situation, particularly in terms of its symbolic aspects—personal or collective values, meaning references, identities, beliefs, desires, and so forth—which reveal the nature of the links between the different social, local actors. Agroecological analyzer–mobilizers are actions that recreate situations which question and analyze the links and interconnections within local society, and between it and the farming sector, focusing on sustainability. This is to then mobilize the local society, particularly in terms of its symbolic aspects. These actions represent moments constructed by the researcher around intermediary objects, and through the materiality of these objects, they reconfigure the relationships between local actors, displaying a highly performative effect on reality (Daniel 2011; Dirksmeier and Helbrecht 2008). Local food fairs, crop varieties contests, public debate events, or traditional dances and games, applied within a participatory process, act as devices that can be useful in fostering agroecological transition, as they open up a symbolic environment necessary for agroecological innovation.

PSC techniques applied to agroecological transition are useful to elaborate the participatory diagnosis, to mobilize the social actors, and to promote subjective and symbolic transformations in the local community. They also allow the evaluation of progress achieved in the social–economic–cultural and sociopolitical dimensions.

9.2.8 Phases of the Agroecological Transition Process

In our proposal for participatory action research, the methodologies described are also organized in time, adhering to a sequential five-step system (Table 9.2), according to the sociopraxical, community-based approach (Villasante 2006). Depending on the context, some phases may be removed and others may be parallel or overlapping. Nevertheless, the linear outline of phases is suitable as it explains and structures the process.

The objective of Phase I (preliminary) is to estimate ex ante the "local agroecological potential," meaning the social, ecological, economical, and cultural resources present in the area that can be used to support an agroecological transition. In this phase, mutual trust and relationships between researchers and social agents is a key. Interviews and participant observation are research techniques that are often used (Table 9.2). Likewise, MESMIS and social metabolism applied in Phases I and II enable the diagnosis in Phase II and the evaluation of sustainability improvement in Phase V.

In Phase II (participatory diagnosis) we hope to examine the situation from a holistic perspective, to obtain objective data about the local situation, as well as subjective views of people with whom we are working. Following Chambers' principle of "optimal ignorance" (1992 p. 14), we do not want to know everything; rather, we want to know what is necessary in each moment of the process to act and transform selected items in which we have decided to intervene. In this phase, formal participation and process monitoring spaces are set up, from which we differentiate two theoretical types: the Driving Force Group, as an operational entity that drives the process, composed of the

Table 9.2   Phases and Most Relevant Methods and Techniques Applied in the Agroecological Processes

PHASES	Phase I Preliminary	Phase II Diagnostic	Phase III Research	Phase IV Action	Phase V Evaluation and Adjustment
Matters to solve	Is there potential for the agroecological process?	How do we explain the situation? Who can we rely on?	How can we determine what is possible? How do we create the plan? How do we generate useful information?	How should we act every day?	How can we further the transformation process?
Toolbox	Secondary information analysis Interview Participant Observation MESMIS Agroecological analyzers Social metabolism	Interview Participant Observation MESMIS Discussion groups Sociogram SWOT analysis Flowcharts PRA techniques Social metabolism	Flowcharts PRA techniques FPR Social metabolism	FPR From peasant to peasant Agroecological-mobilizers	Interview MESMIS Sociogram Deliberative polling Social metabolism

FPR, farmer participatory research; MESMIS, framework for the evaluation of natural resource management systems incorporating sustainability indicators; PRA techniques, participatory rural appraisal techniques; SWOT analysis, strengths, weaknesses, opportunities, and threats analysis.

local "base" population; and the Steering Committee, for the formal supervision, legitimization and consensus regarding the process, which would bring together the economic, social, and political associations representing the region.

Discussion groups, sociograms, situational flowcharts, SWOT (strengths, weaknesses, opportunities, and threats) analysis and, in general, several techniques from the PRA are used in this phase. To start the discussion, it is important to use "analyzers," uttering within the interviews and workshops past or present events significant for local people, on searching for positions and interrelations between the different stakeholders to emerge. These techniques and others listed in Table 9.2 are applied in participatory workshops (Chambers 1992), with different objectives, as follows: (1) "feedback workshops" to provide feedback on the information obtained or actions carried out to conduct an in-depth analysis of the items dealt with; (2) "social creativity workshops" to assess the situation, to plan future scenarios or actions, and to innovate on technical or organizational solutions; and (3) "evaluation workshops" to evaluate the participatory process.

Phase III (participatory research) converts the diagnosis into an action plan, the drafting of which involves all the local actors. It must be as legitimate as possible and must adapt to the most pressing needs and to those actions in which the local population is willing to become involved. This plan includes activities to generate information that reinforces the agroecological transition process. Adopting the FPR framework is an important part of this stage.

In Phase IV (participatory action) the researcher's task is to promote the development of actions included in the action plan, which are organized into sectorial work groups (SWG). Outreach activities are essential in this phase. The transmission of the knowledge generated at farm scale is done "from peasant-to-peasant" and the application of "agroecological analyzer–mobilizers" is particularly effective to achieve social mobilization. Traditional knowledge and local varieties are very powerful agroecological analyzer–mobilizers, which concern all of local society, due to their link

Table 9.3 Case Studies' Characteristics Related to the Local Agroecological Potential

	Sierra de Yeguas (Málaga)	Morata de Tajuña (Madrid)	Alpujarra (Granada)	Vega (Granada)
Spatial scale of agroecological transition	Farm	Town	*Comarca*	*Comarca*
Social importance of agriculture in the area	High	Low	High	Low
Experiences in organic farming in the area	None	Low	High	Low
Experiences in short food supply chains in the area	Medium	Low	Low	Low
Farmers' agrarian experience	Low	High	High	High
Farmers' traditional agrarian knowledge	Low	Low	High	Medium
Farmers' social organization	High	Low	Medium	Low
Farmers' political organization	High	Low	Low	Low
Connecting with other social movements	High	Low	Low	Low

with cultural identity. The activities in which this heritage is revealed displays those types of managements that "exist but are not named," such as peasant-like management and other nonindustrial alternatives, which can mobilize compromised social actors to begin reclaiming them.

Finally, Phase V (evaluation and adjustment) brings the process to a close. It considers the subjective and material aspects of the results obtained, as well as the evolution of the local social map. This phase must encapsulate the construction and accompaniment of new group leadership along the lines of agroecology, developed interactively throughout the process, in view of a new PAR cycle. The use of a sociogram in evaluation workshops and deliberative polling are other useful techniques (Table 9.2) to evaluate the evolution of the local social map and subjective aspects. To evaluate progress in material aspects, MESMIS and social metabolism are useful.

To show the utility of this approach to facilitate agroecological transition in the European context, we chose four case studies with very different characteristics with regard to local agroecological potential: spatial scale of agroecological transition, social validity of agriculture, farmers' agrarian knowledge, farmers' political and social organization, and so forth (see Table 9.3).

9.3 DESCRIPTION OF THE STUDY SITES

The four following case studies were developed and analyzed. Table 9.4 summarizes the characteristics of the areas in which the case studies are located.

9.3.1 El Romeral Cooperative

The cooperative, created in 1991 by 10 laborers from the Farm Laborers' Union (SOC in Spanish), gained access to a public farm of 103 ha in the town of Sierra de Yeguas, Malaga (Table 9.4). The SOC led the fight for agrarian reform in Andalusia during the democratic transition of the 1980s–1990s and the creation of agrarian cooperatives in various towns. A group of these cooperatives defended a production model based on traditional rural knowledge and autonomy, which led them to productive diversification and recycling, not using biocides, creating employment, and developing SFSC. The researchers, who subsequently developed the process of PAR, were invited to participate in discussions with the cooperatives in the years before the formal start of the agroecological transition. What is summarized here forms part of the PAR process (1993–1999), which

Table 9.4 Characteristics of the Areas in Which the Case Studies Are Located

Location	Sierra de Yeguas (Málaga)	Morata de Tajuña (Madrid)	Alpujarra (Granada)	Vega (Granada)
Population (no. of inhabitants)	3,206	6,548	24,750	151,469
Population density (inhab./km²)	37.5	145.5	21.38	415.9
Total farms (no.)	269	285	4249	4027
Farmers participating in the PAR process	10 Families	53	330	8 Farmers with on-farm trials. Other actions: 780
Useful agricultural area (ha)	7,561	2340	39,242	26,494
Rainfall (mm)	492.2	386	559.74	382.7
Climate	Temperate with dry or hot summer	Temperate with dry or hot summer	Temperate with dry or temperate summer	Dry climate. Cold steppe
Predominant watering regime	Irrigated/Rain-fed	Irrigated/Rain-fed	Rain-fed	Irrigated
Main crops	Olives, wheat, horticultural crops	Garlic, olives, horticultural crops	Olives, almonds, vineyard, figs	Horticultural crops, olive
Main livestock	Goat	Unimportant	Sheep, cattle	Dairy cattle
Farmers as first activity (%)	86	22	68	70
Farmers >54 years old (%)	24	61	53	52

Source: Instituto Nacional de Estadística (INE), 1999. *Censo Agrario.* www.ine.es.

began with the decision of the members of the El Romeral Cooperative to adopt an agroecological production and marketing model.

In this case, the study focused on transition of the farm scale, with a small group of farm laborers in a typical rural context in Spain, where agriculture is a very important economic activity. At the beginning of this process, there was no experience about OF in the area. However, the agroecological potential of the group was high, especially for its strong social and political organization. The main shortcomings were due to inexperience of the cooperative members as farmers (Table 9.3).

9.3.2 Town of Morata de Tajuña

The proximity of this town to Madrid (37 km) creates a strong periurban influence in its social dynamic, leading to urban prices in the local land market. The town has suffered from a severe deagriculturalization process, which can be observed in the degradation of agrarian institutions and infrastructures, the dismantling and poor mobilization ability of the agricultural sector, the lack of agroindustry, and the concentration of sales in the hands of those landowners with more land. At the start of the project, only three farms were organic (olive groves), but the product was not sold as such. Nevertheless, dense social networks and the restaurant sector proved to be sensitive toward conserving the landscape, the local agrarian activity. and the potential to boost the local economy, especially to mitigate the threat of further urbanization.

The rapprochement for the case study took place before this research project began. The researchers were participating in another agroecological project in the municipality in 2001. This fact led to first contact with the person who, from 2002 onward, would be the municipal counselor for agriculture. This counselor expressed a strong desire to revitalize agrarian activity in the municipality and the local administration explicitly accepted the proposals of OF and participatory methodologies as the central elements of the project, allowing us to design a comprehensive process of participatory rural development based on agroecological transition processes (López 2012).

In short, this case study focused on an intermediate spatial scale (town), where agriculture is socially invisible, but in which there were some experiences in OF to support the process of agroecological transition. The agroecological potential of Morata de Tajuña was initially very low, mainly due to disorganization and weakness of the agricultural sector (Table 9.3).

9.3.3 Alpujarra de Granada *Comarca*

This *comarca* (local administrative district with common territorial features and agricultural services, and consisting of several towns) is a very mountainous region, located in the southeast of Spain between two important mountain ranges (Sierra Nevada and Sierra de la Contraviesa). Meltwater, channeled by means of irrigation channels or underground corridors, facilitates the irrigation of small plots, generating a clear variety of spatial and temporal arrangements of crops, in which altitude plays an important role. The use of the mountain for livestock complements family farms. The precipitous relief prevented farming from becoming intense and enabled the development of rich, traditional agrarian knowledge, which has survived and is very relevant in the European context. The population has decreased in the last century; it now has an older population, with a higher ratio of males, and is concentrated in fewer groups. La Alpujarra is still an agrarian *comarca*, but the active agrarian population has decreased.

The researchers' involvement in the Alpujarra was invited by an association of 54 farmers, named "Contraviesa Ecológica," in 2005. They requested that we identify, describe, and evaluate the local varieties of fig (*Ficus carica* L.) and the traditional knowledge associated with their management, with consideration to introducing them into the organic market. This evaluation involved identifying conjointly those elements of varieties or knowledge that continued to be valid in the present circumstances and those which farmers needed to adapt to achieve their goal: to protect their jobs in farming and conserve their cultural heritage. Subsequently, from 2008, the participatory project was extended to other types of production (livestock, fruit, vegetables, etc.) at the request of other farmer groups and local government. This article focuses on the PAR process between 2008 and 2010.

In this case, we covered more area and population (Table 9.4). Furthermore, agriculture in La Alpujarra has a greater social significance than in the other cases. At the beginning of this process, there were many experiences of OF in the area, so the intervention took place in an advanced stage of the agroecological transition. Therefore, the role of researchers focused mainly on those aspects that farmers identified as major deficiencies (Table 9.3) such as the organization of farmer groups so that they could identify their needs, prioritize them and find solutions, the identification of solutions for new technical problems (for those that they had not yet found an agroecological alternative, as some pest control), and finally build SFSC.

9.3.4 Vega de Granada *Comarca*

This *comarca* is located in the southeast of Spain, around the city of Granada. It is flat, fertile land with abundant irrigation water. High urban and population growth has caused the degradation of the territory and social disarticulation. Despite its high agricultural potential, very few of the inhabitants farm the land (Table 9.4). Small farms predominate and agroindustry is weak. At the beginning of the process, there were 19 companies (only 10 cooperatives) that marketed their products mainly through long channels, earning a low return.

The PAR process began in 2003 after the foundation of the Centro de Investigación y Formación de Agricultura Ecológica y Desarrollo Rural de Granada (CIFAED) (*"Granadan Organic Agriculture and Rural Development Research and Training Centre"*) in a town in the Vega *comarca*. This center was the result of an agreement between two political parties, the socialist party (PSOE) and the

Greens, and was funded by the regional government and local councils of the province. The aim of the CIFAED was to promote agroecological transition in the province of Granada, especially of the Vega *comarca*, because of its high agricultural potential and the complex problems it faced.

The initial agroecological potential in the Vega de Granada was very low. As well as the situation described in the case of Morata de Tajuña (high urban influence, reduction of farming activity, degradation of agricultural infrastructure, and institutions, etc.), there was also the fact that in the last two centuries, the Vega *comarca* has devoted most of its crop area to agroindustrial crops with prices guaranteed by the administration, first by the Spanish administration and then by the European Union. The farmers of the Vega have been unable to react to the progressive dismantling of this system in recent decades. An appropriate reaction would have required prior organizational experience, which does not exist in the *comarca*.

Nevertheless, dense social networks proved to be sensitive towards conserving the landscape, the local agrarian activity, and the agroecological potential.

9.4 RESULTS AND DISCUSSION

9.4.1 El Romeral Cooperative

The evident agroecological potential of the group and the previous relationship between the cooperative and the researchers rendered Phase I unnecessary. The other Phases proceeded at two levels: (1) the discussion on the global producer–supplier model and how to develop alternatives to this, in which nine cooperatives participated in social creativity workshops, coordinated by members of the SOC; and (2) the planning and carrying out of the agroecological transition process at the farm scale with the El Romeral Cooperative, which included the participatory redesign of the agroecosystems and the development of SFSC (Guzmán and Alonso 2000).

The planning of the transition process on-farm took place in Phase II (1993–1994) in three stages: (1) we analyzed secondary information regarding the transition process to OF, since there was little experience in Spain in 1993 and (2) we diagnosed the initial situation of the farms and cooperative from agronomical, socioeconomical, and technological perspectives, using interviews, participatory observation, and MESMIS. The selection of sustainability indicators used in the MESMIS was performed in social creativity workshops. We drew up a number of indicators related to the objectives of the members of the cooperative (e.g., employment creation, crop diversification, development of SFSC, etc.), their perception of the risks (e.g., invasion of weeds or pests) and, in general, the sustainability of the process (e.g., land improvement, planting hedges, etc.); (3) the cooperative members proposed the "ideal" situation that they wanted to reach and the transition strategy by means of information feedback workshops and social creativity workshops.

Phases III and IV (1994–1999) took place at the same time. We carried out the plan as anticipated, starting with the evaluation and the production of *in situ* information, which would enable the management to be modified if necessary. We started trials to optimize the operation (e.g., fertilization and weed management) and to recover the traditional agricultural knowledge and the local horticultural varieties that were incorporated into the farm. MESMIS and FPR were the most important techniques used.

Phase IV consisted of setting the cooperative strategy in motion at the commercial and training levels. The SFSC were developed through consumer associations and organic food producers in various Andalusian cities, linking in with urban social movements. On a secondary level, these were also developed through local business.

Phase V, the evaluation phase (2001), was based, on the one hand, on follow-up sustainability indicators drawn up during previous phases and, on the other hand, on the development of new attitudes and possible redefinitions of the values and objectives both of the group and the environment.

The results are summarized in Table 9.5. The agroecosystem was redesigned with high crop diversi-fication, introduction of livestock for home consumption (pigs, chickens), planting of trees along the edges of vegetable plots, and composting of organic waste. We solved problems with pests (*Agriotes lineatus*) and weeds, maintaining good yields. Success led to easier adoption of agroecological management practices by neighboring farmers. Also economic viability and social objectives were achieved (employment creation and development of SFSC), which are still running. However, the process required much effort and some cooperative members did not agree, so the cooperative was divided into two in 1996. Currently, the new agroecological cooperative is visited by many groups of farmers and technicians interested in OF.

9.4.2 Town of Morata de Tajuña

The research was conducted in three intermittent periods between 2006 and 2009, determined by local public financing. Phase I was based on the analysis of secondary information and semis-tructured interviews with key informants. Furthermore, the researcher negotiated the objectives and limits of the project with the city government, which sponsored the project by binding the proposals and providing funding.

Phase II continued with interviews using a relational approach to create the initial sociogram. The results obtained were used in feedback workshops for the local agrarian sector, applying PRA techniques. During the process, nine farmers from different spheres made up the Driving Force Group, which put into operation the information collected in the preparation of the participatory diagnosis of the local agrarian sector. To achieve this, social creativity techniques were applied (SWOT analysis, sociogram, flowcharts, etc.). Moreover, the entity of municipal participation in agriculture and the environment, a local, official advisory council composed of all the interested political parties and associations, assumed the role of Steering Committee for the project.

In Phase III, we carried out diagnostic feedback workshops with the population and the research focused on the recovery of traditional knowledge. The Driving Force Group (nine farmers from dif-ferent social and productive conditions), together with the researcher and the population involved (more than 50 farmers and other local actors involved in the different workshops), created the action plan of the project, applying a new flowchart to specify and prioritize the actions proposed that were encouraged by the population.

In Phase IV, we put into action the seven SWG that emerged from the action plan, comprising farmers and other nonagrarian actors, including the local government, interested in each topic. Each SWG carried out a social creativity workshop to draw up detailed sector diagnoses and to prioritize actions. In parallel, different "agroecological mobilizers" were implemented (e.g., local food mar-kets and radio programs) to provide the project with visibility, improve the social value of agrarian activity, and make agroecological management alternatives visible.

Finally, in Phase V, a final round of interviews was conducted, as well as deliberative polling, to estimate the subjective transformations obtained locally. Moreover, two evaluation workshops took place. The aim of these workshops was to carry out a participatory assessment of the results obtained and, furthermore, reconsider objectives for a possible continuation of the process.

The impact—material and symbolic—generated on the local agrarian sector was high (Table 9.5). Particularly important was the transformation of the pessimist views of many farmers that changed into motivation to adopt management methods along the lines of an agroecological transition. For this purpose, a cooperative action objective was developed throughout the project, demanded by the farmers in Phase II, to improve the social appraisal of agrarian activity in the town. Our strategy was to make this visible in all local public places and to seek support from all types of entities. With respect to the steps in the conversion process (Gliessman 2010), in this case study, success was higher in level 4 than in level 3. The agroecosystem improved only in those cases in which local varieties were added. Other farmers were able to reach level 2 of input substitution.

Table 9.5 Summary of Results of the Three Case Studies

	El Romeral Cooperative (1993–1999)	Morata de Tajuña (2006–2009)	Alpujarra de Granada (2008–2010)	Vega de Granada (2003–2009)
Short food supply chains (SFSCs) development	By means of consumer associations, local shops, bio trade fairs, and school dining halls. It required too much work, causing the cooperative to divide. Half of the members followed the agroecological model.	Marketing local organic food in 70% of local businesses and in three local restaurants. Creation of community supported agriculture. Creation of distribution cooperative supplying organic foods through SFSC. 300% income increase to organic olive farmers by manufacturing the first organic certified oil in the region, and selling it through SFSC.	The first olive oil mill to have an organic product was established. Its oil is sold through SFSC. The meat produced by Livestock Breeders Association is sold in a neighboring comarca by means of short trade fairs.	Creation of El Vergel: an association of 10 farmers in the Vega comarca and six farmers from nearby districts who market their products locally: box scheme (for approximately 100 families), shops, farmers markets, and so forth. Other 10 producers grouped together in the Association of Organic Producers of the Province of Granada and opened their own shop. Marketing through institutional consumption channels (school canteens, and hospitals) was discontinued as a result of a change of policy by the regional government.
Organic farming promotion	In 2001, Sierra de Yeguas brought together 28% of the organic horticultural farmers from the province, when it only represents 0.45% of the farms. It is a regional benchmark in organic agroindustry, of which the cooperative is a member.	The organic surface area was multiplied by 3. Organic holdings was multiplied by 2, and kept on growing after field work. Creation of an organic winery.	The number of organic farmers increased from 109 to 205.	The organic crop area multiplied by 4.65 (from 74 to 343 ha) and the number of farmers by 3.6 (from 8 to 29).
Solving management problems	Regional impact establishing it as an agroecological leader of horticultural production. We solved problems with pests (Agriotes lineatus) and improved weed management.	Problems with olive fly (Bactrocera oleae) were reduced.	Problems with fig tree disease were solved. We adapted the organic olive farming techniques to the comarca.	Regional impact establishing it as an agroecological leader of irrigated fruit production (pears, persimmon, etc.). Progress in pest control in corn.
Use of local resources	Mainly organic matter. Recovery of local horticultural varieties. The cultivation of local varieties was not consolidated.	Mainly recovery of local horticultural varieties: seven farmers begin to cultivate local varieties at their farms. Input substitution in four farms: agrochemicals are changed by plant extracts for pest control.	Reassessment of traditional agriculture, especially local varieties of fig trees and migratory herding routes.	High diversification of crops. Planting of hedges and plant cover on farms. Recovery of local horticultural varieties. The cultivation of local varieties was consolidated. The need became evident for the care and treatment of water supplies so as not to prejudice the agricultural future of the comarca.

(Continued)

Table 9.5 (*Continued*)   Summary of Results of the Three Case Studies

	El Romeral Cooperative (1993–1999)	Morata de Tajuña (2006–2009)	Alpujarra de Granada (2008–2010)	Vega de Granada (2003–2009)
Social impact	High; creating employment on the farm and in agroindustry, in a region with high unemployment levels. Employment has tripled in the farm.	The social problem of employing immigrant workers was dealt with. Legal contracts went from 0 to 22. Strengthening of local farmers association. Incorporation into farming of three young local people. Introduction of agriculture into public life locally: festivities, radio, and so forth.	Formation of two associations: of livestock and organic fruit and vegetables. Strengthening of the three preexisting associations: organic olive oil, organic vegetables, and organic figs. Training local agents in PAR techniques to continue the process.	Generation of a fabric of associations. The associations promoted during the process were the Andalusian Network of Women Promoters of Responsible Consumption and Organic Food and the Granadan Association for the Defense and Promotion of Organic Agriculture, which are now promoting organic agriculture.
Institutional impact		High in educational centers. All the political parties supported the action plan. Parts of the public local budget were destined to support the proposals from the participatory process.	High in public rural development institutions. The technicians currently accompany the SWG.	Suspension of activities by public institutions devoted to agricultural research and training. These institutions have partially continued to train technical experts and to assist organic farmers.

However, level 4 was widely developed. The organizational effort made by farmers and their belief in the need to change the marketing model became the creation of infrastructure to transform their products (wine and olive oil) and sell them through local trade and SFSC (Table 9.5).

9.4.3 Alpujarra de Granada *Comarca*

The PAR developed between 2008 and 2010, covering all the phases and is shown in Table 9.2. At the start of the process, there was already a reportable area of OF in the *comarca* (1395 ha), concentrated in particular towns. In these cases, intervention was aimed at improving, from an agroecological perspective, a process that was already underway.

Phase I consisted of a preliminary definition of the problems and claims of the farmers and the local agro industry, as well as of the local sociogram. The Driving Force Group was made up of government agents, farmers, and the researcher. We defined an SWG, with actors that were related to OF, including livestock, fig trees, oliviculture, horticulture, artisan agribusiness, SFSC, and viticulture. We drew up a draft of the work plan for each sector, with a first indicative schedule, thus including steps to be conducted with the people, groups, and institutions involved in kick-starting the process, the dates expected for the next steps, and the resources available for the undertaking.

Participatory workshops were carried out, in which the researcher introduced herself, supported the local actors' demands about OF, and started assessing the local agroecological potential. Besides the sociogram, participant observation was important.

In Phase II, we established a typology of farms, through interviews and visits to 51 farms. In addition, we carried out sector diagnoses, and reflection spaces were opened to discuss local agro-ecological knowledge, highlighting the discourses present in relation to OF, the future of the activity, and the possible means of strengthening the sector. The SWOT analysis and PRA techniques were applied in social creativity workshops.

During Phase III, we established critical problems prioritized per sector and action plans and annual schedules were drawn up for the 2010–2013 period. SWG in all the sectors identified was also established, except in the wine sector, which was already an established sector. Particularly relevant were the feedback workshops and the social creativity workshops in which we set out the work priorities.

In Phase IV, we began to implement the action plans. The effort focused partially on the agro-ecological transition processes on the farm, developing actions outlined in the FPR. This facilitated the methodological and technical transfer between the researcher and the farmers and between the farmers themselves (from peasant-to-peasant). Good qualitative and quantitative results were obtained (Table 9.5).

In Phase V, we reviewed and assessed the 2010–2013 Action Plan, and we reinforced the transfer of group leadership that had been previously started, with the complementary training of farmers involved in PAR techniques. We carried out a participatory qualitative assessment of the process (achievements, challenges, and reorganization of networks) using a sociogram and a quantitative evaluation (increase in OF).

Results are summarized in Table 9.5. Implementation of traditional farming practices, such as transhumance (seasonal migration of livestock) and the cultivation of local fig tree varieties, was the largest contribution to the agroecosystem redesign. This contribution was invaluable from two perspectives: in its symbolic nature through the reinforcement of farmers' identity and the appreciation of traditional knowledge, and in its material nature through the use of local resources, landscape improvement, biodiversity increase, fire risk reduction, and so forth. Technical problems were solved by redesigning the agroecosystem (e.g., biodiversity introduction) and, sometimes, by input substitution (e.g., pest mass trapping). Finally, the best farmer organization progressed in the establishment of a cooperative agroindustry and food marketing through SFSC.

9.4.4 Vega de Granada *Comarca*

Phase I (2003–2005) consisted of the evaluation of the sustainability of agriculture in the Vega *comarca* in the years 1750–2000 applying social metabolism and MESMIS. This study revealed the keys to the current crisis in the Vega *comarca* and allowed proposals for change to be elaborated to increase agricultural sustainability. The diagnosis and the resulting proposals were discussed with farmers and other social actors during Phase II.

In parallel, unstructured interviews were held with key persons and participatory observation allowed an in-depth examination of the discourse of the social actors regarding the agricultural problems facing the *comarca*. As a result of this interaction, the researchers participated in numerous debate forums in which it became evident that the vast majority of the attendees were deeply concerned about the destruction of the Vega *comarca*. There was an evident need for the strengthening of the fabric of social associations to make the problems more visible and to produce solutions that would bring about a global change of strategy in production and marketing, going beyond sectorial solutions. A result of this process was the creation of two associations, which later played a crucial role in the agroecological transition process (Table 9.5).

In Phase II (2006–2007), we interviewed 20 representatives of agroindustry, which gave us in-depth knowledge of the strategies they were using to overcome the agricultural crisis. Subsequently, we used the discussion group to analyze the approaches of the farmers in two thematic areas: the production and marketing strategies they were developing and, secondly, organic agriculture as an integrated strategy. Eighty farmers participated, mostly members of different cooperatives. There were farmers with different production (organic/nonorganic) and marketing strategies (long chain/ short chain). The results were discussed in the feedback workshops, which debated the different strategies and the facilitating and limiting factors affecting the agroecological transition, both with regard to production and marketing, especially the development of short channels. The final stage of this process was the elaboration of the Vega de Granada Organic Agriculture Plan, which included measures and the budget necessary to facilitate the agroecological transition of the *comarca*. It was signed by the four organizations representing farmers and agroindustry operating in the Vega *comarca* and by the three ecological and consumer organizations involved in the process. The regional government committed to cofinance the measures of the Plan for 3 years (2008–2010).

Phases III and IV (2008–2009) ran in parallel. The implementation of the measures was undertaken mainly by the social organizations, which were signatories to the Plan and the CIFAED. FPR techniques played a central role with fruit farmers and extensive irrigation crop farmers (corn, alfalfa) who decided to initiate the agroecological transition. Some of the traditional agricultural knowledge gathered in Phase I was applied here, introducing local horticultural varieties on the farms being converted. One measure was the evaluation of chemical contamination in irrigation water, to allay farmers' fears that it would prevent the organic certification of their produce. Finally, measures were designed for the development of SFSC.

These Phases lasted only 2 years. The political coalition broke down and the regional government, now made up only of the socialist party, withdrew its support and closed the CIFAED in 2009.

Table 9.5 summarizes the results obtained from the evaluation during Phase V (2011). Organic production was consolidated in horticulture and irrigated fruit production. There was a sharp increase in biodiversity on the farms as a result of the diversification of crops, the planting of hedges, and the use of plant cover and local varieties. The cultivation of local horticultural varieties was consolidated with high demand from farmers (including nonecological farmers) and consumers. The associations promoted during the process continue to defend the agroecological transition of the Vega *comarca*, and they have been joined by other initiatives and associations (e.g., for the defense of organic food in schools, secondary school teachers in defense of the Vega, among others).

Historical analyzers were very useful to discuss and refute certain subjective views found in the agricultural sector in the Vega de Granada, for example, the quest for a "miracle crop," which would save them from decline. The historical analysis showed how monoculture was related to institutional frameworks, which were very different from that which exists today and that, in the medium term, it led to the destruction of natural resources and a serious loss of autonomy for the agricultural sector.

9.5 CONCLUSIONS

PAR has developed a plethora of tools for achieving social change through social action–reflection processes (Freire 1972) in very different fields. Applied to agroecology, participatory methods have been developed mostly at the farm scale (e.g., MESMIS, FPR). However, many of the constraints for agroecological transition can only be solved at higher research scales (Dalgaard et al. 2003), including ecological, social, and economic complexity, which are at the core of the farming sector crisis. Community-based participatory approaches, which focus on social networks, can help to link different scales of agroecological research and can confront the asymmetrical relations between actors within the food system.

The methodology applied has combined different research techniques inside a participatory process, following the sequence proposed by Villasante (2006) for sociopraxis. PRA techniques proved to be the most effective in facilitating participation in the workshops and FPR techniques proved to be the most effective in the generation of appropriate technologies and in technological diffusion. Meanwhile, PSC proved to be useful to design and monitor the process by researchers, and showed more limitations for participatory workshops with farmers. MESMIS and, above all, social metabolism were useful to aid comprehension of the processes which undermine the sustainability of agroecosystems and to contribute to their redesign.

Although Gliessman (2010) defines agroecological transition in 4 progressive stages, in some of our case studies, Phase I was not present and Phase IV was developed in parallel to the redesign of the agroecosystem. In fact, SFSC development became the driving force for the change in the management of the agroecosystem in Morata de Tajuña.

Considering the marginal nature of agriculture in the European context, especially in periurban areas, the simple act of opening communication spaces between farmers, in which they are the protagonists, fostered initial interest in further participation. However, obtaining specific results regarding farmers' claims and problems was an essential incentive to widen and further increase participation later. The reduced number of people representing the local community of farmers was often insufficient to develop the proposals set out, and, as a result, collaboration from other nonagrarian actors. Thus, widening the territorial intervention scale—to include the whole region—proved to be useful in obtaining a sufficient critical mass. The integration in the project of other actors and the connection with other networks beyond the limits of study site were essential to develop the SFSC.

Hardly any progress was made in the redesign of the agroecosystems when this required intervention in the territory at a higher level than on the farm scale. An example of this is the fact that virtually no progress was made in the relocalization of energy and nutrient flows or the recuperation of the quality of the water. To overcome these difficulties will require greater organization and influence on public policies at a higher scale.

The PAR implementation period was assessed as short by participants, except in El Romeral. Stable financing is required for long time periods in order for these processes to be correctly carried out—something that is very difficult to obtain.

Despite these limitations, a PAR methodological approach and associated research techniques have been successful in initiating and accompanying agroecological transition processes, in involving farmers in the redesign of their farms to increase their sustainability, and in building local organic food networks by the wider society.

REFERENCES

Bell, M., S.E. Lloyd, and C. Vatovec. "Activating the countryside: Rural power, the power of the rural and the making of rural politics." *Sociologia Ruralis* 50 no. 3 (2010): 205–224.

Best, H. "Organic agriculture and the conventionalization hypothesis: A case study from West Germany." *Agriculture and Human Values* 25 (2008): 95–106.

Buck, D., C. Getz, and J. Guthman. "From farm to table: The organic vegetable commodity chain of Northern California." *Sociologia Ruralis* 37 no. 1 (1997): 3–20.

Chambers, R. *Farmers First. Farmer Innovation and Agricultural Research.* London: Intermediate Technology Publications, 1989.

Chambers, R. *Rural Appraisal: Rapid, Relaxed and Participatory.* IDS Discussion Paper 311. Brighton, United Kingdom: Institute for Development Studies, 1992.

Cuéllar, M. and A. Calle. "Can we find solutions with people? Participatory action research with small organic producers in Andalusia." *Journal of Rural Studies* 27 (2011): 372–383.

Dalgaard, T., N. Hutchings, and J. Porter. "Agroecology, scaling and interdisciplinarity." *Agriculture, Ecosystems & Environment* 100 (2003): 39–51.

Daniel, J.F. "Action research and performativity: How sociology shaped a farmers' movement in the Netherlands." *Sociologia Ruralis* 51 no. 1 (2011): 17–34.

Darnhofer, I., T. Lindenthal, R. Bartel-Kratochvil, and W. Zollitsch. "Conventionalisation of organic farming practices: From structural criteria towards an assessment based on organic principles. A review." *Agronomy for Sustainable Development* 30 (2010): 67–81.

De Wit, J. and H. Verhoog. "Organic values and the conventionalization of organic agriculture." *NJAS— Wageningen Journal of Life Sciences Articles* 54 no. 4 (2007): 449–462.

Dirksmeier, P. and I. Helbrecht. "Time, non-representational theory and the 'performative turn'—Towards a new methodology in qualitative social research." *Forum Qualitative Sozialforschung/Forum Qualitative Social Research* 9 no. 2 (2008): 1–15.

Eds. O. Fals Borda and M. A. Rahman, O. "Algunos Ingredientes Básicos." In: *Acción y Conocimiento. Como Romper el Monopolio con Investigación-Acción Participativa.* Bogotá, Colombia: CINEP, 1991, 7–19.

Farrington, J. and A. Martin. "Farmer participatory research: A review of concepts and practices." *Agricultural Administration Network* Discussion Paper 19 (1987).

Francis, C., G. Lieblein, S. Gliessman, et al. "Agroecology: The ecology of food systems." *Journal of Sustainable Agriculture* 22 no. 3 (2003): 99–118.

Freire, P. *Extensión o Comunicación.* Santiago de Chile, Chile: Instituto de Capacitación e Investigación en Reforma Agraria, 1969.

Freire, P. *Pedagogy of the Oppressed.* London: Penguin, 1972.

García-Ruiz, R., M. González de Molina, G.I. Guzmán, D. Soto, and J. Infante. "Guidelines for constructing nitrogen, phosphorus and potassium balances in historical agricultural systems." *Journal of Sustainable Agriculture* 36 (2012): 650–682.

Geilfus, F. *80 Tools for Participatory Development.* Costa Rica: IICA, 1997.

Gliessman, S.R. "The framework for conversion." In: *The Conversion to Sustainable Agriculture: Principles, Processes, and Practices. Advances in Agroecology.* S.R. Gliessman and M. Rosemeyer (Eds.). Boca Raton, FL: CRC, Taylor & Francis Group, 2010, 3–14.

Goldberger, J.R. "Conventionalization, civic engagement, and the sustainability of organic agriculture." *Journal of Rural Studies* 27 (2011): 288–296.

González de Molina, M. and V. Toledo. *The Social Metabolism: A Socio-ecological Theory of Historical Change.* New York: Springer, 2014. Guzmán, G.I. and A.M. Alonso. "The European Union: Key roles for institutional support and economic factors." In: *The Conversion to Sustainable Agriculture: Principles, Processes, and Practices. Advances in Agroecology.* S.R. Gliessman and M. Rosemeyer (Eds.). Boca Raton, FL: CRC, Taylor & Francis Group, 2010, 239–272.

Guzmán, G.I. and A.M. Alonso, "Transición Agroecológica en Finca." In *Introducción a la Agroecología como Desarrollo Rural Sostenible.* G. I. Guzmán, M. González de Molina, and E. Sevilla Guzmán (Eds.). Madrid, Spain: Mundi-Prensa, 2000, 199–226.

Guzmán, G.I. and M. González de Molina. "Preindustrial agriculture versus organic agriculture: The land cost of sustainability." *Land Use Policy* 26 no. 2 (2009): 502–510.

Guzmán, G.I., M. González de Molina, and A.M. Alonso. "The land cost of agrarian sustainability. An assessment." *Land Use Policy* 28 (2011): 825–835.

Holt Giménez, E. *Campesino a Campesino: Voces de Latinoamérica Movimiento Campesino para la Agricultura Sustentable*. Managua, Nicaragua: SIMAS, 2008.

Instituto Nacional de Estadística (INE), 1999. *Censo Agrario*. www.ine.es.

Kindon, S., R. Pain, and M. Kesby. (Eds.). *Participatory Action Research Approaches and Methods*. Routledge Series in Human Geography. Oxon, Oxfordshire: Routledge, 2012.

Lapassade, G. *L'Analyseur et l'Analyste*. Paris: Gauthier Villars, 1971.

Lobley, M., A. Butler, and M. Reed. "The contribution of organic farming to rural development: An exploration of the socio-economic linkages of organic and non-organic farms in England." *Land Use Policy* 26 (2009): 723–735.

López, D. *Hacia un Modelo Europeo de Extensión Rural Agroecológica. Praxis Participativas para la Transición Agroecológica. Un Estudio de Caso en Morata de Tajuña, Madrid*. PhD Thesis, Baeza, Spain: Universidad Internacional de Andalucía, 2012.

López-Ridaura, M., O. Masera, and M. Astier. "Evaluating the sustainability of complex socio-environmental systems. The MESMIS framework." *Ecological Indicators* 2 no. 1–2 (2002): 135–148.

Milestad, R., R. Bartel-Kratochvil, H. Leitner, and P. Axmann. "Being close: The quality of social relationship in a local organic cereal and bread network in Lower Austria." *Journal of Rural Studies*, 26 no. 3 (2010): 228–240.

Park, P., M. Brydon-Miller, B.L. Hall, and T. Jackson (Eds.). *Voices of Change. Participatory Research in the United States and Canada*. Toronto, ON: OISE Press, 1993.

Reed, M. "The rural arena: The diversity of protest in rural England." *Journal of Rural Studies* 24 no. 2 (2008): 209–218.

Rhoades, R. and R. Booth. "Farmer-back-to-farmer: A model for generating acceptable agricultural technology." *Agriculture Administration* 11 (1982): 127–137.

Scoones, I. and J. Thompson. "Knowledge, power and agriculture-towards a theoretical understanding." In: *Beyond Farmers First: Rural People's Knowledge, Agricultural Research and Extension Practices*. I. Scoones and J. Thompson (Eds.). London: Intermediate Technology Publications Ltd., 1994, 16–31.

Selener, D. "Farmer participatory research." In: *Participatory Action Research and Social Change*. D. Selener (Ed.). New York: Cornell University Press, 1997, 149–188.

Sevilla Guzmán, E. *De la Sociología Rural a la Agroecología*. Icaria, Barcelona, 2006.

Tandon, R. "Civil society. Adult learning and action in India." *Convergence* 33 (2000): 1–2.

Tello, E., E. Galán, V. Sacristán, et al. "*A Proposal for a Workable Analysis of Energy Return On Investment (EROI) in Agroecosystems*." Social Ecology Working Paper, Institute of Social Ecology at the Alpen-Adria Klagenfurt University in Vienna, 2014. http://www.uni-klu.ac.at/socec/inhalt/1818.html (Accessed January 15, 2015).

Villasante, T.R. *Desbordes Creativos*. Madrid: Los libros de la Catarata, 2006.

Villasante, T.R., M. Montañés, and J. Martí. *La Investigación Social Participativa. Construyendo Ciudadanía I*. Madrid, Spain: El Viejo Topo, 2000.

Wezel, A., S. Bellon, T. Doré, C. Francis, D. Vallod, and C. David. "Agroecology as a science, a movement and a practice. A review." *Agronomy for Sustainable Development* 29 (2009): 503–515.

Willer, H., J. Lernoud, and L. Kilcher. "*The World of Organic Agriculture. Statistics and Emerging Trends 2013*." Bonn: FiBL, Frick, and, IFOAM, 2013.

CHAPTER 10

Agroecology, Food Sovereignty, and Urban Agriculture in the United States

Margarita Fernandez, V. Ernesto Mendez, Teresa Mares, and Rachel Schattman

CONTENTS

10.1 INTRODUCTION

In the last 15 years, movements for just and sustainable food systems in the United States have burst into the national stage. Local action on sustainable and organic agriculture, community food security, food justice, food sovereignty, urban agriculture, local food policy, childhood obesity, local foodsheds, and direct farmer to consumer marketing continues to expand across the country (Holt-Giménez and Shattuck 2011; Allen 2004; Mares and Alkon 2011). Most practitioners in US alternative agrifood movements do not use the term "agroecology," but share and are guided by similar ecological and social principles and a vision for transforming local and global agrifood systems. While "agroecology" in the United States is a term most often used in association with the academic literature, university research, and educational institutions, this approach has also played a role in the evolution of alternative agrifood movements (Buttel 2004; Wezel et al. 2009). The field of agroecology has evolved from an early focus on integrating ecology into agriculture at the farm scale toward a more integrative study of the ecology of food systems (Francis et al. 2003). This evolution takes the field beyond a technological approach to one that actively pursues sustainability in agriculture and food systems using a systems-based transdisciplinary, participatory, and action-oriented approach (Gliessman 2010; Mendez et al. 2013).

With growing recognition that agroecology is the key agricultural approach to confronting the multiple crises of climate change, global hunger, and the unsustainability of the corporate agrifood system (Food and Agriculture Organization [FAO] 2014; International Assessment of Agricultural Knowledge, Science and Technology for Development [IAASTD] 2009; Chappell and LaValle 2011; Horlings and Marsden 2011; de Schutter 2010; de Schutter and Vanloqueren, 2011), we believe it is important to examine the current state of agroecology in the United States, and specifically assess its role both in academia and alternative agrifood movements. Since an examination of all alternative agrifood movements is out of the scope of this paper, we chose to focus on two important initiatives—urban agriculture and food sovereignty. We chose these for two main reasons: (1) food sovereignty has mobilized one of the largest global social movements today and identifies agroecology as one of its key strategies; however, food sovereignty in the United States is a nascent movement and does not so obviously identify with agroecology; and (2) urban agriculture is an essential venue for a global transformation of our agrifood system due to the multiple social, ecological, and economic benefits it provides to a growing global urban population. Thus, the objective of this paper is to explore how agroecology, food sovereignty, and urban agriculture have evolved in the United States and identify opportunities for a better integration between the three in order to advance overlapping goals of creating sustainable agrifood systems. We believe that a greater integration between academia and alternative agrifood movements, and in particular between agroecology, urban agriculture, and food sovereignty, can help facilitate scaled-up change toward more ecologically resilient, socially just, and economically viable agrifood systems.

10.2 EVOLUTION AND SCOPE OF AGROECOLOGY IN THE UNITED STATES

10.2.1 Overview

Agroecology emerged as a response to the negative environmental, social, and economic externalities of the agro-industrial system (Gliessman 1990; Altieri 1987; Rosset and Altieri 1997; Vandermeer 2010), proposing that ecological concepts and principles were needed in order to design and manage sustainable agroecosystems (Gliessman 1998). Although many pioneers of the field (i.e., Altieri, Gliessman, Vandermeer, Perfecto, and Sevilla-Guzmán) worked mostly in the tropics, they were predominantly based in United States and European academic institutions. Susanna Hecht (1995) traces the intellectual lineage of agroecology through influences from tropical ecology, studies of indigenous agriculture systems, ecological methods, rural development, geography, and anthropology. This evolution of a more interdisciplinary approach stems in part from an understanding that in order to analyze the interactions between ecology and agriculture, agroecology must also analyze the interactions between human systems and natural systems (Hecht 1995).

One of the most widely used definitions of agroecology today comes from Francis et al. (2003: 100) who described agroecology as "the integrative study of the ecology of the entire food system, encompassing ecological, social and economic dimensions." While this perspective expands the focus of agroecology to an interdisciplinary perspective, it was Wezel et al. (2009) who proposed that agroecology is expressed not only as a science, but also a practice and a movement. This evolution in the meanings and applications of agroecology paralleled the rise of alternative agrifood movements in the United States, which were motivated by concerns not only about on-farm sustainability, but also community food security, food safety, labor, environmental health, and broader sustainability issues of the agrifood system (Allen 2004). Although some interaction between agroecology and US agrifood movements can be seen in the 1990s (see e.g., Allen et al. 1991), there seems to have been little integration since.

In the 1970s, the science of agroecology influenced the emergence of the concept of sustainable agriculture as a practice and movement (Wezel et al. 2009). Simultaneously, the environmental and sustainable agriculture movements and the practice of sustainable agriculture influenced agroecology as a science (Hecht 1995). As described by Allen (2004), the growth of academic programs with a focus on sustainable agriculture and community food security issues reflected an institutionalization of social movement agendas. For example, social movement work, with leadership from the sustainable agriculture coalition,[*] was instrumental in passing the US Department of Agriculture (USDA) Low Input Sustainable Agriculture program (now known as Sustainable Agriculture Research and Education [SARE]). The SARE program and other programs under the USDA National Institute for Food and Agriculture have contributed significantly to the growth of agroecology-based programs in universities across the country. As a result, many academic programs promoting the study and application of agroecology benefited from the social advocacy work around food and sustainable agriculture in the 1960s, 1970s, and 1980s.

10.2.2 Higher Education and Research in Agroecology

Agroecology in the United States has been most prominently used and advanced by academics in US universities (Francis et al. 2003; Gliessman 2007). Agroecology courses were initially offered within environmental studies or agriculture programs, with one of the first to be offered by the Environmental Studies Program at the University of California, Santa Cruz, in 1981 (Francis et al. 2003). The late 1980s and early 1990s saw a boom in sustainable agriculture programs in research universities, including the University of California Davis (1986), the University of Maine (1986), Iowa State University (ISU) (1987), the University of Illinois (1988), the University of Wisconsin Madison (1989), the University of Minnesota (1991), Washington State University (1991), and the Center for Agroecology and Sustainable Food Systems at the University of California Santa Cruz (1993). These remain major institutional centers for both sustainable agriculture and increasingly transdisciplinary agroecological research and education. Today, there are more than 55 land grant and private colleges and universities offering undergraduate and graduate degrees in sustainable agriculture and food system studies with 12 of those offering programs and degrees specifically in agroecology.

Higher education institutions have played three roles in the agroecology movement: (1) conducting research on innovative agroecological methods; (2) providing practical educational initiatives for producers through extension programs and outreach; and (3) training students in agroecological approaches through undergraduate and graduate education. The land grant university system, established by the Morrill Act of 1862 and funded by public tax dollars, has a long history of research into agricultural technologies and the subsequent transfer of those technologies to farmers via extension programs (Warner 2007). Initially, this model was seen as a success, as it improved farm productivity and provided abundant food for the nation. However, following the publication of Rachel Carson's *Silent Spring* and James Hightower's *Hard Tomatoes, Hard Times*, the industrial agricultural science produced by land grant institutions came into question, as well as the extension model that granted "expert" status to scientists and relegated farmers to the role of their "clients," which curtailed farmer-initiated innovation (Gliessman 2010; Warner 2006, 2007). Therefore, sustainable agriculture approaches began to be integrated in other pedagogical arenas and stand-alone programs were founded to meet the needs of the growing sustainable agriculture community.

Agroecological knowledge cannot be "transferred" from scientist to farmer as easily as new chemical or mechanical technologies. Ecological farming practices, such as biological insect control or use of cover crops, tend to be information intensive, adapted to a specific site, and labor intensive to implement and monitor (Warner 2007). Additionally, agroecological farming requires not just the insertion of one or two key technologies into an existing agroecosystem, but an entire system redesign (Gliessman 2010). Thus, agroecological research and education are necessarily

---

[*] Today known as the National Campaign for Sustainable Agriculture.

transdisciplinary in nature, incorporating elements of plant science, soil science, ecology, economics, political science, sociology, geography, and anthropology, as well as farmer knowledge (Méndez et al. 2013). Since land grant universities have traditionally been disciplinary in nature (Parr et al. 2007), it has been challenging for them to incorporate the new perspectives brought by agroecological education, research, and extension. However, faculty, students, and farmers at universities throughout the United States are increasingly demanding and designing innovative programs and partnerships grounded in that transdisciplinarity and participatory research and extension.

One example of such a partnership is the Organic Agriculture Program at ISU. In 1996, and in response to demand from producers, industry, and citizens, ISU established an organic agriculture and research program housed in the departments of horticulture and agronomy. This program was directed by a new faculty position specializing in organic agriculture, with the goal of addressing the gap in field-tested research and education in organic production (Delate 2002). At the heart of the program was the establishment of four Long-Term Agroecological Research (LTAR) sites, one in each of the four agroecological zones of Iowa. Research at these sites is complemented by on-farm research projects in collaboration with Iowa farmers. Most of these activities are funded by grants from the USDA-SARE and USDA-IFAFS (Initiative for Future Agriculture and Food Systems).

The ISU program is characterized by an interdisciplinary focus, a strong commitment to farmer participation, and a grounding in whole-systems research. Over 10 academic departments are involved in the activities of the ISU Organic Agriculture Program (Delate 2002), and ISU offers an interdepartmental graduate program in sustainable agriculture. The Leopold Center for Sustainable Agriculture, a publicly funded research and education center established in 1987 and housed at ISU, has offered logistical and financial support to the Organic Agriculture Program. The Leopold Center has also encouraged program staff to allow the priorities of organic and transitioning farmers to guide the research agenda, and cohosted a series of focus groups during the establishment of the program to determine those priorities. Farmer groups such as Practical Farmers of Iowa comprised a substantial part of these focus groups, as farmer networking has historically been the primary means of education on organic farming in Iowa (Delate 2002). Immediate application of the research on Iowa farms is fostered by conducting research on organic farms, despite the additional time involved. This research approach recognizes the complexity of agroecological systems by utilizing and implementing systems theory, which recognizes the emergent properties of agroecosystems, with new relations and phenomena arising at certain junctures within the system (Delate 2002).

ISU's partnership with Practical Farmers of Iowa is a key feature of agroecological research and education. Farmer-to-farmer networks are one of the first routes where many farmers begin to learn about and advocate for a more sustainable form of agriculture, and these social networks have long been the backbone of the agroecology movement globally (Warner 2007; Holt-Giménez 2006; Rosset et al. 2011; Rosset and Martinez-Torres 2012). In many cases, these networks and organizations formed as a response to the absence of research or education in sustainable methods within the land-grant university system. Such was the case, for example, with dairy farmers interested in rotational grazing in Wisconsin, who developed the Southwestern Wisconsin Farmers Research Network, in 1986, to conduct on-farm research and share knowledge about a form of agriculture that was not supported by the University of Wisconsin extension system. By partnering with nongovernmental rural organizations, the network was eventually able to secure some state funding, and developed into the Sustainable Agriculture Program that provided funding for farmer groups to conduct their own research and extension (Warner 2007).

Despite the fact that agroecology in the United States is strongest in the academic sphere, and to a certain extent, in extension programs (see Darby et al., Chapter 11 this edition), agroecology as a practice and movement has not taken as strong a hold in the United States as it has internationally (Wezel et al. 2009). As agroecology-based academic programs increasingly offer courses and curriculum that focus on agroecology as the study of the ecology of food systems (Francis et al. 2003), incorporating participatory, transdisciplinary, and action-based research (Méndez et al. 2013), there will be more opportunities for interactions between agroecology and alternative agrifood movements in the United States. An increased connection between the science of agroecology and movements aligned with its principles can help contribute to systemic policy changes. Leading agroecologists contend that ecological change in agriculture and food systems cannot happen without social, economic, and policy change (Altieri 2009). In order for agroecological change to happen, partnerships between agroecology and alternative agrifood movements are critical. In the next section, we summarize alternative agrifood movements in the United States, using a political lens. We then highlight advancements of two important movements—food sovereignty and urban agriculture—and discuss synergies between these movements and agroecology.

10.2.3 Alternative Agrifood Movements in the United States

In the United States, various food movements have developed under a variety of terms and from different origins (Allen 2004; Mares and Alkon 2011). These include sustainable agriculture, ecological agriculture, organic agriculture, permaculture, multifunctional agriculture, low input agriculture, conservation agriculture, community food security, food justice, food sovereignty, and sustainable/local food systems. These are not monolithic concepts, and different actors in the US agrifood system have differing views and uses for each one of these terms, which can vary according to their social, political, environmental, and economic values (National Research Council [NRC] 2010; Gliessman 2010). As Mares and Alkon argue, "both 'localized food systems' and the 'corporate food regime' are complex and multidimensional social, political, and economic formations necessitating multidisciplinary and multisectoral research approaches" (2011: 69). Some areas of the food movement, such as community food security and local food, are primarily oriented toward improving access to foods through market-based approaches, and changing consumption behaviors at the individual and household level. Food justice and food sovereignty discourses, on the other hand, point to deeper critiques of the class and race based inequalities that pervade the food system, from production through disposal. When endorsing market-based strategies, food justice and food sovereignty approaches both emphasize that these strategies should be created by and within the control of those most affected by food inequalities, whether that is low-income communities of color or small resource-poor farmers.

In a related vein, Holt Gimenéz and Shattuck (2011) provide a political characterization of these different perspectives according to the degree of transformation from the current agrifood system that they propose (either explicitly or implicitly), ranging from neoliberal/reform to progressive/radical. Based on this work, we summarize here salient characteristics of those agrifood initiatives that have explicitly engaged with sustainable agriculture and agroecology issues in the United States.

The neoliberal/reform view represents the perspectives of predominantly corporate, global trade, and development actors. This standpoint advocates the expansion of trade liberalization, increased production, certification schemes, genetically modified organisms (GMOs), and agrofuels as solutions to the environmental and social challenges facing the current agrifood system. However, these policies and actions tend to approach sustainability through a patchwork approach (often based primarily upon market mechanisms) rather than a whole systems approach (NRC 2010). Stakeholders in this domain have appropriated "sustainable" to include monocultures, GM crops, unjust labor systems, and other practices that do not align with agroecological principles (Rosset and Altieri 1997; Altieri and Toledo 2011). The corporatization of organic agriculture demonstrates how the concept of sustainability can be stripped of its social and environmental values, as well as its holistic approach to fit the agroindustrial model (Guthman 2004; Jaffee and Howard 2010; Thompson 2001). Large agribusinesses, pharmaceuticals, and food processors now produce many of the inputs approved by the USDA National Organic Program for production and processing, have acquired many of the organic food processors, and sell through a few large retailers (Howard 2009). The increasing consolidation of the agrifood system severely undermines decision-making power at the farm level (NRC 2010; Pimbert et al. 2001). Although the corporate organic food system has made some strides in improving the environmental and health impacts of food production, processing, and nutrition, it still falls short of directly addressing structural issues inherent to the agroindustrial food model (Guthman 2004).

Actors espousing the progressive/radical view are farmers, distributors, processors, consumers, nonprofits, researchers, and local governments who are working toward a radically alternative agrifood system. This perspective views the relocalization of food production, distribution, consumption, and waste management as a key strategy to transforming the agrifood system. This vision seeks to enhance and expand diversified farming, food cooperatives, community-supported agriculture (CSA), farmer's markets, food hubs, regional food system plans, and food sovereignty laws. Many in this domain call for an overhaul of the corporate-dominated agrifood system through structural changes—land reform, shifts in research, credit and subsidies, and increased regulatory pressure on corporate actors (Holt-Giménez and Shattuck 2011; Rosset 2009). These structural changes pursue the reversal of the neoliberal trends affecting the United States and global agrifood system, most notably, increasing market consolidation, which has resulted in oligopsonies, oligopolies, and land consolidation exemplified by the land-grabbing phenomena (de Schutter 2010; Brent and Kerssen 2014; Zoomers 2010). Actors in this domain build on the environmental and health concerns that spurred the organic and sustainable agriculture movement of the 1960s and 1970s, but integrate more recent thinking in terms of food sovereignty, food justice, and human rights. The progressive/radical domains align with the agroecological perspective that seeks a transdisciplinary, participatory, and action-oriented approach, which is the focus of this volume (Mendez et al. 2013; Mendez et al., Chapter 1 in this volume). In the United States, this progressive/radical approach is primarily grounded in projects that seek to empower farmers and farmworkers and support diversified small-scale farms. Perhaps more importantly, actors in this domain move away from a focus on productivity and equally value the social and environmental effects of agrifood systems.

The trajectory of food and agriculture movements in the United States has widened the political and economic spectrum that promotes sustainable agriculture (Holt-Giménez and Shattuck 2011; Kloppenburg et al. 2000). Central to this trajectory has been a dynamic relationship between grassroots movement actors, policymakers, and educators in shaping and reshaping the United States' agrifood system. This dynamic is particularly active in today's political climate, where issues deeply connected to agrifood systems—climate change, peak oil, global food crises, and loss of biodiversity—are prompting a diversity of actors to prescribe, negotiate, compromise, and/or resist varying strategies for a more sustainable agrifood system (Gliessman 2010). As with most social change processes, actions and policies aimed at reshaping the current agrifood system are part of a dialectical process where civil society advocacy, private sector lobbying, and policy making engage and negotiate through different social, economic, environmental, and political interests to develop new alternatives. This produces unique challenges and opportunities for agroecology to contribute in these efforts toward agrifood system transformation. As Altieri and Toledo (2011, 597) state, "the new agroecological scientific and technological paradigm is being built in constant reciprocity with social movements and political processes."

In a review of organizations funded by the three top US funders of sustainable agriculture and food systems initiatives—the USDA Community Food Program, SARE, and the W.K. Kellogg Foundation (Sustainable Agriculture and Food Systems Funders [SAFSF] 2006)—and a Web-based search, we found that very few organizations working on alternative agrifood systems use the term agroecology to describe their work. However, a review of a sample of these organizations' missions and objectives shows a large majority promoting strategies in line with the agroecological principles of systems based, participatory, action-oriented, and transdisciplinary work for agrifood system change (see Mendez et al., in this issue, Chapter 1, and www.agroecology.org for detailed principles). The organizations that do use the term agroecology, including Food First, Pesticide Action Network, Oxfam America, Heifer International, Institute for Agriculture and Trade Policy, Family Farm Defenders, and National Family Farm Coalition, engage in both domestic and international work. These organizations are connected to international food and agriculture movements that advocate for agroecology as a key strategy to further their goals, including Via Campesina, the Landless Peasant Movement of Brazil, and the Campesino a Campesino Movement.

Holt-Giménez and Altieri (2013) demonstrate that agroecology as a social movement is constantly changing, and ownership over the very name of the discipline is contested. In their recent article, they point out that the anticipated need to increase worldwide food production to feed a population that is growing at a dramatic rate has put steam behind a renewed Green Revolution. They call on agroecologists to align themselves in reform or radical camps, implying that those who choose the reformist route are aiding in the cooptation of agroecology. This suggests that the culture of agroecology is both strengthened and weakened by its inclusion of multiple disciplines and foci under a single conceptual umbrella. To better address food system sustainability, the participation of greater diversity of actors is required, along with mediating institutions and public policies that support collaboration. However, the diffusion of focus can make agroecology's goals difficult to communicate and understand.

While much of the recent international focus of agroecology has concentrated on the global south, it is also applicable to agroecosystems in the global north. The northeastern United States is of particular interest when using an agroecological frame because of the relatively small size of many northeastern farms, the recent focus on local and regional food systems, and the high level of access to local and state political processes. There is a notable lack of organizing in the United States under the banner of agroecology, with the exception of a relatively small network of researchers and extension professionals. Rather, we see a more fluid integration of the term "agroecology" throughout literature that pertains to alternative production systems, including organic agriculture (Goodman 2000), sustainable agriculture (Dover and Talbot 1987), and ecological agriculture (Magdoff 2007). We also see organizations devoted to maintaining access to farmland and farming as a livelihood option. Some of these organizations (the National Family Farm Coalition, the Rural Coalition, Farmworkers Association of Florida, and the Borders Farmworkers Project) have allied themselves with international networks primarily representative of small landholders in the global south (La ViaCampesina), demonstrating the overlap of issues faced by landholders and farmworkers in the global north and south. Many principles are shared between these frameworks, the strongest being resistance to the corporate consolidation of agriculture and the food supply and the natural resource depletion associated with industrial agriculture. In the global north, there has been less attention paid to agroecology as a social movement, with social justice and empowerment for farmers and farmworkers finding homes in other frameworks and contexts.

10.3 FOOD SOVEREIGNTY AND URBAN AGRICULTURE INITIATIVES IN THE UNITED STATES

Over the past five decades different social and environmental movements around food and agricultural issues have fueled changes in higher education and government policies, which have laid a foundation for a growing agroecology movement. These movements have also built a strong knowledge base among farmers and activists in terms of ecological farm management, community organizing, and political advocacy. Despite the fact that agroecology in the United States is strongest in the academic sphere, the broadening of the agroecological approach (Francis et al. 2003; Mendez et al. 2013) has made it easier for agroecology and alternative agrifood movements in the United States to interact, as it has opened more spaces for conceptual and applied overlap. This closer interaction can also be facilitated through the participatory action research (PAR) approach. In the following sections, we focus our attention toward initiatives working on food sovereignty and urban agriculture in the United States.

10.3.1 Food Sovereignty

The concept of food sovereignty can serve as both a policy framework with a strong social and political movement behind it and a conceptual framework that can be implemented by researchers to better understand and address agrifood system inequalities. Born out of farmers' movements protesting the economic, social, and environmental impacts of the neoliberal trade system, food sovereignty seeks to link local progressive actions to a larger political agenda in order to make structural changes to local and global agrifood systems. The concept of food sovereignty was coined at a Via Campesina meeting in the mid-'90s, but its definition has evolved through an iterative process characteristic of the movement's dynamism (Martinez-Torres and Rosset 2010). The food sovereignty paradigm is guided by the following key principles: (1) food as a basic human right; (2) gender equality; (3) genuine agrarian reform; (4) protecting natural resources; (5) reorganizing food trade; (6) ending the globalization of hunger; (7) social peace; and (8) democratic control of food (Wittman 2011; Pimbert 2008). The most recent definition from Via Campesina states that food sovereignty is "The right of peoples to healthy and culturally appropriate food produced through ecologically sound and sustainable methods, and their right to define their own food and agriculture systems. It puts those who produce, distribute, and consume food at the heart of food systems and policies rather than the demands of markets and corporations" (Via Campesina 2007). Leaders in both the agroecology and international food sovereignty movements emphasize that the application of agroecology within agrifood systems is a key strategy to achieving food sovereignty (Altieri and Toledo 2011; Martinez-Torres and Rosset 2010; Cohn et al. 2006). La Via Campesina has explicitly adopted agroecology as its guiding approach for agricultural and farm management (Rosset and Martinez-Torres 2012).

Many principles of agroecology are directly linked to the goals of food sovereignty (Altieri and Toledo 2011). For example, agroecology advocates for farmer autonomy by relying on local, renewable resources and minimizing external inputs linked to industrialized agrifood structures (synthetic fertilizers and pesticides, commercial seed, machinery, etc.) (Rosset and Altieri 1997). In addition, a respect and value for the knowledge and priorities of farmers aligns with food sovereignty principles of autonomy, equity, and a relocalization of food systems (Altieri 2009). Agroecology's focus on farmer self-sufficiency can be perceived as a "subversive act" by those with a neoliberal view. Ultimately, the redesign of agricultural systems, by harnessing ecological processes inherent in natural systems, enables independence from the agroindustrial system (Coleman and Damrosch 2010). For these reasons, agroecology is the essential foundation for food sovereignty processes and goals. We agree with many other authors who advise against a strict definition of food sovereignty (Wittman 2011; Boyer 2010; Jarosz 2014), as we view it as a process, a vision, a means, and an end at the same time and because it is a multidimensional, context-dependent approach. Hence, food sovereignty requires flexibility to be adapted to unique situations. In this sense, it is similar to the concept of agroecology, which is guided by a number of key principles that can be adapted to distinct contexts (Altieri and Toledo 2011; Gliessman 2007). The ongoing challenge has been how to connect local forms of resistance grounded in food sovereignty and agroecology to larger social and political movements for structural change. In the United States, a growing number of alternative agrifood movements are identifying with and applying the food sovereignty framework to their unique struggles (Schiavoni 2012; Ayres and Bosia 2014; Block et al. 2011; US Food Sovereignty Alliance [USFSA] 2014), but few explicitly link their conceptualization of food sovereignty with agroecology as has been done in international movements.

A salient example of this is Brazil's Landless Workers Movement (Movimiento Sin Tierra [MST]), a member of Via Campesina who takes as its primary action the occupation of land so that it can be used by landless rural workers. The MST has embraced agroecology to guide management strategies in lands that have been occupied and transferred to landless families. In 2005, the MST founded the Latin American School of Agroecology, which was designed to train members of the MST and Via Campesina in the principles of agroecology, as grounded in a social and political movement to change the global agrifood system (MST 2014). Although the integration of agroecology into rural social movements like the MST and Via Campesina is not without struggle (Delgado 2008; Rossett and Martinez-Torres 2012), the successes of these international movements are significant in reshaping rural food systems, particularly in the global south.

In March 2011, Sedgwick, Maine, became the first US town to pass a food sovereignty ordinance. Within 6 months of Sedgwick's ordinance, food sovereignty ordinances had been passed in Maine, Vermont, Massachusetts, Georgia, North Carolina, Utah, Wyoming, and Montana. These ordinances are meant to protect the rights of local producers, in particular meat and dairy farmers, so that they can produce artisanal products without the financially prohibitive laws that are meant to control quality and safety in large meat and dairy operations—quality and safety issues that are often irrelevant in small operations. These ordinances have been framed, in part, as a response to the Food Safety Modernization Act that was signed by President Obama in January 2011 and that increases the power of the federal government in the control of food safety. Food sovereignty in the United States represents a politicization of the sustainable food movement at the local level defending the values of self-reliance, self-provisioning, and autonomy. Just as with other alternative agrifood movements in the United States, the concept and discourse of agroecology is not employed. As a movement at the national level, the USFSA was formed, in 2010, "to end poverty, rebuild local food economies, and assert democratic control over the food system" (USFSA 2014). The USFSA serves as a venue for catalyzing food sovereignty movements in the United States and connecting them to similar movements abroad. Members of the USFSA are diverse, representing rural farm groups, urban farm groups, migrant farmworker groups, and policy/advocacy groups.

US food sovereignty initiatives have interacted with agroecology much less than initiatives on the international stage. As Ayres and Bosia (2011) noted, food sovereignty articulates in different ways in different locales with distinct historical, political, and cultural contexts. For example, in the state of Vermont (as in other states), food sovereignty often unfolds as an autonomous, antiregulatory approach eschewing the dominant agrobusiness corporate model and embracing localism, rather than embracing a broader and more deeply integrated framework provided (at least in part) by agroecology. Social, economic, and political changes needed to address issues related to food sovereignty cannot happen without ecological change. Agroecology provides the framework with which to make that ecological change without losing sight of greater systemic forces affecting the sustainability of this change. While food sovereignty movements in the United States and abroad have emerged predominantly from rural landscapes, urban agriculture, globally and in the United States, is a ripe and dynamic venue for food sovereignty (Schiavoni 2009; Block et al. 2011).

10.3.2 Urban Agriculture

While local and slow food movements get much of the media attention, the politics of urban food justice and the practices connected to urban agriculture are among the most dynamic alternative agrifood initiatives in the United States. Many organizations working on sustainable food systems in urban areas are based in and led by low-income communities of color, drawing significant inspiration from environmental justice action and theory. Both urban agriculture and food justice have long independent histories. The history of urban gardening in the United States reflects a cyclical process of urban garden creation and destruction that moves in conjunction with economic crisis and recovery. Urban gardening in the United States dates back to the economic depression of the mid-1890s, when the city of Detroit allotted 455 acres of land and seed potatoes for planting, to 945 families. The temporary leasing by the city of abandoned land spread to more than 20 cities in the United States, but with the increase in real estate development these gardens were short-lived (Hynes 1996). The next revival of urban gardening came with the "liberty gardens" of World War I and then the postwar "victory gardens," which were part of a national campaign to supplement food shortages and "maintain morale on the homefront" (Kurtz 2001). The war gardens were part of a collective effort that reflected the current cultural and national ideals with "an estimated five million gardeners rallying to such slogans as 'plant for freedom' and 'hoe for liberty' (Hynes 1996). However, once the immediate need to produce food subsided so did governmental support. With the economic crisis felt in cities across the country in the 1970s came a new wave of urban gardens, many of which still exist today. This new wave of urban agriculture brought with it a more explicit framework of racial and economic justice (Mares 2014). Over the past decade, hundreds of urban gardens and nonprofit organizations, such as Just Food, The Food Project, Rooted in Community, Food What!, and Community Harvest, have emerged as part of the local food, food justice, and youth empowerment movements. As importantly, urban agriculture projects have also found significant support from municipal governments, such as Seattle's P-Patch Program coordinated by the city's Department of Neighborhoods (Mares 2014).

Today there are over 16,000 community gardens and urban farms across the country (American Community Gardening Association [ACGA] 2011). Community gardens provide a host of ecological, social, and economic benefits. Extensive research on urban farming, including community gardens, greenbelt gardens, and personal gardens, assert that growing food within city limits significantly contributes to an increased quality of life by building social capital, improving access to food, providing jobs, improving mental and physical health, and providing a multitude of environmental benefits, such as reducing a community's carbon footprint (Fernandez 2006; United Nations Development Program [UNDP] 1996; Blair et al. 1991; Brown and Jameton 2000; Glover 2004; Pinderhughes 2003; Saldivar-Tanaka and Krasny 2004). As urban populations continue to increase globally, policies that promote urban agriculture and its associated benefits are ever more pressing.

The most recent economic crisis has brought interesting imperatives and opportunities for urban agriculture, especially in cities that have experienced significant postindustrial decline. For example, between 2005 and 2009, tens of thousands of properties were left vacant and abandoned in the City of Cleveland and Cuyahoga County, Ohio, after a dramatic increase in foreclosures. The local government created a land bank to turn foreclosed properties back to productive use, including urban agriculture. Thanks in part to recommendations by the Cleveland-Cuyahoga Food Policy Council, the city now has one of the most progressive urban agriculture zoning policies in the country. In addition to supportive zoning policies, the city created a program to make available publicly owned urban properties from the land bank for food production on both single and 5-year leases, is piloting an irrigation program to reduce water costs to urban farmers, and is helping negotiate other obstacles to entrepreneurial urban farms like liability insurance. As of August 2011, Cleveland has leased some 60 parcels to entrepreneurial urban farms and community gardens (Walsh et al. 2015).

There is a wide variety of political expressions in current urban agriculture movements, some of which overlap with more overtly political calls for food justice. Food justice connects efforts to expand access to healthy food with a critique of historical patterns of racism (Alkon and Agyeman 2011; Alkon and Norgaard 2009). The concept emerged from historical struggles over racial and environmental justice in urban centers in the United States and tends to take a social, rather than consumer perspective on food systems change, and in many ways is among the strongest movements for food system change in the urban United States (Holt-Giménez and Shattuck 2011). While food justice is a more recently defined term, its roots can be traced to civil rights efforts like the school breakfast program initiated by the Black Panther Party in Oakland, California, in January 1969, as part of the group's militant struggle for racial and economic justice (Patel 2011). Groups organizing under the food justice banner work primarily on creating access to healthy food in low-income communities (both in terms of improving quality, convenience, and affordability in the "food environment" and increasing purchasing power). But food justice organizations are not limited to this framework—organizations under the banner of food justice are also creating farmworker-owned cooperative businesses, organizing and advocating for better working conditions across the food chain, opening locally or cooperatively owned grocery stores in low-income communities, establishing buying clubs, addressing health disparities through education, connecting low-income consumers with fresh produce from local farmers through CSA systems, and developing youth leadership. The Detroit Black Community Food Security Network, for example, uses urban gardening as part of a broader agenda including addressing structural racism and lack of black ownership in the food system. This broader agenda includes moving beyond looking only at food security and food justice toward food sovereignty by prioritizing rights to land and food and linking this to a contestation of the dominant agrifood regime.

Just as with food sovereignty in the United States, there has been little interaction between agroecology and urban agriculture. Agroecology can be an important tool for urban agriculture in the United States and abroad, but very little research has looked into the concept and practice of "urban agroecology."

Agroecology's key principles provide an essential toolkit for optimizing urban agriculture's multiple benefits. Most notably is the principle of diversified farming systems, which internalizes soil fertility and pest management without the need for toxic chemicals, many of which have restricted use in urban areas. Diversified farming systems can also produce a variety of food crops that can contribute to improved nutrition. Recent research from agroecologists Philpott et al. (2014) look at factors that drive species richness and abundance of arthropod populations in urban landscapes, including community gardens. Arthropods are essential species for the functioning of ecosystems, including urban ecosystems, and are especially important for the pollination of food crops. Partnerships between urban agriculture movement actors and agroecologists are not common, but have great potential to be mutually beneficial. Given urban agriculture's diverse goals that span social, ecological, and economic spheres, agroecology's transdisciplinary, participatory, and action-oriented approach can help the scaling-up of urban agriculture with urban farmers as key protagonists.

10.4 INTEGRATING AGROECOLOGY AND ALTERNATIVE AGRIFOOD MOVEMENTS IN THE UNITED STATES: CHALLENGES AND OPPORTUNITIES

Across the United States, there is a growth in food policy councils, food sovereignty ordinances, new farmers, the urban food justice movement, and educational institutions offering agroecology-based programs. Collectively, this reflects a growing influence of transformational and transdisciplinary approaches in alternative agrifood movements both within society and academia (Allen 2004). A marked rise in youth, women, and minority groups farming and politically organizing around food systems brings a new dynamism to agrarian movements, many of whom may be interested in, or potentially already incorporating agroecological principles to their work. The growing links between concerns about the environment, health, food security, poverty, and social justice reflect an emerging systemic understanding of agriculture as a social and ecological activity in addition to an economic one.

One key challenge to creating sustainable agrifood systems is to connect progressive local actions to a larger political agenda in order to remove structural barriers to the scaling-up of these systems (Holt-Giménez and Shattuck 2011; Mares and Alkon 2011). Federal policy that perpetuates the agro-industrial model, market concentration, and the orientation of research and extension toward these sectors, are central barriers to the scaling-up of sustainable agrifood systems (Reganold et al. 2011). Alternative agriculture receives comparatively little state support for extension services, storage, distribution and processing facilities, affordable credit, and insurance policies (Carolan 2005). Furthermore, land values in the United States are divorced from their productive uses (USDA, Economic Research Service [ERS] 2011) and over half of US cropland is rented, often on single-year leases where incentives are low for agroecological innovation (Carolan 2005). Until producers have access to land and infrastructure and are consistently paid a better price for both their product and the environmental services they steward, sustainable agrifood systems will be on tenuous footing (Robertson and Swinton 2005).

On the consumer side, economic justice is a challenge for the movement. With nearly 15% of Americans on food stamps, purchasing power in low and middle income communities is often insufficient to purchase enough food, much less food from alternative networks (Food Research and Action Center [FRAC] 2011). Although food justice movements are making strides to increase accessibility to sustainable products, systemic change in federal policy is necessary to reorient monies that currently support the production of abundant, cheap and nutritiously deficient food toward diversified farming systems that produce nutritious diets at an affordable price.

Urban agriculture and food sovereignty movements in the United States are crucial to the advancement of alternative agrifood systems. Agroecology can contribute to this process by partnering with social movements and local food system actors through PAR. As Patricia Allen points out, there is a dearth of studies of alternative agrifood movements and great potential for further collaboration between academia and agrifood movements (Allen 2004, 2008). Agroecology can complement other research and action frameworks (e.g., rural sociology and political ecology) in order to better understand and analyze strengths and weaknesses of agrifood system strategies and identify solutions for ecological, social, and political action. Because agroecology espouses participatory and transdisciplinary approaches it dovetails with the democratic, multistakeholder, systems-based approaches embraced by many agrifood movements (Mares and Alkon 2011). Furthermore, with its use of PAR it aims to empower people to become well-informed agents of change for themselves and their communities. Likewise, agrifood movement actors can enrich agroecology students and researchers by helping them remain grounded in analysis of real problems and real solutions. Social, economic, and political changes needed to address issues of food justice, food sovereignty, and food security cannot happen without ecological change. Likewise, ecological change cannot happen without social, economic, and political change. Agroecology provides technological, scientific, and methodological tools that can contribute to facilitate this change (Altieri and Toledo 2011). Hence, we believe that a deeper interaction between agroecology and alternative agrifood movements could provide additional impetus to transforming current agrifood systems to more sustainable ones.

REFERENCES

Alkon, A.H., and J. Agyeman. *Cultivating Food Justice: Race, Class, and Sustainability.* Boston: The MIT Press, 2011.

Alkon, A.H., and K.M. Norgaard. "Breaking the food chains: An investigation of food justice activism." *Sociological Inquiry* 79 no. 3 (2009): 289–305.

Allen, P. *Together at the Table: Sustainability and Sustenance in the American Agrifood System.* University Park, PA: The Pennsylvania State University Press, 2004.

Allen, P. "Mining for justice in the food system: Perceptions, practices, and possibilities." *Agriculture and Human Values* 25 no. 2 (2008): 157–161.

Allen, P., D.V. Dusen, J. Lundy, and S. Gliessman. "Integrating social, environmental, and economic issues in sustainable agriculture." *American Journal of Alternative Agriculture* 6 no. 1 (1991): 34–39.

Altieri, M. "Agroecology, small farms, and food sovereignty." *Monthly Review* 61 no. 3 (2009): 102–113.

Altieri, M.A. *Agroecology: The Scientific Basis of Alternative Agriculture.* Boulder, CO: Westview Press, 1987.

Altieri, M.A., and V.M. Toledo. "The agroecological revolution in Latin America: Rescuing nature, ensuring food sovereignty and empowering peasants." *Journal of Peasant Studies* 38 no. 3 (2011): 587–612.

American Community Gardening Association. 2011. http://www.communitygarden.org/learn/faq.php (Accessed August, 2011).

Ayres, J., and M.J. Bosia. "Beyond global summitry: Food sovereignty as localized resistance to globalization." *Globalizations* 8 no. 1 (2011): 47–63.

Blair, D., C. Giesecke, and S. Sherman. "A dietary, social, and economic evaluation of the Philadelphia urban gardening project." *Journal of Nutrition Education* 23 (1991): 161–167.

Block, D.R., N. Chávez, E. Allen, and D. Ramirez. "Food sovereignty, urban food access, and food activism: Contemplating the connections through examples from Chicago." *Agriculture and Human Values* 29 no. 2 (2011): 203–215.

Boyer, J. "Food security, food sovereignty, and local challenges for transnational agrarian movements: The Honduras case." *Journal of Peasant Studies*, 37 no. 2 (2010): 319–351.

Brent, Z., and T.M. Kerssen. *Land and Resource Grabs in the United States: Five Sites of Struggle and Potential Transformation*. Land & Sovereignty in the Americas Series, No. 7. Oakland, CA: Food First/Institute for Food and Development Policy and Transnational Institute, 2014.

Brown, K., and A. Jameton. "Public health implications of urban agriculture," *Journal of Public Health Policy* 21 no. 1 (2000): 20–39.

Buttel, F.H. "Envisioning the future development of farming in the USA: Agroecology between extinction and multifunctionality?" In W.L. Bland and F.H. Buttel (Eds.). *New Directions in Agroecology Research and Education*. Madison, WI: Center for Integrated Agricultural Systems, University of Wisconsin, 2004.

Carolan, M.S. "Barriers to the adoption of sustainable agriculture on rented land: An examination of contesting social fields." *Rural Sociology* 70 (2005): 387–413.

Chappell, M.J., and L.A. LaValle. "Food security and biodiversity: Can we have both? An agroecological analysis. *Agriculture and Human Values* 28 no. 1 (2011): 3–26.

Cohn, A., J. Cook, M. Fernández, R. Reider, and C. Steward. (Eds.). *Agroecology and the Struggle for Food Sovereignty in the Americas*. New Haven, CT: IIED, IUCN-CEESP, and Yale F & ES Publication Series, 2006.

Coleman, E., and B. Damrosch. Paper prepared for the *Agrarian Studies Colloquium*, November 9, 2010. Yale University.

Delgado, A. "Opening up for participation in agro-biodiversity conservation: The expert-lay interplay in a Brazilian social movement." *Journal of Agricultural & Environmental Ethics* 21 no. 6 (2008): 559–577.

deSchutter, O. *Report submitted by the Special Rapporteur on the right to food*. UN General Assembly. Human Rights Council Sixteenth Session, Agenda item 3A/HRC/ 16/49, 2010.

deSchutter, O., and G. Vanloqueren. "The new Green Revolution: How twenty-first-century science can feed the world." *Solutions* 2 no. 4 (2011): 33–44.

Delate, K. "Using an agroecological approach to farming systems research." *Horttechnology* 12 no. 3 (2002): 345–354.

Dover, M.J., and L.M. Talbot. *To Feed the Earth: Agro-Ecology for Sustainable Development*. Washington, D.C.: World Resources Institute, 1987.

FAO. *International Symposium on Agroecology for Food and Nutrition Security*, 2014. http://www.fao.org /about/meetings/afns/en/.

Fernandez, M. "Cultivating community, food and empowerment: Urban gardens in Havana and New York." In *Agroecology and the Struggle for Food Sovereignty in the Americas*. A. Cohn, J. Cook, M. Fernández, R. Reider, and C. Steward (Eds.). New Haven, CT: IIED, IUCN-CEESP, and Yale F&ES Publication Series, 2006.

Food Research and Action Center. *SNAP/Food Stamp Monthly Participation Data*, 2011. http://frac.org/reports -and-resources/snapfood-stamp-monthly-participation-data/ (Accessed December 3, 2011).

Francis, C. et al. "Agroecology: The ecology of food systems." *Journal of Sustainable Agriculture* 22 no. 3 (2003): 99–118.

Gliessman, S. *Agroecology: Ecological Processes in Sustainable Agriculture*. Ann Arbor, MI: Ann Arbor Press, 1998.

Gliessman, S. *Agroecology: The Ecology of Sustainable Food Systems, Second Edition*. CRC Press, 2007.

Gliessman, S.R. (Ed.). *Agroecology: Researching the Ecological Basis for Sustainable Agriculture*. New York, NY: Springer-Verlag, 1990.

Gliessman, S.R. *The Conversion to Sustainable Agriculture: Principles, Processes, and Practices.* New York, NY: CRC Press, 2010.

Glover, T. "Social capital in the lived experiences of community gardeners." *Leisure Sciences* 26 (2004): 143–162.

Goodman, D. "Organic and conventional agriculture: Materializing discourse and agro-ecological managerialism." *Agriculture and Human Values* 17 (2000): 215–219.

Guthman, J. *Agrarian Dreams: The Paradox of Organic Farming in California.* Berkeley: University of California Press, 2004.

Hecht, S.B. "The evolution of agroecological thought." In *Agroecology: The Science of Sustainable Agriculture.* M.A. Altieri (Ed.). Boulder, CO: Westview Press, 1995.

Holt-Giménez, E. "*Campesino a Campesino: Voices from Latin America's Farmer to Farmer Movement for Sustainable Agriculture.*" Oakland, CA: Food First Books, 2006.

Holt-Giménez, E., and M.A. Altieri. "Agroecology, food sovereignty and the new green revolution." *Agroecology and Sustainable Food Systems* 37 no. 1 (2013): 90–102.

Holt-Giménez, E., and A. Shattuck. "Food crises, food regimes, and food movements: Rumblings of reform or tides of transformation?" *Journal of Peasant Studies* 38 no. 1 (2011): 109–144.

Horlings, L.G., and T.K. Marsden. "Towards the real Green Revolution? Exploring the conceptual dimensions of a new ecological modernisation of agriculture that could feed the world." *Global Environmental Change* 21 no. 2 (2011): 441–452.

Howard, P.H. "Consolidation in the North American organic food processing sector, 1997 to 2007." *International Journal of Sociology of Agriculture and Food* 16 no. 1 (2009): 13–30.

Hynes, P.J. *A Patch of Eden: America's Inner-City Gardeners.* White River Junction, VT: Chelsea Green, 1996.

International Assessment of Agricultural Knowledge, Science and Technology for Development. *Global Report.* Washington, WA: Island Press, 2009.

Jaffee, D., and P.H. Howard. 2010. "Corporate cooptation of organic and fair trade standards." *Agriculture and Human Values* 27 no. 4 (2010): 387–399.

Jarosz, L. "Comparing food security and food sovereignty discourses." *Dialogues in Human Geography* 4 no. 2 (2014): 168–181.

Kloppenburg, J., S. Lezberg, K. De Master, G.W. Stevenson, J. Hendrickson, M. Mead, et al. "Tasting food, tasting sustainability : Defining the attributes of an alternative food system with competent, ordinary people." *Human Organization* 59 no. 2 (2000): 177–186.

Kurtz, H. "Differentiating multiple meanings of garden and community." *Urban Geography* 22 no. 7 (2001): 656–670.

Magdoff, F. "Ecological agriculture: Principles, practices, and constraints." *Renewable Agriculture and Food Systems* 22 no. 2 (2007): 109–117.

Mares, T. "Engaging Latino immigrants in Seattle food activism." In *Food Activism: Agency, Democracy, and Economy.* V. Siniscalchi and C. Counihan (Eds.). New York, NY: Bloomsbury Press, 2014, 31–46.

Mares, T.M., and A.H. Alkon. "Mapping the food movement: Addressing inequality and neoliberalism." *Environment and Society: Advances in Research* 2 (2011): 68–86.

Martinez-Torres, M.E., and P. Rosset. "La ViaCampesina: The birth and evolution of a transnational social movement." *Journal of Peasant Studies* 37 no. 1 (2010): 149–175.

Méndez, V.E., C.M. Bacon, and R. Cohen. "Agroecology as a transdisciplinary, participatory, and action-oriented approach." *Agroecology and Sustainable Food Systems* 37 no. 1 (2013): 3–18.

Movimiento Sin Tierra. 2014. http://www.mstbrazil.org/?q=LAschoolofagroecology

National Research Council. *Towards Sustainable Agricultural Systems in the 21st Century.* Washington, DC: National Academies Press, 2010.

Parr, D.M., C.J. Trexler, N.R. Khanna, and B.T. Battisti. "Designing sustainable agriculture education: Academics' suggestions for an undergraduate curriculum at a land grant university." *Agriculture and Human Values* 24 (2007): 523–533.

Patel, R. *Survival pending revolution: What the Black Panther Party can teach the U.S. Food movement.* In E. Holt-Giménez (Ed.), Food Movements Unite! (pp. 115–135). Oakland, California: Food First Books, 2011.

Philpott, S.M., J. Cotton, P. Bichier, R.L. Friedrich, L.C. Moorhead, S. Uno, et al. "Local and landscape drivers of arthropod abundance, richness, and tropic composition in urban habitats." *Urban Ecosystems* 17 no. 2 (2014): 513–532.

Pimbert, M. *Towards Food Sovereignty: Reclaiming Autonomous Food Systems.* London: IIED, 2008.

Pimbert, M.P., J. Thompson, W.T. Vorley, T. Fox, N. Kanji, and C. Tacoli. *Global Restructuring, Agri-Food Systems and Livelihoods* (Gatekeep Series no. 100). London: IIED: Sustainable Agriculture and Rural Livelihoods Program, 2001.

Pinderhughes, R. "Poverty and the environment: The urban agriculture connection." In J.K. Boyce and B.G. Shelley (Eds.). *Natural Assets: Democratizing Environmental Ownership.* Washington, DC: Island Press, 2003.

Reganold, J.P., D. Jackson-Smith, S.S. Batie, R.R. Harwood, J.L. Kornegay, D. Bucks, et al. "Transforming U.S. agriculture." *Science* 332 (2011): 670–671.

Robertson, G.P., and S.M. Swinton. "Reconciling agricultural productivity and environmental integrity: A grand challenge for agriculture." *Frontiers in Ecology and the Environment* 3 no. 1 (2005): 38–46.

Rosset, P. "Fixing our global food system: Food sovereignty and redistributive land reform." *Monthly Review* 61 no. 3 (2009): 114–128.

Rosset, P., and M.A. Altieri. "Agroecology versus input substitution: A fundamental contradiction in sustainable agriculture." *Society and Natural Resources* 10 (1997): 283–295.

Rosset, P.M., and M.E. Martínez-Torres. "Rural social movements and agroecology: Context, theory, and process." *Ecology and Society* 17 no. 3 (2012).

Rosset, P.M., B.M. Sosa, A.M.R. Jaime, and D.R.Á. Lozano. "The Campesino-to-Campesino agroecology movement of ANAP in Cuba: Social process methodology in the construction of sustainable peasant agriculture and food sovereignty." *Journal of Peasant Studies* 38 no. 1 (2011): 161–191.

Saldivar-Tanaka, L., and M.E. Krasny. "Culturing community development, neighborhood open space, and civic agriculture: The case of Latino Community Gardens in New York City." *Agriculture and Human Values* 21 (2004): 399–412.

Schiavoni, C. "The global struggle for food sovereignty: From Nyeleni to New York." *Journal of Peasant Studies* 36 (2009): 663–706.

Sustainable Agriculture and Food Systems Funders. *Trends in Sustainable Agriculture and Food Systems Funding 2003–2006.* The Headwaters Group Philanthropic Services, 2006.

Thompson, P.B. "The reshaping of conventional farming: A North American perspective." *Journal of Agricultural and Environmental Ethics* (2001): 217–229.

United Nations Development Program. *Urban Agriculture: Food, Jobs, and Sustainable Cities.* New York, NY: United Nations Development Program, 1996.

US Department of Agriculture, Economic Research Service. *Land Use, Value, and Management: Agricultural Land Values.* Economic Research Service. United States Department of Agriculture, 2011. http://www.ers.usda.gov/Briefing/landuse/aglandvaluechapter.htm (Accessed October, 2011).

US Food Sovereignty Alliance. 2014. http://www.usfoodsovereigntyalliance.org/home (Accessed October, 2011).

Vandermeer, J. *The Ecology of Agroecosystems.* Burlington, MA: Jones & Bartlett Publishers, 2010.

Via Campesina. *Declaration of Nyeleni*, 2007. http://www.viacampesina.org/en/index.php?option=com_content&task=view&id=282&Itemid=38 (Accessed May, 2011).

Walsh, C.C., M. Taggart, D.A. Freedman, E.S. Trapl, and E.A. Borawski. "The Cleveland–Cuyahoga County Food Policy Coalition: 'we have evolved.'" *Preventing Chronic Disease* 12 (2015) 140538. DOI: http://dx.doi.org/10.5888/pcd12.140538.

Warner, K.D. "Extending agroecology: Grower participation in partnerships is key to social learning." *Renewable Agriculture and Food Systems* 21 no. 2 (2006): 84–94.

Warner, K.D. *Agroecology in Action: Extending Alternative Agriculture through Social Networks.* Cambridge, MA: MIT Press, 2007.

Wezel, A., S. Bellon, T. Dore, C. Francis, D. Vallod, and C. David. "Agroecology as a science, a movement and a practice. A review." *Agronomy for Sustainable Development* 29 no. 5 (2009).

Wittman, H. "Food sovereignty: A new rights framework for food and nature?" *Environment and Society: Advances in Research* 2 no. 1 (2011): 87–105.

Zoomers, A. "Globalisation and the foreignisation of space: Seven processes driving the current global land grab." *Journal of Peasant Studies* 37 no. 2 (2010): 429–447.

CHAPTER 11

On the Ground
Putting Agroecology to Work through Applied Research and Extension in Vermont

Debra Heleba, Vern Grubinger, and Heather Darby

CONTENTS

11.1 INTRODUCTION

The purpose of this chapter is to describe concrete examples of how farmer adoption of agroecological practices can be supported by researchers, extension personnel, and other professionals that work directly with farmers. We explore how the concepts of participatory action research (PAR) and outcomes-based outreach of agroecology and sustainable agriculture are applied at the community level through programs conducted by cooperative extension in the U.S. state of Vermont. This chapter illustrates the practical application of theoretical frameworks provided in other sections of this book.

11.1.1 Sustainable Agriculture and Agroecology

The underpinnings of the case study examples that follow come from our collective experience of and knowledge about sustainable agriculture. Currently, the term "agroecology" is not widely used in extension or among farmers, although as this field of study evolves and interest in food system work increases, more extension educators are adopting the term and agroecological practices in their work with the farming community (Cecil 2004; Getz and Warner 2006; Ikerd 1993; Perez and Howard 2007). On the other hand, "sustainable agriculture" is a broadly used and accepted term among farmers, extension educators, and other agricultural service providers. "Sustainable agriculture," as it is legally defined (U.S. Code Title 7, Section 3103), means an "integrated system of plant and animal production practices having a site-specific application that will over the long term:

- Satisfy human food and fiber needs.
- Enhance environmental quality and the natural resource base upon which the agriculture economy depends.
- Make the most efficient use of nonrenewable resources and on-farm resources and integrate, where appropriate, natural biological cycles and controls.
- Sustain the economic viability of farm operations.
- Enhance the quality of life for farmers and society as a whole (U.S. Code 2011).

While there may be debate about the terminology that is most appropriate to use when referring to whole farm systems research and outreach work with farmers, "sustainable agriculture" and "agroecology" share many principles, values, and goals—especially those around environmental stewardship, community engagement, and systems approaches to addressing root causes of problems versus treatment of symptoms. Therefore, for the purposes of this chapter, we choose to use the terms interchangeably. The ultimate goals of the work described here are captured in the outcome statement of the United States Department of Agriculture (USDA) Northeast Sustainable Agriculture Research and Education program, a competitive grants program responsible for funding much agroecological research and outreach conducted in our region over the past 25 years: "Agriculture in the Northeast will be diversified and profitable, providing healthful products to its customers; it will be conducted by farmers who manage resources wisely, who are satisfied with their lifestyles, and have a positive influence on their communities and the environment" (NE-SARE 2014).

11.2 THE ROLE OF EXTENSION IN AGRICULTURE

To set the stage, it seems appropriate to provide a brief history of cooperative extension and the U.S. land grant university (LGU) system. In many ways, the twenty-first century case studies presented here exemplify efforts to *modernize* the traditional LGU/extension role, which started as an expert system designed to deliver scientific findings to better inform farmers and the general public. Today, although the importance of scientific knowledge is not diminished, there is an added commitment among a growing number of personnel within LGUs and extension to empower farmers by honoring their knowledge and experiences as part of the educational process, giving them a voice in determining the research agenda, and including them as partners in on-farm research and testing of new ideas. This expanded approach is critical when it comes to addressing farming systems because of their complexity and variability across the geographical, economical, and cultural landscape. In other words, while the principles of agroecology can be widely applied, in practice, farming is a place-based activity that requires each farmer to act on a unique synthesis of information in order to make the best decisions. Having access to different kinds of information can assist this synthesis. Thus, the usefulness of applied research results can be enhanced by related knowledge derived from the experiences and observations of farmers that share common operating features and/or environmental conditions.

To explore extension and LGU efforts to promote the adoption of agroecological practices, it seems fitting to examine case studies in the U.S. state of Vermont because the entire LGU system got its start through the work of a U.S. senator from Vermont, Justin Morrill. The Morrill Acts of

1862 and 1890 established and then expanded the land grant system (Mayeske 1990; Schuh 1986). These acts allocated federal funds to establish public land grant colleges and universities throughout the United States, providing education to "ordinary people" (farmers and others in the working class) who, before that time, were largely left out of higher education. The original mission of the LGU was to focus on the teaching of practical agriculture, science, and engineering. This served to expand higher education opportunities from solely *classical* forms of knowledge to *applied* knowledge that could be linked to problem solving in the real world (Jones 1986).

Cooperative extension was created as the outreach arm of the LGU system through the Smith–Lever Act of 1914. The act stated, "Cooperative agricultural extension work shall consist of the development of practical applications of research knowledge and giving of instruction and practical demonstrations of existing or improved practices or technologies in agriculture" (U.S. Congress 1914).

From its roots, then, extension's mission was "… to aid in diffusing among the people of the United States useful and practical information on subjects relating to agriculture" (U.S. Congress 1914). Interestingly, the bill itself focused on an educational purpose rather than that of "technology transfer" (i.e., moving publically funded research ideas, inventions, and technologies into the private sector), a role for which extension has been both credited and roundly criticized (Jones 1986; Röling 1988).

The Extension Workers Creed, developed in the 1930s (Bliss 1952), expresses the core values and guiding principles of extension educators. The Creed is still in use today; its emphasis on education and client-centered community development has driven extension programming over the past century.

Criticisms of extension and the LGU system (Jones 1986; Warner 2006, 2008)—which, not surprisingly, coincided with the emergence of "corporate agribusiness" starting in the late 1930s through the 1980s (Kleiman 2010; Raup 1973)—have centered on concerns about extension and LGU researchers serving a corporate-driven agenda rather than a "public good" agenda, as well as an inability to effectively serve the local citizenry (due to lack of adequate funding, capacity, conflicts of interest, and/or not staying current with agricultural trends). That is, at some point,

EXTENSION WORKERS CREED

I BELIEVE in people and their hopes, their aspirations, and their faith; in their right to make their own plans and arrive at their own decisions; in their ability and power to enlarge their lives and plan for the happiness of those they love.

I BELIEVE that education, of which extension work is an essential part, is basic in stimulating individual initiative, self-determination, and leadership, that these are keys to democracy and that people, when given facts they understand, will act not only in their self-interest but also in the interest of society.

I BELIEVE that education is a lifelong process and the greatest university is the home; that my success as a teacher is proportional to those qualities of mind and spirit that give me welcome entrance to the homes of the families I serve.

I BELIEVE in intellectual freedom to search for the present, the truth without bias and with courteous tolerance toward the views of others.

I BELIEVE that the extension service is a link between the people and the ever-changing discoveries in the laboratories.

I BELIEVE in the public institutions of which I am a part.

I BELIEVE in my own work and in the opportunity I have to make my life useful to mankind.

Because I BELIEVE these things, I am an extension worker.

extension and the LGUs started to stray from their missions. The role of money is often cited as the root of this problem; as public funding levels have diminished relative to external grant funding, it has been claimed that research and extension outreach have been conducted for the benefit of private industry (Hightower 1972; Rausser et al. 2008; Schuh 1986; Scott 2006).

While there is merit to these criticisms, they did not and do not apply to all personnel working at these public institutions. However, because these actions have served to damage citizen trust in extension/LGUs, they have, at times, created credibility challenges for a growing cadre of extension educators and land grant researchers dedicated to addressing the needs of their local farm communities in ways that improve agroecological health and long-term sustainability.

11.2.1 Effective Extension Education in the Twenty-First Century: A Farmer-First, Outcomes Approach

Today, "extension defines itself as a 'nationwide educational network' that is a vital link to the food and agricultural system—meeting the needs of both producer and consumer," (Mayeske 1990: 50). To successfully meet both agroecological and effective educational goals, we contend that extension work must use farmer-centered and outcomes-oriented approaches.

First, research and extension work needs to have a farmer- or client-first focus. Several scholarly works (Anderson and Feder 2007; Eckert and Bell 2006; Jones 1986; Mayeske 1990; Peters et al. 2006) recognize that to be effective, programming must clearly identify what is important to the end user and then develop approaches that specifically address those needs, that is, meet farmers where they are. Wyman says, "You have to be able to bring ecological principles to people's everyday work, their businesses, how they live their lives, and how they make the choices that they make" (Peters et al. 2006: 101). There are, of course, cases where farmers may not recognize certain needs or ask for the information to address them, especially when it comes to meeting regulatory requirements aimed at serving the greater "public good." For example, protecting water quality, reducing food safety risks, and managing pesticides in a safe manner are some extension program and applied research goals that were not originally farmer-driven. However, over time, farmers have become involved in the development and testing of practical approaches to these issues (Getz and Warner 2006; Harden et al. 2013; Pretty and Ward 2001). As an example, see the case study on the Vermont Farmers' Watershed Alliance in Section 13.4.4.

The farmer-first model popularized by Chambers and Ghildyal (1985) provides a sharp contrast with the technology transfer model. The transfer of technology model starts with ideas generated from the research community with results imposed upon (and presumably adopted by) the public. Chambers and Ghildyal suggest that this model is ineffective among most farmer audiences, not because of lack of experience, knowledge, or capacity to adopt the technology but rather because the technology at hand may simply not fit farmer needs nor their physical, social, and/or economic conditions. Therefore, these researchers suggest a "farmer-first-and-last" model as a fundamentally different approach to research exploration and farmer adoption. Here, farmers are recognized as experts of their own agroecosystems (farms) and for their spirit of discovery and innovation. Further, they drive the research and education agenda, that is, using participatory methodologies, farmers identify the needs within their own community and resource base, and researchers and educators work in collaboration with farmers to conduct the research and deliver the results. The researcher-educator position in the relationship, then, becomes what King (1993) describes as moving from a "sage on the stage" to a "guide on the side." Emphasis in these models for the researcher–educator is placed on the role of facilitator or catalyst of information transfer (Morse et al. 2006; Peters et al. 2006).

Married to the farmer-first concept is an outcomes-based approach to research and education. Covey's (1989) idiom, "start with the end in mind," is a common mantra in extension work as many units across the United States have adopted a focus on farmer behavior change. That is, for extension to be effective and relevant in the twenty-first century, it is no longer sufficient to provide just the one-way knowledge transfer of the traditional expert model. Further, calls for increased accountability of extension programming have necessitated the adoption of outcome-based models to demonstrate changes in knowledge and behaviors as a result of extension interventions. There are many outcome models used in the public sector (Penna and Phillips 2004); in extension, popular outcome models include the logic model (Braverman and Engle 2009; University of Wisconsin Extension 2014; Workman and Scheer 2012) and the outcome funding framework (Penna and Phillips 2004). These outcomes models assist with program planning and evaluation but, more importantly, measure accountability in serving the public. As such, the inclusion of outcomes models, particularly the logic model, is often a requirement in receiving federal funds to conduct research and extension programming (Hoffman and Grabowski 2004). Interestingly, outcome models focus on what actions the farmers take, such as the acquisition of new knowledge, skills, intentions, and resulting changes in behavior, rather than focusing on the activities of researchers and extension personnel. The latter are determined by what is needed to achieve the former.

11.2.2 Focus on Adult Learning Techniques

Effective extension work must also be based on sound principles of adult learning (Bell and McAllister 2012; Eckert and Bell 2005; Franz et al. 2010b). For the most part, emphasis of extension programs is placed on nonformal adult education. Ota et al. (2006) suggest that the learning needs of this audience—adults—differ from those of younger students. They suggest that specific educational approaches be used to address adult learners' unique characteristics of (1) need to know, (2) self-concept, (3) prior experience, (4) readiness to learn, (5) learning orientation, and (6) motivation to learn.

Steele (1995) suggests that adults who already know something about a topic are more likely to voluntarily seek out venues for further learning. Today, adults are able to more easily access practical and technical information from a variety of sources. They seldom start from base zero. It is important, therefore, to give adults opportunities to indicate where changes in their knowledge may occur. The application of case studies, discussion formats, and problem-based learning opportunities are cited as particularly effective instructional strategies to accomplish this goal (Ota et al. 2006). They suggest that adult learning may be enhanced using a combination of these educational techniques.

Further, Bell and McAllister (2012) outline four principles for effective adult learning:

1. A safe environment for learning must be provided (one where learners feel physically comfortable as well as psychologically and emotionally safe).
2. Learners' assumptions, expectations, and motivations are understood, acknowledged, and addressed.
3. Opportunities are given to relate the content delivered to learners' prior experiences as well as to experiment with the content.
4. Opportunities are given to learners to contribute to their own learning through control over the specific content delivered as well as the process in which the content is delivered.

Eckert and Bell (2005) suggest that these approaches are especially important when approaching farmer education because their research has shown that farmer "mental models" (their values, beliefs, knowledge, and skills, and ways of processing information and applying skills) influence the degree of acceptance and application of the education conferred.

Franz et al. (2010a) go one step further to suggest that extension educators "need to not only be experts in a particular subject matter but also be architects of relationships, learning processes, and environments that directly meet farmers' needs to catalyze transformative learning." They suggest that "to design successful educational programs, agricultural educators must understand farmers' needs and struggles and design programs to address them."

In sum, effective extension education needs to consider the learning preferences and motivations of farmers as well as to incorporate their expressed needs into the design, implementation, and evaluation of these educational programs (Bell and McAllister 2012; Eckert and Bell 2006; Franz et al. 2010a; Getz and Warner 2006). Left unstated but essential to successful long-term engagement of farmers is the need for mutual trust and respect. Many extension personnel spend most of their careers in a single geographic location, serving the evolving needs of a farming community. Integrity, interpersonal skills, and a willingness to become a part of the community, not just an advisor or expert, are keys to gaining the trust and participation of farmers in the exploration of new practices, especially when they push the boundaries of "conventional" thinking, challenging farmers to deviate from what have become familiar, traditional practices.

11.3 EXAMPLES OF AGROECOLOGY IN THE FIELD

11.3.1 The Vermont Context

Vermont is a small state (9623 square miles) in the Northeast United States with a total population of about 626,000 people (State of Vermont 2014). Agriculture is a critical component to Vermont's economy, landscape, and culture. According to the U.S. Census of Agriculture (2012), the 7338 farms in Vermont generated US$776 million in farm gate cash receipts in 2012; dairy was the top commodity, contributing 72% of total cash receipts to the state. Estimates indicate that farming has a US$4 billion impact on the state's economy through direct sales and indirect impacts like tourism (Altendorfer et al. 2010). Interestingly, Vermont leads the nation in both the direct sales per capita at farmstands, farmers' markets, and CSAs (US$36.80 per capita), as well as the value of organic sales as a percentage of total farm sales (11%) (NASS 2014).

In Vermont, extension and the LGU seem to have avoided, to a large extent, strong ties with corporate agribusiness that some other states have experienced. That the University of Vermont (UVM) is a relatively small LGU with about 15,000 students, housed in a liberal New England state with predominantly small-scale farms, has undoubtedly reduced the potential for engagement with corporate agribusiness. The state's libertarian traditions of self-sufficiency and frugality may have also helped steer people in its institutions to emphasize ecology, not just technological inputs, to answer many agricultural challenges.

11.3.2 Partnerships for Successful Applied Research

As described in previous chapters of this book, PAR "combines local insights of community members with the technical expertise of researchers to explore mutual interests and issues through a democratic and collaborative exchange" (Franz et al. 2010b). Such an approach can also yield a wealth of information which cannot be obtained by either traditional research methods or community dialogue alone.

The work of the UVM Extension Northwest Crops and Soils program is an example of PAR that has led to successful stakeholder–researcher partnerships. Dr. Heather Darby, an agronomist, created the program 10 years ago when she joined the UVM faculty. At that time, farmers were facing increased scrutiny of their nutrient and soil management practices (especially related to water quality regulations), livestock feed prices were rising (particularly grain prices for certified organic producers), and notions of "localvores" and "food systems" were just starting to gel in the state (Grubinger et al. 2005).

Because Darby grew up on a dairy farm in northern Vermont (where she farms today), she had an understanding of the cultural and physical landscapes as well as the social networks of farmers in the state. However, it was her vision, tenacity, and aptitude to turn farmer-based needs into actionable research that has helped create a successful applied research and extension program.

Her philosophy mirrors the original land grant mission, as expressed by the following statements:

> One of the most important reasons for conducting research at a LGU is to answer the questions that are coming from the community around you. And the reason we are doing this work in the first place is because of the farmers and the end users saying, "look, we want to improve the quality of our agricultural products so that we can have better markets and become more viable," so if our research is effective, the end users are happy as well (Darby 2011).

> Building a relationship with farmers that are willing to help answer these questions is really important because there's no better place to conduct research than on an actual farm, in their conditions, on their soil types. It's difficult, it's time consuming, and it takes resources to do it and so to be able to work with a farmer ... that's willing and interested in helping to answer those questions is just the kind of partnership that we look for at the university because we couldn't do it without them (Darby 2011).

An example of this type of partnership is the Northern Grain Growers Association (NGGA). Small grains had not been grown in Vermont on a large scale since the nineteenth century, when wheat was a large part of the state's agricultural economy (Bushnell 2010; Podhaizer 2008). Vermont, in fact, had been dubbed the "breadbasket of New England" in the 1880s because of its high production of wheat. During that heyday of wheat growing, a UVM botanist, Dr. Cyrus Pringle, conducted numerous plant breeding trials on small grains, some of which resulted in commercial varieties that have been used by farmers for decades, and which provided the genetic material for many cultivars still grown throughout the United States today (Burns Davis 1936).

The interest in on-farm livestock feed production and farm product diversification to meet a growing consumer demand for local products prompted an informal network of farmers to gather in 2004 to exchange ideas, seeds, and equipment related to cereal grains. At that time, there were only a handful of farmers growing grains commercially in Vermont, and they were challenged by inconsistent crop yields and quality. Darby provided a facilitation role in these first gatherings, which were primarily driven by agronomic concerns. A project initiated by Darby and an organic dairy farmer, Jack Lazor, early in the network's development focused on participatory plant breeding (Lazor 2008). The project sought to rediscover on-farm traditional plant breeding knowledge and skills to develop varieties appropriate to Vermont's growing conditions (Figure 11.1).

Another concern was product acceptance in the local marketplace. In response, Darby and the farmer network invited several bakers to the table. By learning more about quality parameters needed by this group of end users and, further, by inviting them to become full partners of the network, the NGGA essentially birthed a functional, whole systems local grains research and outreach system (NGGA 2014). This partnership of bakers, farmers, and researchers/educators has achieved the following to date.

- Funding from local foundations has been secured to purchase research equipment such as a plot combine and grain drill, allowing the expansion of a grain research program.
- Fifty-two on-farm cereal grain research trials have been conducted over a 4-year period. Most of the trials have investigated agronomic aspects, including varietal compatibility with local growing conditions, weed and pest management under organic production, organic fertility management, plant breeding, and optimal planting and harvest timing.
- A university-based cereal grains quality testing laboratory has been established that provides analyses on quality parameters used by the baking industry as well as food safety screens.
- Culinary milling and baking trials have been performed as well as public tasting surveys to assess consumer preferences of local wheat varieties (Figure 11.2).

Figure 11.1   Participatory plant breeding is one example of farmer–university relationships. Here, grain variety crosses using nineteenth-century germplasm are grown in field conditions to be evaluated by farmers for desirable production characteristics. (Photo courtesy of Debra Heleba, University of Vermont Extension.)

- Participatory breeding of five wheat crosses—some using the germplasm from nineteenth-century varieties bred by Cyrus Pringle—have been implemented. These crosses have been grown out on-farm for more than 6 years where farmers have been improving them through annually selecting for favorable traits. It is expected that a public release of these varieties will occur in 2015—the first wheat varieties developed in Vermont since 1901.
- The network has been formalized into an association of growers, bakers, millers, researchers, extension educators, and others that meet annually through conferences, field days, and workshops.

NGGA members have also sought out the experiences of other small-scale growers and millers by traveling together to locations like Canada and Denmark—this has not only resulted in effective farmer-to-farmer knowledge exchange but by traveling together, has strengthened the social network of this group.

Franz et al. (2010b) argued that "building relationships and trust are crucial to meeting the educational needs of farmers." The example of the NGGA illustrates how building trust and ongoing relationships among a diversity of stakeholders can play central roles in forming and sustaining strong local food systems.

Darby has facilitated a similar partnership around an emerging hops industry in the Northeast. Again, grower and processors (in this case, beer brewers) have come together to co-learn about production and quality challenges (UVM Extension Northwest Crops and Soils Program 2014). This partnership has taken a systems approach to hop production, including overcoming equipment and other production barriers; assessing hop quality challenges; and identifying, encouraging, and protecting important predatory arthropods in hopyards.

The approaches described above differ from a more traditional research approach, where the results of university research is disseminated, leaving farmers to put the "big picture" together. Instead, by engaging farmers and the end users of their products (e.g., bakers and brewers), knowledge is developed in a systems context that accounts for differences across farms as well as quality concerns that may impact the market for new products that farmers seek to produce. These cases exemplify the

Figure 11.2 The Northern Grain Growers Association is an example of how building relationships among extension educators, researchers, farmers, bakers, and consumers can lead food system improvements. Here, extension agronomist Heather Darby and farmer Jack Lazor conduct a consumer taste test of Vermont-grown wheat breads. (Photo courtesy of the University of Vermont Extension Northwest Crops and Soils Program.)

agroecological approach promoted in this book, which seeks to integrate a diversity of stakeholders in efforts to improve the entirety of the food system, from farm to table.

11.3.3 Facilitating Farmer-based Learning Communities and Knowledge Exchange

As mentioned previously, a central role of the extension educator is as facilitator and catalyst of farmer-to-farmer knowledge exchange. Research on how farmers learn best has revealed that contextual and peer-based information exchanges are very effective (Bell and McAllister 2012; Franz 2007; Kilpatrick et al. 1999)

The UVM Vegetable and Berry Program is an example where farmer-to-farmer knowledge exchange is used to improve agricultural sustainability. Dr. Vern Grubinger has led the program for 24 years. Though Grubinger has and continues to work with all types of growers—large and small, new and established, conventional and organic—his early career work with sustainable and organic farmers served to build trust among this marginalized group, who felt they had not been well served by extension and LGUs (Figure 11.3).

Throughout its history, the UVM Vegetable and Berry Program has relied on extension education offerings that allow for farmer-to-farmer exchanges of information, such as farmer panels at educational conferences, on-farm "twilight meetings," and individual farm visits where farmers express their needs and interests as well as ask for technical assistance. As new information and communication technologies emerged, Grubinger recognized the potential for capturing and sharing much more of the knowledge and experience held by farmers across the state and region. To that end, he has produced videos that focus on farmer experiences, facilitated information exchanges among farmers through a listserv and e-mail newsletter, and produced farm case studies of innovative production methods (UVM Extension Vermont Vegetable and Berry Program 2014).

The videos were intended to address complex agroecological questions not suited to formulaic answers, and not easily solved by applied research conducted in a single location. What are the best cover crops to grow? Which tillage practices optimize soil health? What types of cultivation equipment are effective for mechanical weed control? To help farmers answer these questions for themselves, the videos addressed each question using six to eight case studies of experienced farmers from across the Northeast. The farmers explained their approaches to the questions, rationale behind their practices, and their experiences to date including what worked and what did not. Video footage demonstrated how sustainable agriculture practices on these farms were implemented (UVM Extension Vermont Vegetable and Berry Program 2014).

Grubinger has been a strong proponent of information exchange among farmers, as illustrated by the following statement:

Figure 11.3 New and aspiring farmers, pictured here, are among a number of audiences with whom Vern Grubinger works as the University of Vermont Extension's Vegetable and Berry Specialist. Grubinger (on right) describes equipment needs for successful vegetable production systems at an on-farm field day. (Photo courtesy of Debra Heleba, University of Vermont Extension.)

I realized very quickly when I came to work for extension that the old model of information dissemination was no longer optimal. There were simply too many complex questions from too many different
types of growers for a handful of research and outreach folks to address them in a satisfactory manner.
It was critical that we start to do a better job of capturing credible information held by the farmers so
that we could focus limited resources on "the gaps" that farmers could not address on their own. For
example, if you ask a couple of dozen experienced growers about their favorite carrot cultivars, you
often get a lot of agreement and from there it is reasonable to identify a "top 10" list without doing
variety trials. But if you ask about the optimal storage facility, you'll probably find a lot of uncertainty,
suggesting the need for applied research (Grubinger 2014).

To help farmers benefit from the knowledge held within their community, in 2010, Grubinger
worked with the Vermont Vegetable and Berry Growers Association, a farmer membership organization, to establish an e-mail listserv that includes more than 450 growers. A key purpose of the
listserv is to help farmers tap into their collective wealth of information and share it among themselves. A protocol has been established where once a farmer asks a question, all responses are sent
to that individual, who compiles the answers and then distributes them back to the list in a single
message. This method reveals where there is agreement among farmers, and where there are differences of opinion based on a variety of experiences. The questions addressed have ranged from preferred cultivars and farm employment policies to experiences with permanent cover crops planted
between vegetable beds. The listserv has become a top resource for commercial vegetable and berry
growers in the state. A 2013 survey of listserv members revealed that 83% of those who responded
improved farm production practices as a result of exchanges on the listserv. These improvements
included pest and disease management strategies and soil fertility practices, including cover cropping and water management. One grower said, "We enjoy reading all the experiences farmers have
had when faced with an issue … it is like having an amazing bunch of life experiences at your
disposal." Another said, "The listserv is one of my go-to places to source useful information about
specific crop production (cold tolerances, best practices, pest and disease management, etc.). I find
the farmer-to-farmer information sharing is an excellent balance to the crop production information
readily available online" (Grubinger 2014).

As the listserv manager, Grubinger views his role as "light" facilitation, interjecting when factually incorrect information is posted or when there is relevant research or regulatory information
to share with the audience. Grubinger also uses the listserv as an efficient method to gather critical research needs identified by his constituency and to track the severity of certain pest problems
around the state. Because the listserv is a part of the membership to the Vermont Vegetable and
Berry Growers Association, it requires paying a nominal annual fee per farm. This also assures that
the list is primarily made up of commercial growers with a keen interest in the content.

Another communication tool that harnesses farmer knowledge exchange is the biweekly newsletter "Vermont Vegetable and Berry News," which is compiled and distributed by Grubinger (UVM
EXT VVBP 2014). The newsletter is available to anyone on request and currently has more than
750 subscribers. The vast majority are farmers; two-thirds are located Vermont and the remaining
members are spread across the Northeast United States. Unlike traditional extension newsletters
that tend to provide unidirectional information developed by "experts," Grubinger solicits information from farmers through an e-mail invitation to submit "Reports from the Field" several days
before the newsletter is compiled. These reports are intended to share current crop conditions,
production activities, recent pest observations, comments on markets and labor, and other items of
interest to growers. Submissions are compiled, lightly edited (for spelling, grammar, and to revise
factually incorrect information), and then distributed to the newsletter subscribers. This type of
farmer-to-farmer exchange has received highly favorable responses from growers, which is not surprising, as scholarly works have suggested that farmers prefer learning from peers and experts who
have experience with their situation (Franz 2007; Kilpatrick et al. 1999). As with the listserv, the
newsletter also serves to identify emerging issues in the farming community from market trends

(e.g., strength of sales at farmers' markets) and emerging pests (e.g., which farms have noted the arrival of spotted wing drosophila) to infrastructure management (e.g., how to prepare greenhouses for storm events). Grubinger is able to immediately respond by adding supplemental information to the newsletter and/or by steering readers to links so they can learn more about issues raised by the field reports. From a 2013 grower survey, the newsletter is the most frequently referenced resource used in the program and has been particularly useful in making pest management and equipment decisions (Grubinger 2014).

One grower said, "We love the 'Reports from the Field' and gather great info and things to trial, seed varieties, tips and techniques year-round from the reports." Another farmer described the importance of the facilitated exchange: "You have done an amazing job of fostering a culture of ownership by the farmers themselves here in Vermont. The partnership that exists between agricultural service providers (especially extension) and farmers is a strong one. A continuation of this approach will only serve to extend University knowledge to the entire growing community. Having established and even newer growers spreading the word for you has been very effective" (Grubinger 2014).

Grubinger has also developed applied research efforts aimed at improved farm sustainability using farmer-generated knowledge. He worked with an agricultural engineer to engage 14 growers in evaluating renewable energy systems for greenhouse heating (Callahan and Grubinger 2010). Grubinger obtained funds to provide a cost share project for farmers to implement these systems, contingent upon the growers committing to document installation costs, fuel use, and furnace performance for use in a report distributed to other growers. These growers also agreed to hold on-farm meetings so other farmers could see their systems and evaluate their potential for adoption.

The examples of the Vermont Vegetable and Berry Program support scholarly work on the effectiveness of farmer-to-farmer approaches (Chambers and Ghildyal 1985; Franz et al. 2010b; Morse et al. 2006; Peters et al. 2006) (Figure 11.4).

Figure 11.4 Extension specialist Vern Grubinger describes renewable energy options to a group of farmers at an on-farm biodiesel facility. Grubinger helps facilitate farmer-to-farmer information exchanges through on-farm meetings and other methods. (Photo courtesy of Debra Heleba, University of Vermont Extension.)

11.3.4 Empowering Farmers to Protect Public Interests

Grubinger and Darby have also helped facilitate farmer groups in taking proactive steps to protect water quality, reduce food safety risks, and manage other environmental risks to the general public. Often, policies regarding public health are made with little input from growers and the agricultural community, resulting in defensive and reactive responses on the part of farmers and other rural stakeholders (Falconer 2000). However, when farmers are given a voice and the power to help craft the policies that affect them, creative and productive "win–win" solutions may be achieved to benefit both the farmer and the public. The Franklin and Grand Isle County Farmer's Watershed Alliance (FWA) is an example.

The FWA is a farmer–member nonprofit organization that aims to use peer networks to improve water quality in Lake Champlain, Vermont's major water body (FWA 2014). Members are primarily dairy producers from the state's two northernmost counties (both of which border the lake and have significant agricultural acreage). Darby helped a team of farmers form the FWA in 2006 as a proactive response to ongoing concerns from the public about nonpoint source pollution, especially excessive phosphorus loads from farms into Lake Champlain. At that time, state regulators had introduced a medium farm operation (MFO) program as a response to changes in federal regulations for concentrated animal feeding operations. Essentially, the state expanded its regulatory arm to require both large farm operations and MFOs—all operations with 200 to 699 mature dairy cows—to obtain permits from the state to regulate animal wastes, follow accepted agricultural practices, and develop and implement nutrient management plans.

Darby met with this group of farmers to first help them better understand the new regulations and learn how to comply with them. She also responded by developing a multisession course that teaches farmers about soil and crop management and potential impacts on water quality. Based on farmer feedback, course materials were developed—first came a workbook, then an Excel spreadsheet, and most recently, a mobile application (app) called "goCrop™." All of these materials were designed to assist farmers in the development of nutrient management plans used to better meet state and federal regulations (UVM Extension Northwest Crops and Soils Program 2014).

The farmer group, however, was interested in moving beyond just regulatory compliance; they also wanted to make strides in improving the soil, air, and water of the Lake Champlain Watershed while promoting the benefits of dairy agriculture to the community and the state's economy. Therefore, Darby worked with the FWA to help the group provide farmer-to-farmer nonregulatory technical assistance to implement environmentally friendly production practices, as well as deliver farmer feedback to policymakers and the public that shows a positive image of agriculture and its influence on the environment.

A major program of the FWA is a water quality assessment where farmers can receive a confidential farm visit to look at methods to improve water diversion, soil erosion, and livestock exclusion from water bodies, among other practices (FWA 2014). Working closely with Darby, the FWA has developed innovative solutions to minimizing environmental impact while protecting the farmer's bottom line. Practices—including soil aeration, cover cropping, reduced tillage, and manure injection—have all been farm tested in this local area and, as a result, have led to broadscale adoption across the entire state. State and federal cost share programs once largely untapped have become competitive as an increasing number of producers have become interested in implementing these practices. To date, the FWA has implemented practices on over 250 farms in their target area, covering more than 50,000 acres in the Lake Champlain Basin.

To acknowledge their work, in 2014, the New England office of the U.S. Environmental Protection Agency recognized the FWA with an Environmental Merit Award for its efforts "to promote good environmental stewardship practices and improve water quality in the Lake Champlain Basin," (EPA New England 2014). Darby and FWA members continue to work together to conduct on-farm research, refine the nutrient management planning mobile application, and provide farmers with the education and support they need to make on-farm changes to protect water quality.

11.4 CONCLUDING THOUGHTS

The transformation of agriculture based on the tenets of agroecology will not happen without the intensive involvement of farmers. Research done in isolation or outreach provided to people with little "skin in the game" will not yield the extent of change necessary to improve the sustainability of our food system. Rather, meaningful and trusting partnerships among scientists, educators, and farmers are needed. These partnerships must function throughout the continuum of inquiry and application, from the identification of needs and goals to the implementation of research and testing of practices, and the refinement and widespread adoption of improved approaches. Such partnerships serve extension and the land grant institutions well by reaffirming their commitment to improving the well-being of farmers, their communities, and food consumers across the nation.

The agroecological approach underscored in this volume (see Chapter 1) suggests the need to improve our food systems through the integration of diverse forms of knowledge systems, including localized knowledge of farmers, the more general knowledge of scientists, and the hybrid knowledge of extension personnel that work closely with farmers and rural communities. This chapter elucidates successful examples of how strong relationships and farmer-to-farmer exchanges can improve agricultural management not only for the benefit of farmers, but also for the environment and the general public. It is not always easy for researchers and educators to build relationships with farmers and other stakeholders that honor a wide range of knowledge and experiences. It requires the ability to relinquish a presumed role as "expert" while also viewing farmers not merely as students or research subjects, but as collaborators who are profoundly knowledgeable about their own local ecosystems. Giving farmers a voice through full participation in the development of applied research and education programs is an idea that has gained greater acceptance in recent years. In future years, it must become a mandate for extension and the land grant universities—for their good as well as for the nation's agriculture.

REFERENCES

Altendorfer, I., D. Holland, and A. Isaacson. *Economic Impact of Agriculture in Vermont*. University of Vermont Legislative Research Service, Burlington, VT, 2010.

Anderson, J.R., and G. Feder. "Agricultural extension." *Handbook of Agricultural Economics* 3 (2007): 2343–2378.

Bell, S., and J. McAllister. "*Sustainable Agriculture through Sustainable Learning: Applying Principles of Adult Learning to Improve Educational Outcomes*." Northeast SARE, Storrs, CT, 2012.

Bliss, R.K. (Ed.) *The Spirit and Philosophy of Extension Work*. Washington, DC: Epsilon Sigma Phi, 1952.

Braverman, M.T., and M. Engle. "Theory and rigor in extension program evaluation planning." *Journal of Extension* 47 no. 3 (2009): 1–10.

Bushnell, M. "In Early Vermont, Grain was More than Food." *Times Argus* (Barre-Montpelier) January 31, 2010.

Burns Davis, H. "Life and Work of Cyrus Guernsey Pringle." *The Pringle Herbarium*. 1936. http://www.uvm.edu/~plantbio/pringle/index.html (Accessed June 13, 2014).

Callahan, C., and V. Grubinger. *Biomass Furnaces for Greenhouse Vegetable Growers*. University of Vermont Extension, Brattleboro, VT, 2010.

Cecil, K. "Integrating ecology and relating natural systems to agriculture: An increased priority for extension agricultural programming." *Journal of Extension* 42 no. 5 (2004).

Chambers, R., and B.P. Ghildyal. "Agricultural research for resource-poor farmers: The farmer-first-and-last model." *Agricultural Administration* 20 no. 1 (1985): 1–30.

Code, U.S. "Title 7 – Agriculture." Chapter 64 – Agricultural Extension, and Teaching, Section 3013 Definitions (2011).

Congress, U.S. "Smith-Lever Act, Agricultural Extension Work Act. 63 PL 95; 63 Cong." Chapter 79 (1914): 38.

Covey, S.R. *The 7 Habits of Highly Effective People.* New York, NY: Simon & Schuster, 1989.

Darby, H. "Sustainable Agriculture Research Education (SARE) Program Grants in the Northeast." *Across the Fence.* Burlington, VT: University of Vermont Extension, October 7, 2011.

Eckert, E., and A. Bell. "Continuity and change: Themes of mental model development among small-scale farmers." *Journal of Extension* 44 no. 1 (2006).

EPA New England. "EPA Awards 3 Environmental Merit Awards to Vermont Recipients." *United States Environmental Protection Agency Region 1 Press Release* (April 22, 2014).

Falconer, K. "Farm-level constraints on agri-environmental scheme participation: A transactional perspective." *Journal of Rural Studies* 16 no. 3 (2000): 379–394.

Farmer's Watershed Alliance (FWA). "Farmer's Watershed Alliance." 2014. http://farmerswatershedalliance. com (Accessed June 13, 2014).

Franz, N. "Adult education theories: Informing cooperative extension's transformation." *Journal of Extension* 45 no. 1 (2007).

Franz, N.K., F. Piercy, J. Donaldson, J. Westbrook, and R. Richard. "Farmer, agent, and specialist perspectives on preferences for learning among today's farmers." *Journal of Extension* 48 no. 3 (2010a): 3RIB1.

Franz, N., F. Piercy, J. Donaldson, R. Richard, and J. Westbrook. "How farmers learn: Implications for agricultural educators." *Journal of Rural Social Sciences* 25 no. 1 (2010b): 37–59.

Getz, C., and K.D. Warner. "Integrated farming systems and pollution prevention initiatives stimulate co-learning extension strategies." *Journal of Extension* 44 no. 5 (2006).

Grubinger, V. *Evaluation of UVM Vegetable and Berry Program.* Unpublished survey. University of Vermont Extension, Brattleboro, VT, 2014).

Grubinger, V., K. Mulder, and D. Timmons. *Vermont's Agriculture: Generating Wealth from the Land.* Vermont Sustainable Agriculture Council. 2005 http://www.uvm.edu/~susagctr/Documents/SAC2005Report.PDF (Accessed June 13, 2014).

Harden, N.M., L.L. Ashwood, W.L. Bland, and M.M. Bell. "For the public good: Weaving a multifunctional landscape in the corn belt." *Agriculture and Human Values* 30 no. 4 (2013): 525–537.

Hightower, J. "Hard tomatoes, hard times: Failure of the land grant college complex." *Society* 10 no. 1 (1972): 10–22.

Hoffman, B., and B. Grabowski. "Smith lever 3(d) extension evaluation and outcome reporting—A scorecard to assist federal program leaders." *Journal of Extension* 42 no. 6 (2004): 1–12.

Ikerd, J.E. "The need for a system approach to sustainable agriculture." *Agriculture, Ecosystems & Environment* 46 no. 1 (1993): 147–160.

Jones, G.E. (Ed.). *Investing in Rural Extension: Strategies and Goals.* London: Elsevier Applied Science Publishers, 1986.

Kilpatrick, S., S. Johns, R. Murray-Prior, and D. Hart. "Managing farming: How farmers learn." Kingston, Australia: Rural Industries Research and Development Corporation, 1999.

King, A. "From sage on the stage to guide on the side." *College Teaching* 41 no. 1 (1993): 30–35.

Kleiman, J. "The rise of agribusiness and the demise of the new deal order: Shane Hamilton, trucking country: The road to America's Wal-Mart economy." *Technology and Culture* 51 no. 1 (2010): 216–219.

Lazor, J. *"A Vermont Farmers Breeding Club: Developing Varieties That Work For Us!" NE-SARE Farmer Grant Report.* 2008. http://www.uvm.edu/Extension/cropsoil/wp-content/uploads/jack-lazor-report.pdf (Accessed June 13, 2014).

Mayeske, G.W. *"An Assessment of the Value of a Nationwide Extension System." Accountability and Evaluation Reporting System* (USA) (1990).

Morse, R.S., P.W. Brown, and J.E. Warning. "Catalytic leadership: Reconsidering the nature of extension's leadership role." *Journal of Extension* 44 no. 2 (2006): 6.

NASS, USDA. "2012 Census Volume 1, Chapter 1: State Level Data—Vermont." *Census of Agriculture* (2014).

NE-SARE. *"USDA Northeast Sustainable Agriculture Research and Education Program."* 2014. http://www .nesare.org (Accessed June 6, 2014).

Northern Grain Growers Association (NGGA). *"Northern Grain Growers Association."* 2014. http://northerngraingrowers.org (Accessed June 13, 2014).

Ota, C., C.F. DiCarlo, D.C. Burts, R. Laird, and C. Gioe. "Training and the needs of adult learners." *Journal of Extension* 44 no. 6 (2006).

Penna, R.M., and W.J. Phillips. *Outcome Frameworks: An Overview for Practitioners.* Center for Outcomes, Rensselaerville Institute, Delmar, New York, 2004.

Perez, J., and P. Howard. "Consumer interest in food systems topics: Implications for educators." *Journal of Extension* 45 no. 4 (2007).

Peters, S.J., D.J. O'Connell, T.R. Alter, and A.L. Jack. *Catalyzing Change: Profiles of Cornell Cooperative Extension Educators from Greene, Tompkins, and Erie Counties, New York.* Ithaca, New York: Cornell University, 2006.

Podhaizer, S. *"No Grain, No Gain." Seven Days* (Burlington). 2008. http://www.sevendaysvt.com/vermont/no-grain-no-gain/Content?oid=2135384 (Accessed June 4, 2014).

Pretty, J., and H. Ward. "Social capital and the environment." *World Development* 29, no. 2 (2001): 209–227.

Raup, P.M. "Corporate farming in the United States." *The Journal of Economic History* 33 no. 1 (1973): 274–290.

Rausser, G., L. Simon, and R. Stevens. "Public vs. private good research at land-grant universities." *Journal of Agricultural & Food Industrial Organization* 6 no. 2 (2008).

Röling, N.G. *Extension Science: Information Systems in Agricultural Development.* University of Cambridge, New York, 1988.

Schuh, G.E. "Revitalizing land grant universities." *Choices* 1 no. 2 (1986): 6.

Scott, J.C. "The mission of the University: Medieval to postmodern transformations." *Journal of Higher Education* 77 no. 1 (2006): 1–39.

State of Vermont. *"About Vermont: Facts and Statistics."* http://www.vermont.gov/portal/vermont/index.php?id=253 (Accessed June 13, 2014).

Steele, S.M. "Looking for more than new knowledge." *Journal of Extension* 33 no. 3 (1995).

University of Wisconsin Extension. *"Program Development and Evaluation."* http://www.uwex.edu/ces/pdande/evaluation/evallogicmodel.html (Accessed June 13, 2014).

University of Vermont Extension Northwest Crops and Soils Program (UVM EXT NWCS). *"Hops."* http://www.uvm.edu/Extension/cropsoil/hops (Accessed June 13, 2014).

University of Vermont Extension Vermont Vegetable and Berry Program (UVM EXT VVBP). *"Vermont Vegetable and Berry Grower Pages."* 2014a. http://www.uvm.edu/vtvegandberry (Accessed June 13, 2014).

University of Vermont Extension Vermont Vegetable and Berry Program (UVM EXT VVBP). *"Videos for Vegetable and Berry Growers."* 2014b. http://www.uvm.edu/vtvegandberry/?Page=videos.html (Accessed June 13, 2014).

Warner, K.D. "Agroecology as participatory science emerging alternatives to technology transfer extension practice." *Science, Technology & Human Values* 33 no. 6 (2008): 754–777.

Warner, K.D. "Extending agroecology: Grower participation in partnerships is key to social learning." *Renewable Agriculture and Food Systems* 21 no. 2 (2006): 84–94.

Workman, J.D., and S.D. Scheer. "Evidence of impact: Examination of evaluation studies published in the *Journal of Extension.*" *Journal of Extension* 50 no. 2 (2012).

CHAPTER 12

Agroecology as a Food Security and Sovereignty Strategy in Coffee-Growing Communities
Opportunities and Challenges in San Ramon, Nicaragua

Heather Putnam, Roseann Cohen, and Roberta Jaffe

CONTENTS

12.1 INTRODUCTION

Smallholder coffee farmers produce an estimated 70% of the world's coffee (Eakin et al. 2009), yet are especially vulnerable to food insecurity in the form of seasonal hunger. Agroecology is recognized and promoted as critical to improving household food security and sovereignty (FSS) in rural areas by the global food sovereignty movement led by *La Via Campesina*, and is promoted by grassroots farmer organizations like *Campesino-a-Campesino* throughout Latin America. At the same time, many Latin American governments have enacted "Right to Food" laws, which often include commitments supporting agroecology as a key strategy for household, community, and national food security. Similarly, international aid agencies are increasingly promoting sustainable agriculture and agroecology as food security strategies. However, even with this increased focus on agroecology as an FSS strategy in smallholder coffee farming communities, little is understood about the best approaches for doing so within a donor-funded and international development agency project context.

Through a 3-year case study of a donor funded multistakeholder FSS project in San Ramon, Nicaragua, this chapter explores the feasibility, challenges, and opportunities of promoting agroecology as an FSS strategy in smallholder coffee cooperatives in an international development context. This case study project is a long-term collaboration between a second-level cooperative in Nicaragua (Unión de Cooperativas Agropecuarias Augusto César Sandino, UCA San Ramón hereafter), eight of its affiliated primary level member cooperatives, and a US-based nonprofit organization, Community Agroecology Network (CAN). These stakeholders have been engaged in a long-term project funded by Keurig Green Mountain with the goal of building community youth- and women-led household food security, community food sovereignty, and reducing seasonal hunger. The main strategies utilized are production diversification and promotion of agroecological food production (primarily through homegardens and fruit production in coffee shade environments), income diversification aimed at women, and relocalizing of control over the availability and access to staple foods in the cooperatives. The project utilizes established models of farmer-to-farmer exchange and capacity building to promote household scale production diversification. We find that farmer-to-farmer exchange and horizontal capacity building not only lead to early and widespread farmer adaptation of best practices to improve food security and promote long-term food sovereignty, but also foment farmer leadership of these processes (Figure 12.1).

Our analysis of this experience finds that these methodologies for promoting agroecology as a food security strategy are the most effective. Building on these established models, our analysis of this case demonstrates how horizontal exchange and capacity building not only impacts farmers, as has been widely recognized, but *shapes* agroecological approaches to building food sovereignty. That is, through a combination of collective and individual practice, experimentation, and reflection, all stakeholders developed a mutual understanding of agroecology, identified project shortcomings/gaps, and allowed project facilitators (CAN and UCA San Ramon) to integrate ongoing or emergent community practices, as well as introduced innovations, into key FSS strategies. Ultimately, the case study demonstrates that successful outcomes relied on ongoing reflection and practice to achieve a shared understanding of agroecology relevant to the local context, ultimately shaping the agroecological approach of the entire project to include various components of the food system.

Three specific lessons emerged from the case study experience. First, the emphasis on making more food available through homegarden diversification must be placed within the larger production context of staple crops, cash crops, soil fertility, and access to water and land, as well as within the larger food systems context of markets and distribution, and food usage and consumption. This ensures a focus on transformation of the food system, rather than only in the production system. Agroecology must be promoted not only in homegarden production, but also simultaneously in coffee and staple crops production, to positively impact household food security and resilience to economic and ecological shocks. Second, strategies utilizing agroecology must especially embrace and highlight existing local and ancestral knowledge that positively impact FSS at the household and community levels, while also introducing new technologies and fostering a spirit of experimentation and innovation. Third, food production must be linked to food consumption and marketing

Figure 12.1 Farmers in San Ramon municipality demonstrating pedal pump technology during farmer-to-farmer exchange (2012).

and must be structured to favor women as the primary agents of these activities. These critical elements favor transformation of the entire food system (rather than only the food production system) toward one in which local communities are food sovereign and not just food secure, going beyond increased availability of food.

In the next sections, we explore the dynamics of food insecurity in the coffee lands and existing approaches to agroecology as an FSS strategy, before presenting the case study experience and findings.

12.2 FOOD SECURITY IN THE COFFEE LANDS

Seasonal hunger is the most common manifestation of food insecurity in coffee lands, although chronic hunger and malnutrition also affect families and especially children under the age of five in the most vulnerable communities. Caswell et al. ask why the reality of persistent food insecurity among coffee smallholders has "remained hidden for so long," analyzing the existing body of knowledge to identify causes and possible solutions to this reality (2012, p. 1). Their research demonstrates that persistent food insecurity remained hidden due to a single-minded focus among the coffee industry, fair trade movement, and international development complex on increasing access to specialty markets for smallholders through alternative market certifications and developing smallholder organizations' business capacities. There is a growing understanding among these actors that the rural areas where some of the world's best specialty coffee is grown are vulnerable not only to overdependence on coffee as a single cash crop, but also to various "food security risk factors" as listed by Caswell et al.: (1) depletion of natural resources from which the population makes its living; (2) environmental degradation; (3) shocks such as natural disasters and conflict; and (4) seasonal changes in food production and food prices" (p. 4, citing FIVIMS 2012). Six studies reviewed by Caswell et al. (2012) in their research brief support this finding: in studies ranging from $n = 25$ to $n = 469$, at least 44% and up to 100% of coffee smallholder families surveyed experience "thin months" or periods where they are unable to meet basic food needs. These periods of seasonal hunger are experienced in coincidence with three other phenomena: the rainy season between May and November, the period after income from the previous coffee harvest has been spent and cash for purchasing basic foods is scarce, and the period after basic grains have been harvested. Caswell et al. assert that the combination of overdependence on coffee as a cash crop at the expense of less food production, insecure land tenure, and the overarching vulnerability of being subject to volatile coffee prices and cyclical food prices contributes to insecurity (2012, p. 6). However, our knowledge is still scarce on the specific causal dynamics beyond those listed above, as much of our knowledge is based on "anecdotal evidence or outcomes generated by organizations as part of internal evaluations" (Caswell et al. 2012, p. 5).

The volatility of coffee as a commodity deserves special note here. First, rural populations dependent on coffee either as plantation workers, smallholder farmers, or service providers to the coffee industry are subject not only to global economic crisis, but to the fact that coffee, like petroleum, is an extremely volatile commodity, experiencing often-violent swings in prices every 20 years or so. The last major price dip lasted approximately 4 years (1999–2005, known as the "coffee crisis") and resulted in mass out-migration from rural coffee-growing communities, as well as the disruption of local, national, and regional economies and entire farming systems (International Coffee Organization 2002, p. 2). For smallholders, who depend primarily on coffee for cash income to supplement their subsistence farming, the coffee crisis was especially harmful. The loss of income meant less access to food, medicine, education, transport, communications, and other basic needs. It also left smallholders more heavily in debt primarily because of short-term loans to meet basic needs, forcing them to abandon farms and look for work in cities or other countries, according to the International Coffee Organization (2002, p. 2).

The second dynamic is that coffee is a perennial crop, which is problematic in a volatile market. When prices swing upward for more than 1 year, smallholders are often motivated to plant more of their land in coffee, often replacing basic food crops and in effect "putting their eggs in one basket" so to speak. When the price drops again, farmers find themselves with less land available to plant food crops or other cash crops to make up for the gap left by low-priced coffee. They are left particularly vulnerable to seasonal hunger, malnutrition, and less preferred (cheaper, but often less healthy) food choices. Over-dedication of land to coffee thus leads to less resilience to economical or ecological shocks (including climate change) over time.

During the coffee crisis, strategies to address this overdependence promoted by the International Coffee Organization as well as international development agencies in coffee-growing regions included diversification into additional or alternative economic activities among smallholder coffee farmers. Farmer organizations looked to diversification into coffee product segmentation, namely, specialty and alternative coffee markets like fair trade and other certifications that purported to promote social justice or environmental sustainability (see International Coffee Organization 2002, p. 4; Jaffee 2007, pp. 55–57). Fair trade certification brought higher and more stable prices that would ostensibly reduce vulnerability among smallholder families. There was little effort during the crisis to diversify smallholder production into food crops for subsistence. Thus little attention was paid globally to improving farming families' access to food or increasing local food availability except by increasing income available to purchase food. Research on the impacts of fair trade and organic certified markets on smallholder families revealed that higher prices were simply not adequate by themselves as solutions to the systemic vulnerability faced by smallholders. Méndez et al. state that "certifications did not have a discernible effect on … livelihood-related variables, such as education and incidence of migration at the household level… and contributions derived from [certified markets] premiums has limited effects on household livelihoods" (2010, p. 236). Likewise, Bacon's study of a coffee cooperative organization that sells to alternative markets noted a decline in their quality of life during the period of the crisis. However, Bacon points out that membership in a coffee cooperative organization that sells to alternative markets does reduce the risk of smallholders losing their land titles during periods of low prices by four times (2008, p. 168), suggesting that vulnerability is reduced by selling through a cooperative to a certified market due to the support services provided through the cooperative structure.

Ultimately, vulnerability and instability among smallholder coffee farming families cannot be simply addressed by income-centered approaches. They must be complemented by efforts to "increase access to land, build stronger producer organizations, promote access to alternative markets, increase government investments in rural health and education, and diversify income sources" Bacon (2008, p. 170). As the shortcomings of income-centered strategies have become clearer since the coffee crisis that ended in 2005, research on strengthening coffee farmer livelihoods and vulnerability has shifted to focus more heavily on household and community food security and food resilience.

12.3 THE RELATIONSHIP BETWEEN AGROECOLOGY AND FOOD SECURITY AND SOVEREIGNTY

12.3.1 Food Security and Food Sovereignty: Mainstream Approaches

The global agroindustrial food system model has served to further marginalize rural communities and deepen the process of "depeasantization," defined both as "the erosion of an agrarian way of life" (Vanhaute 2012, p. 6) and as "the phasing out of a mode of production to make the countryside a more congenial site for intensive capital accumulation" (Bello and Baviera 2009, p. 27, citing Bryceson 2000). According to McMichael (2008), depeasantization results from "the combined pressures of evaporation of public support for peasant agriculture, the *second* green revolution (privatized biotechnologies and export agricultures to supply global consumer classes), market-led land reform, and World Trade Organization trade rules that facilitate targeting southern markets with artificially cheapened food surplus exports from the North" (p. 209). Related to these factors are other challenges to maintaining traditional food and production cultures, which include a shift away from the use of traditional crop varieties to cash crops for export (Ghosh 2010), the associated introduction of genetically modified organisms—either voluntarily or involuntarily (see McAfee 2003 on genetic pollution of native maize varieties in Oaxaca, México)—and the influx of imported foods into local communities that are culturally inappropriate (see Friedmann 2005, p. 257). Furthermore, the standardized approach of the industrial model fails to value the diversity of practices that are reflected in traditional agricultural systems, which are the product of specific cultural traditions adapted to local environments. Finally, the subordination of traditional production systems to the industrial approach serves to weaken the ecological integrity of rural communities. This is particularly important in light of the high level of environmental impact of industrial agroecosystems and the relatively low level of environmental impact of traditional systems (Altieri and Koohafkan 2008).

The Food and Agriculture Organization (FAO) defines food security as "when all people, at all times, have physical, social and economical access to sufficient safe and nutritious food that meets their dietary needs and food preferences for an active and healthy life" and, furthermore, "household food security is the application of this concept to the family level, with individuals in households as the focus of concern" (FAO 2008, 2010a). Within this definition are four dimensions that need to be satisfied simultaneously to achieve food security: availability, access, utilization, and stability. Culture is mentioned by way of the significance of food preferences, noting that these are either "culturally or socially determined" (FAO, 2003, p. 27), though some argue that this reference is quite weak (Schanbacher 2010, p. 30) and requires more nuance to define how food preferences interact with other factors in food security.

The meaning of the term "food security" has become more nuanced over time, expanding beyond balance sheet type calculations of food self-sufficiency at the national scale, toward incorporating regional and community scales. Furthermore, the radical food sovereignty discourse of participation and power used by civil society groups such as La Via Campesina has influenced traditional international bodies like the FAO. Hence, whereas approaches in the 1970s and 1980s focused on redistribution of food and strengthening of markets to primarily increase access to food, current discourse focuses on the participation of poor, smallholder farmers as the drivers of food security, and the goal as one of improved nutrition, not just enough calories. The 2012 FAO Report "State of Food Insecurity in the World," for example, argued that in order for economic growth to enhance the nutrition of the neediest, the poor must participate in the growth process and its benefits, and that agricultural growth is particularly effective in reducing hunger and malnutrition; furthermore, agricultural growth involving smallholders, especially women, will be most effective in reducing extreme poverty and hunger when it increases returns to labor and generates employment for the poor. Since the 2012 FAO report, the organization has moved further in identifying the individual smallholder farmer as a key actor in the alleviation of hunger: Its report, "The State of Food and Agriculture 2014," states, "The world must rely on family farms to grow the food it needs and to do so sustainably" (FAO 2014, p. 7). The report recognizes innovative approaches that value smallholder farmer knowledge by stating, "Combining farmer-led innovation and traditional knowledge with formal research can contribute to sustainable productivity" (FAO 2014, p. 9) and "A supportive environment for producer organizations and other community-based organizations can also help promote innovation among family farms." These findings all support an argument that international development agencies working with local communities of smallholder farmers to alleviate hunger must facilitate spaces to support innovation among smallholder farmers while helping them overcome obstacles to sustainable food production (FAO 2014, p. 8). Essentially, farmers must be supported in the innovation of their own context-specific approaches to alleviate hunger.

This is in stark contrast to traditional state-level approaches to improve food security. These often include providing seed and/or production inputs, and even giving food subsidies offered by many national governments. One example is México's PROCAMPO and *Oportunidades* programs. In the case of México, it has been shown that although food subsidy programs like *Oportunidades* do alleviate immediate hunger, production subsidy programs like PROCAMPO have not actually resulted in diversified production and have not achieved their programmatic goals related to food security (Fox and Haight 2012, pp. 19–20). Meanwhile, new approaches by other governments, such as Nicaragua and Guatemala, have heeded the call for participation by creating municipal level food security committees and other government structures that are part of new legal structures, including "Right to Food" Laws (based on the FAO framework), that attempt to integrate government food security programs and projects into the existing state legal and government structure in varying ways. Their engagement with the smallholder agricultural sector varies widely. In Guatemala, most of the state's food security programs are being led by a joint program called Hambre Cero, but most actions are coordinated by the Ministry of Health, with little coordination with, and very few resources channeled through, the Ministry of Agriculture in that country (Bernal Larrazabal 2012). In contrast, Nicaragua has made the promotion of agroecological practices a central strategy of its Food Security Law and its own Hambre Cero program (speech by Amanda Lorío Arana, Vice-Minister of Agriculture and Forestry, August 12, 2012, Matagalpa, Nicaragua).

The FAO approach to alleviating food insecurity outlined in the 2014 report actually echoes language more often associated with food sovereignty movements. Food sovereignty can be considered a comprehensive, rights-based *approach to achieving* food security, according to Windfuhr and Jonsén (2005, pp. 23–24). This is perhaps the key distinction between food security and food sovereignty. As Pimbert (2009, p. 50) explains, "The mainstream definition of food security ... does not talk about where that food comes from, who produced it, or the conditions under which it was grown," key principles of agroecology as a practice, science, and political movement. Establishing food security must take into account a more complex web of interacting elements that at its core respects the breadth and depth of community participation in defining and shaping their food security. Thus, food sovereignty can be considered an approach to achieve food security that values innovation and smallholder agency in transforming the food system.

A definition of food sovereignty developed at the 2007 Forum for Food Sovereignty defines the concept as "the right of peoples to healthy and culturally appropriate food produced through ecologically sound and sustainable methods, and their right to define their own food and agriculture systems" (Declaration of Nyéléni 2007). Additionally, the food sovereignty discourse emphasizes the rights of indigenous peoples to retain traditional production systems and food cultures (Declaration of Nyéléni 2007; Ruelle et al. 2011). This acknowledgement of the value of culture in attaining food security is of particular importance as it expands the reference to the cultural preferences included in the FAO's definition of food security to that of a right of (indigenous) communities and nations to establish their own food systems that are reflective of cultural values and traditions (Ruelle et al. 2011, p. 164). Additional dimensions of what constitutes food sovereignty, including the rights and roles of women and other disadvantaged peoples, have entered the debate, and various contradictions or issues among the tenets of food security have been explored, including the tension between local self-sufficiency and national self-sufficiency, and between the perceived need to promote food crops, and a farmers' rights to decide what to produce and to what extent to farm in the first place (Agarwal 2014), as well as the not-well-understood role of international trade in food sovereignty (Burnett and Murphy 2014).

12.3.2 Agroecology as a Food Security and Sovereignty Strategy

Despite the negative impacts of petrochemical agroindustry, which include increased pest resistance, reductions in insect pollination, water contamination, and pesticide drift (which has adverse effects on human as well as environmental health), farmers continue to rely on petrochemical pesticides, according to Wilson and Tisdell (2001). Two theories they propose are the low cost of pesticides and the chemical companies' pairing of the purchase of seed and chemicals, creating a cyclical dependence on the part of the farmer on chemicals for higher yields. In contrast, as Altieri and Toledo (2011) explain, "[a]groecological initiatives aim at transforming industrial agriculture partly by transitioning the existing food systems away from fossil fuel-based production largely for agroexport crops and biofuels toward an alternative agricultural paradigm that encourages local/national food production by small and family farmers based on local innovation, resources, and solar energy" (p. 588). So far these initiatives, including those that aim to protect and encourage traditional and local food systems, show much promise as an alternative to agroindustrial systems as they encourage genuine food sovereignty and not just security for the most vulnerable communities. Research shows that agroecological systems can be just as if not more productive than agroindustrial systems (Altieri and Toledo 2011) and have been shown to increase yields (De Shutter

2011, citing Pretty et al. 2006; Pretty 2003). Furthermore, agroecological production is more resilient to climate change and climactic disturbances and disasters (Altieri and Koohafkan 2008; Holt-Giménez 2006a) and more energy efficient (Gomiero et al. 2008), both of which are key factors in the contemporary era of energy and climate change debates. Altieri and Toledo (2011) delve more deeply into the relationship between sovereignty and resiliency by arguing, "Agroecology provides the principles for rural communities to reach food sovereignty but also energy and technological sovereignty within the context of resilience. Agroecology provides the principles to design resilient agroecosystems capable of withstanding variations in climate, markets, and so forth, while ensuring the three broadly but inter-linked sovereignties" (p. 607). Thus, agroecology is a key strategy to promoting community food sovereignty and security because it emphasizes local control over food supply and distribution systems.

Agroecology and food sovereignty also share a focus on empowerment of local peoples over their food systems; agroecology then becomes a critical tool to creating food sovereignty to achieve food security, because it values and integrates local knowledge as an equal part of the toolbox that rural peoples can use to gain control over their food systems, echoing Berkes' statement that "traditional knowledge, especially of the ecological kind, have practical significance for the rest of the world" (2008, p. xiii). In comparing food security and food sovereignty, agroecology makes the physical—and especially the agricultural—landscape an actor in any study of multiscalar interactions that involve food, placing an emphasis on the transformation of the food system (Giessman 2010).

Agroecological methods are increasingly positively linked to strengthening food security (Altieri 2002, 2009; De Shutter 2011). This is in part because of agroecology's multidimensional approach to production that emphasizes environmental health, socioeconomic well-being, and cultural preservation. This is particularly important for rural communities where the bulk of the world's hungry and malnourished reside (FAO 2010b), the majority of which are involved in agriculture and food production.[*] De Shutter (2011) points out that one of the greatest challenges will be achieving food security for the world's poorest, especially small scale farmers in the global South. He argues for the benefits of agroecology as a vehicle to strengthen food security among small-scale farmers, specifically its focus on empowering small farmers by revaluing their knowledge and participation as experts; its potential to increase incomes of rural farmers with less dependence on external inputs, thereby reinvigorating rural economies; diversifying local agricultural production, which leads to more nutritional diversity; and enhancing environmental sustainability "by delinking food production from our reliance on fossil energy (oil and gas) … [and] mitigating climate change, both by increasing carbon sinks in soil organic matter and aboveground biomass, and by reducing greenhouse gases through direct and indirect energy use" (235). Douwe van der Ploeg (2014) adds another dimension to this argument by arguing that peasant farmers practicing agroecology in the form of peasant agriculture are not only made food insecure by factors of global capital, but more importantly can also shape global capital if they can achieve the goal of producing enough food (agricultural growth), and thus food sovereignty through peasant agriculture.

12.4 APPROACHES TO IMPLEMENTING AGROECOLOGY AS AN FSS STRATEGY WITHIN A DONOR-FUNDED PROJECT CONTEXT: A FOCUS ON HOMEGARDENS

Arguably the most common strategy implemented by development organizations to promote increased household food security is production diversification for household consumption and income generation through alternative markets, often in the form of homegardens. Homegardens have long been acknowledged as critical components of rural food security because of their role as sources of

---

[*] See http://www.fao.org/hunger/en/.

food and income (Marsh 1998, p. 4), but are now being adopted as a particularly important and increasingly widespread strategy for supporting household food security in developing countries through their potential for increasing dietary diversity, income generation, and health (Galhena 2013, p. 2). Their value in addressing what the FAO is calling a world food crisis fits within the FAO's call for a reinvigorated focus on household subsistence agriculture as a way for the poor to participate in hunger alleviation (FAO 2012). Furthermore, the United Nations Conference on Trade and Development has stated that "we need to see a move from a linear to a holistic approach in agricultural management, which recognizes that a farmer is not only a producer of goods, but also a manager of an agroecological system that provides a number of public goods and services" (2013, p. 1). Thus the imperative of supporting and strengthening homegardening systems includes not only food security and the preservation of genetic agrobiodiversity, but also for other "services" that such agroecological systems may provide.

Governments and NGOs alike are increasingly promoting homegardens as part of food security projects as a way to increase the availability and accessibility of fresh foods, increase household income, and thus improve nutrition among rural families. We know little about how the growing fashion of promoting homegardens as food security strategies through donor-funded projects affects the form, function, and meaning of homegardens, and relatively little about their limitations in relation to locally available inputs, environmental, and ecological conditions, gender and generational dynamics, cultural preferences, and especially if they positively impact food security in any meaningful way. Limitations of gardens as rural development strategies within a donor-funded project context include poor project design, poor management, and poor monitoring (Marsh 1998, p. 4), which are all familiar critiques of development projects in general. The benefits of homegardens to improving household food production, nutrition, and income have also been established (Cleveland and Soleri 1987, p. 259).

There are various gaps in the homegarden strategy. First, questions remain as to the best ways to promote or implement agroecological garden production among farming families that positively and significantly impact household food security, or even if agroecology as a food systems approach that effects food sovereignty (rather than as a set of production practices) can be promoted within the context of a donor-funded project. Rural development projects have tended to emphasize the "what" (homegardens or production diversification) rather than the "how" of improving household food security. This is generally done through a top–down model in which the sponsoring organization delivers packages of seed (often varieties or species that local farmers are not familiar with), works with farmers to create Western-style homegardens with clearly defined boundaries, and supports farmers with training in garden management through to harvest. The challenge with this approach is that it does not value existing farmer knowledge (Marsh 1998), nor try to disseminate it, nor address why that knowledge is not being applied in the first place if it is not.

However, organizations working within the global food sovereignty movement have focused rather independently from the international intergovernmental organizations promoting food security, on the methodologies for acquiring knowledge and ways of knowing that facilitate food sovereignty in local communities. La Via Campesina's *dialogo de saberes* (knowledge dialogues) among different communities worldwide has indeed led to its food sovereignty framework and has greatly facilitated the dissemination of agroecology among peasants in different parts of the world (Martínez-Torres and Rosset 2014). This focus on horizontal farmer-to-famer exchanges with an emphasis on farmer capacity building and experimentation promotes farmer control and decision-making over which foods she is producing, for what, and how. Campesino a Campesino piloted this methodology among smallholder farmers in Central America in the 1980s and has recently been applying it to promote farmer-based agroecological food production to specifically improve FSS in Nicaragua, as part of the Union Nacional de Agricultores y Ganaderos (National Union of Farmers and Cattlemen, UNAG), a government agency. This case study demonstrates how dissemination of agroecology through farmer-to-farmer exchanges requires the continual adaptation of project design to attain a multifaceted agroecological approach relevant to the particular context of smallholder farmer beneficiaries and the social organization facilitating the process.

The study employed a project evaluation process designed and implemented collaboratively with project stakeholders that employed household surveys, semistructured interviews, and focus groups to collect data on FSS indicators as well as farmer perceptions about changes in their lives resulting from the project process. Evaluations were conducted annually, and this chapter discusses both the 2-year and 3-year evaluations.

12.5 CASE STUDY: YOUTH LEADERSHIP AND FOOD SOVEREIGNTY PROJECT IN SAN RAMON, NICARAGUA

The Youth Leadership and Education for Sustainable Agriculture and Food Sovereignty Project (referred to as the Youth Leadership and Food Sovereignty [YLFS] Project from here) is collaboration between the CAN, the Union of Cooperatives Augusto Cesar Sandino (UCA San Ramon), and eight first-level coffee farmer cooperatives. The primary funder of the project is Keurig-Green Mountain, Inc. CAN is a nonprofit organization based in Santa Cruz, California, working in eight regions of Mexico and Central America. CAN partners with local community-based organizations, farmer cooperatives, nonprofits, and universities in each of the eight regions to generate participatory approaches to sustainable development, food insecurity, and other rural environmental and social problems, through research, education, and action.

The UCA San Ramon, a second-level cooperative organization, was founded in 1992 in response to the liberal return to power in 1990 that threatened to reverse many of the agrarian reforms instituted during the Sandinista Revolution of the 1980s. The new organization quickly started filling the gap in rural support left by the new government by offering technical assistance and access to credit to its members, who produced both coffee and basic grains (corn and beans). Over the years, the UCA San Ramón has developed a strategy that can be described as social and entrepreneurial and has transformed itself into a cooperative business organization that, in addition to having social ends, has also taken actions aimed at increasing and diversifying the incomes of the 1080 members of its 21 first-level cooperatives through initiatives including community-based tourism; access to specialty and alternative markets including fair trade; processing and commercialization of members' milk production; production and income diversification into vegetables and fruits; various economic initiatives aimed at women; and technical assistance and training for members to improve coffee production and yields. Social programs that support those economic initiatives include gender empowerment training for women, men, and youth; a scholarship program for cooperative youth to pursue higher education and technical programs; a youth leadership program to train youth leaders to be cooperative *promotores* in health, food sovereignty, and other cooperative programs and provide them with part-time employment; a sustainable agriculture program to promote appropriate environmentally friendly technologies and practices; and nutrition education associated with its 5-year FSS action plan developed in 2012.

The relationship between CAN, the UCA San Ramon, and the eight first-level cooperatives is multifaceted and long term, having been established first informally through a researcher collaboration in 2000, and evolving as CAN and the UCA San Ramon developed programs in the following decade for student field studies in the cooperatives in San Ramon and sustainable coffee chain models. The long-term relationship has facilitated trust as well as the collective capacity to develop the YLFS project. The project was launched in February of 2011 after being developed collaboratively by CAN and the UCA San Ramon. The project aims to alleviate food insecurity and seasonal hunger among 95 coffee farmer households (124 families) and build sustainable local food systems in eight coffee farmer cooperatives through education and the empowerment of local youth leaders in these communities. The project focuses on capacity building in participating cooperatives, with a specific focus on youth (aged 17–25) and women to promote and implement sustainable food production, usage, and consumption practices at the family and community levels, by impacting rural/agricultural livelihood opportunities for youth and women, knowledge and skills in agro-ecological production, of diverse foods, youth pride in rural culture and livelihoods, and community access to the means of production (water, soil, and seed) and to fresh and locally grown food throughout the year. CAN and the UCA San Ramon engaged in a participatory planning process to define the goals and strategies based on the recognized needs to establish year-round access to healthy food in participating rural communities, preserve and promote local sustainable food cultures, as well as stem the tide of youth outmigration from these communities, where people, especially youth, eat less fresh, healthy, locally grown foods and are turning more to prepackaged chips and other high-fat processed foods available at the local *pulpería* when they do have cash.

Since the project was launched in 2011, CAN has acted as the primary project administrator for the YLFS project, ensuring fulfillment of project vision and goals, participatory planning processes, transparent financial management, the execution of the participatory monitoring and evaluation plan and the continued integration of monitoring data collected every year into project execution, capacity building and organizational strengthening of the stakeholder partnership, and disseminating project results both internally and externally to the CAN network. The UCA San Ramon is the primary organizer of on-the-ground implementation activities, including workshops and capacity building, technical assistance, infrastructure investments, and other follow-through. The eight beneficiary cooperatives themselves also play an important role in facilitating project activities, and each cooperative has a part-time youth leader (*promotor*) paid through the project, and who acts as the primary link between the UCA San Ramon and the individual beneficiary households, providing individual technical assistance and also actively implementing monitoring and evaluation activities by managing data collection and participating in analysis of results. The households themselves also participate in annual monitoring and evaluation through workshops in which collective data analysis occurs and consensus is reached as to current problems and appropriate strategies to integrate to address them the following year.

The partnership between CAN, the UCA San Ramon, and its member cooperatives also includes other programs that complement the YLFS Project. Youth leaders from the project participate in an International Youth Network for Food Sovereignty dedicated to facilitating knowledge and technology exchange related to food sovereignty across the CAN network. An action education program has also facilitated US student field studies and field courses in the cooperatives of the UCA San Ramon since 2009. Finally, through CAN's AgroEco® coffee program, CAN has purchased coffee since 2011 from one of the cooperatives involved in the YLFS Project, and supports the cooperative in its transition towards agroecological coffee production as well as production diversification.

12.5.1 An Approach to Implementing Agroecology as a Food Security and Sovereignty Strategy

Agroecology was a stated foundational principle throughout project visioning, planning and implementation, and monitoring processes. However, initially all stakeholders entered those processes with different conceptions of what agroecology is, and coming to a common understanding of what agroecology is and how it is applied was our first challenge. At the initial project launch meeting in Santa Cruz, California, in which CAN researchers and project managers from Nicaragua participated, it became clear that the project managers did not really have a clear personal understanding of agroecology, but knew that it meant not using chemicals, promoting soil conservation, and not polluting water. CAN researchers entered into the dialogue with an understanding of agroecology that centered on a more global perspective of sustainable food production, distribution, and consumption, but little concrete knowledge about how to apply that to FSS. In the coming months as the initial project vision was shared with cooperative agricultural technicians, youth leaders, and project beneficiary households themselves, other understandings of the term "agroecology" emerged. The agricultural technicians of the UCA San Ramon perceived agroecology as a distinct set of production practices synonymous to organic practices, which involved, as they understood it, not using chemical fertilizers or insecticides. Youth leaders and families, however, experienced the word "agroecology" as a new concept, but as we explored it with them, they increasingly equated the concept with "traditional" or "ancestral" knowledge and practices that they in most cases were no longer practicing in their milpas and patios.

We did not achieve a common definition or understanding of agroecology during that initial visioning process. Rather, the process extended beyond that into the first and second years of the project, in which project funds supported establishment of homegardens, the evaluation of practices, and the identification of new needs and requirements to make the gardens more effectively achieve the goals of increased FSS. Dialogue around the meaning of "agroecology" and how to practice it occurred during facilitated workshops and planning sessions, and in individual conversations among families, youth leaders, field technicians and UCA San Ramon staff, and CAN project managers. Farmers, for instance, initially expressed skepticism of their ability to produce food without applying chemical pesticides or fertilizers; cooperative agricultural technicians focused on compost production as their main "agroecological" solution; CAN researchers focused on promoting the planting of gardens. At the end of the second year, we had come closer to understanding that agroecology meant not just garden production or compost production, but a system that encompasses on-farm and in-garden production practices that favor locally available resources (land, seed, soil, fertilizer, water) and knowledge, food distribution systems that are locally determined and controlled (local markets), and household food usage, preparation, and consumption habits that promote improved human and environmental health, echoing and applying agroecology's global emergence as an approach "grounded in transdisciplinarity, participation, and transformative action" (Méndez et al. 2013).

The YLFS Project initially utilized three interrelated strategies as it set out originally in 2011 (see Table 12.1). The first strategy was to utilize intercultural exchange for youth capacity building and network development around agroecological practices and garden management, and healthy food cultures to foment youth leadership in cooperative processes. The second strategy was education and capacity-building programs aimed at educating women and men farmers and local schoolchildren about agroecological food production, preparation, and consumption practices and benefits. The third strategy was to diversify and extend production through the implementation of the sustainable production practices learned in trainings on individual family farms in homegardens, patios, and coffee fields, the end goal being to increase household access to more diverse foods during the entire year and increase dietary diversity.

A participatory evaluation of the first 2 years of the project yielded some important lessons. The evaluation was conducted over a 3-month period, and included an initial survey designed in conjunction with cooperative staff and youth leaders, who executed the survey among beneficiary families in their communities. Results of the survey analysis were shared in workshops with farmer

Table 12.1   Project Strategies to Link Agroecology and Food Security and Sovereignty

Project Strategies	
Three Original Strategies	1. Intercultural exchange for youth leader capacity building and network development in agroecological practices and garden management
	2. Education and capacity building in agroecological food production, preparation, and consumption practices
	3. Diversification and extension of production practices learned through capacity building and training in homegardens, patios, and coffee fields
Four Strategies Developed As a Result of 2-Year Evaluation	4. Increase household capacity and access to seed and rootstock production
	5. Strengthen staple food production, availability, and access
	6. Build soil fertility in gardens, milpa, and coffee fields
	7. Intensification of initiatives to diversify income with women farmers

beneficiaries, youth leaders, and leaders of the UCA San Ramon, in which participants interpreted the data, performed collective qualitative evaluations, and made recommendations for changes in project strategies.

The first lesson learned was that access to a stable local supply of seed and rootstock for vegetables, fruits, and tubers and roots is critical to the continued success of the homegardens and diversified patios and coffee fields. Building capacity at the household and community scales in seed production and storage became a priority. Second, even with demonstrated increased production diversification in gardens, patios, and coffee fields to increase availability of more diverse foods year-round, the periods of scarcity ("thin months") experienced by beneficiary households had not changed at the end of 2 years of project implementation. They remained at 4.7 months average, partially due to the first year being dedicated to organizational processes and basic capacity building in garden management. While real progress was made in production diversification and nutrition education in the second year, impact on reducing the thin months remained limited because project interventions did not impact local access or availability of staple foods (corn and beans) during the thin months when families have already consumed their stored grains from the previous harvest, and staple foods prices typically soar. This demonstrated the need to include long-term interventions that would positively impact staple foods availability during the thin months. Third, farmers argued during workshops that lack of soil fertility was still a major barrier to production diversification and continued production of staple foods, as well as to maintaining yields of coffee harvest; soil fertility was then a livelihood issue. The first 2 years of the project had included a focus on small-scale compost, vermicomposting, and foliar biofertilizer production for homegardens and patios, but a shift in focus was required to scale up production to include coffee fields in fertilizing strategies. Finally, new research has shown that increased income diversification is correlated to higher levels of household food security (Baca et al. 2013); this supported our finding in 2012 that increasing women's earning power also increased household nutrition levels, and led us to intensify our focus on income diversification strategies focused on women project beneficiaries, expanding from monthly farmers markets where women could take their excess garden produce and fruit production to sell, to the development of collective women's microenterprises for value-added products from their own production.

In response to lessons learned, four more strategies were added to the original three: increase household capacity and access to seed and rootstock production; strengthen staple food production, availability, and access; enhance focus on building soil fertility in gardens, milpa, and coffee fields; and intensify initiatives to diversify income for women.

Five main methodologies developed during the first 3 years to promote these seven main strategies (see Figure 12.2). The first year of the project focused on garden production and youth leadership capacity building utilizing the first methodology of in-garden trainings and

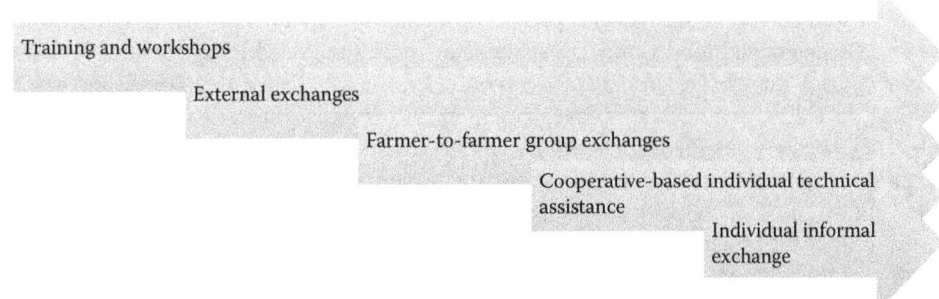

Training and workshops

External exchanges

Farmer-to-farmer group exchanges

Cooperative-based individual technical assistance

Individual informal exchange

Figure 12.2 Five methodologies implemented over 3 years in sequential order.

workshops facilitated by UCA San Ramon staff. As the eight youth leaders and beneficiary households became more proficient at agroecological production practices, and appropriated their own vision of the long-term goals of the project, they became aware of other skills and technologies that they wanted to appropriate but that were not immediately or locally available. This led to the usage of the second methodology, external exchanges, in the second and third year of the project that involved groups of men, women, and youth visiting other communities and experiences to learn from them and exchange knowledge about specific identified topics or practices. It is important to mention that CAN also facilitated annual international exchanges among youth leaders from its two FSS projects in Nicaragua and Mexico, which also included US-based students and food systems activists. These international external exchanges, similar to Via Campesina's *dialogos de saberes*, served to introduce to the youth leaders new knowledge and practices applied in other places to the same problems, in effect catalyzing new tools to experiment with once they returned home. The international exchanges themselves were also critical to San Ramon youth leaders and project managers identifying the gaps in the project's strategies to improve FSS described above.

Upon returning home from these external and international exchanges, farmers and youth leaders would then horizontally disseminate their new knowledge using the third methodology, farmer-to-farmer group exchanges, replicating the Campesino-a-Campesino model vary widely in use in Nicaragua (see Holt-Gimenez 2006b), within and among the eight cooperatives. Farmer-to-farmer group exchanges were also utilized to spread existing knowledge in the cooperatives among more households in production techniques (e.g., the production of seed for a local heirloom tomato called *tomate gallina*), food preparation, and consumption practices.

Following 2 years of these first three methodologies, much of the exchange took place using the last two methodologies. Cooperative-based individual technical assistance involved youth leaders or cooperative technicians visiting individual farms to work with farmers to identify problems and develop solutions; this is distinct from traditional prescriptive technical assistance, in that the youth leader or technician talks with the farmer to draw upon their collective existing knowledge to identify solutions. Finally, individual exchange was the fifth methodology, with women and youth visiting each other's gardens and patios to see what others were doing. These individual exchanges inspired many of the women to start experimenting with new plants and varieties in their own parcels. One key example of this was with the cultivation of amaranth; a few key women farmers planted it in their gardens and took note of where it thrived and where it did not, and what practices the plant seemed to favor. They then saved seed and passed the seed on to other women farmers to try, while also sharing how they had learned to prepare amaranth for different dishes. After a year, approximately 25 women farmers were producing amaranth in their gardens and preparing it for their families.

In a 2013 survey of 24 women project beneficiaries, we found that 41% said they had acquired new knowledge about seed saving through individual exchange (we had only held two formal exchanges on seed saving in the previous year). This demonstrated a major change in the women's roles as experimenters and food systems managers, as well as the need for the project to facilitate more exchanges of knowledge around seed saving. Given that 90% of the women did not have vegetable gardens 2 years before, this development of farmer experimentation was momentous for them all. They began to see themselves differently as farmers and household providers and also as a community of innovators working to improve their own communities, according to interviews conducted among the women in 2013. The UCA San Ramon nurtured this spirit of experimentation in the third year of the project by providing new plants to experiment with, including amaranth and basil. Although the fifth methodology of individual exchanges resulted in the shift in the roles of the youth and women beneficiaries, especially toward leaders of experimentation and innovation, this methodology built upon the foundations established through the first three methodologies. We argue that all five used in conjunction achieved the outcomes and impacts described below (Figure 12.3).

Figure 12.3 Youth leaders and CAN researchers visiting a food pantry in San Jose, California, during the Third Annual International Youth Exchange for Food Sovereignty (2013).

12.5.2 YLFS Project Results: 2011–2013

Another evaluation of the project was performed in 2013, which focused on the results of the four strategies that had been integrated after the 2-year evaluation. The evaluation again included a collaboratively designed survey based on indicators for each strategy determined by CAN and the UCA San Ramon, as well as workshops to collect qualitative perspectives on the project's progress, share interpretations of the data, and offer recommendations on how to improve project strategies. The outcomes are outlined below.

Improved production diversification:

- 45 Homegardens in San Ramon, and 63% of 24 households are now producing more than 12 species of vegetables, tubers and roots, and vine crops in their gardens and patios, up from 5% in 2011.
- 95 Households diversified to include vine crops (e.g., chayote) and fruit trees in San Ramon.
- Six school gardens established in San Ramon in schools attended by 274 schoolchildren through workshops and awareness building.

Improved capacity in agroecological soil fertility practice:

- A strong continued focus on building households' capacity to produce various kinds of fertilizers (worm compost, traditional compost, and foliar biofertilizers) through capacity building and infrastructure improvement, has been brought to a new level with an experimental initiative in one cooperative to produce mycorrhyzae fertilizer at a collective scale that will support coffee renovation in its fields that were affected by the leaf rust in the past 2 years.

Improved agroecological capacities for seed production and water usage:

- Capacity building in seed selection, saving, and storage, with a focus on horizontal exchange of knowledge among participants, especially as pertains to vegetable seeds, which are the most challenging for gardeners to save due to lack of knowledge. Advances were made in cooperative level seed saving as well, with the establishment of one pilot seed bank established in the Ramon Garcia Cooperative, focusing this year on heirloom varieties of beans, and fulfilling our strategy of multiple levels of seed saving (individual and collective).

- 34 Manual pump irrigation and water storage systems were installed on those farms that required it to produce during the dry season, enabling a vegetable harvest during the thin months.

Increased access and availability of food and dietary diversity:

- A nutrition guide and recipe book was developed based on the new knowledge and the women's knowledge of local recipes.
- After 9 months of visioning, planning, and organizing, a pilot Food Storage and Distribution Center (CADA) was established in Ramon Garcia Cooperative. The goal of the CADA is to increase local access to basic foods during the thin months: the cooperative-run CADA purchases farmers' corn and beans during the harvest seasons, and then sells it back to them during the period when household grain stores have been depleted and prices are higher, allowing farmers to purchase food below market price right in their own communities. The cooperative wrote an operational protocol for the CADA, set goals for volumes to collect and sell, set prices for the first purchase, and installed the infrastructure (silos) at the end of the project year.

Enhanced income diversification opportunities for women, and capacity for youth leadership:

- Nine monthly farmers markets were put on annually, in which an average of 35 women came to sell their excess vegetables, roots and tubers, and fruit. Markets are attended by an average of 100 people (see Figure 12.4).
- Business plans were developed for two additional women-focused rural enterprises.
- Youth empowerment: 14 youth involved in the project hold leadership posts in their local cooperative or in the UCA San Ramon, and seven of the eight youth leaders now hold leadership posts in local government or other organizations (Figure 12.5).

12.5.3 Impacts of Promoting Agroecology as an FSS Strategy in the YLFS Project

Monitoring and evaluation of the YLFS Project used 30 indicators to measure changes in food security and sovereignty (access, availability, usage, stability, months of adequate provisioning soil fertility practices); and other indicators related to specific project goals defined by all stakeholders.

Figure 12.4 Women selling produce at monthly farmers market in San Ramon, Nicaragua (2014).

Figure 12.5 Woman sharing her new vermiculure box on her farm in San Ramon municipality.

The indicators presented below related specifically to the relationship between FSS and the food systems approach to agroecology promoted through the project. See Table 12.2 for a summary of indicators.

12.5.3.1 Production Diversification

Households with homegardens have expanded from 24 to 45 since 2011 (the goal after 5 years is 75), and agricultural diversification into fruits, tubers and roots, and vine crops in other production areas (patio and coffee field) has also expanded to 95 households from 24 in 2011. Production

Table 12.2 Summary of Indicators

Indicators	2011 (*n* = 72)	2012 (*n* = 32)	2013 (*n* = 24)
Percent of Households Producing More Than 12 Species of Vegetables	5%	20%	63%
Percent of Households with Over 15 Species of Fruits and Fruit Trees	30%	No data	75%
Percent of Households Producing More Than Three Species of Roots and Tubers	20%	20%	50%
Percent of Households Producing Over 10 Species of Repellent, Aromatic, and Culinary Herbs	30%	30%	87.50%
% Households Saving Over Six Varieties of Seed	No data	10%	30%
Household Dietary Diversity Score	No data	6.61	7.84
Households Consuming More Than Six Food Groups Daily	12%	83%	100%
Households with Access to Water for Irrigation for Production	15%	31%	75%
Households with Access to Clean Drinking Water	no data	33%	47%
Households with Diversified Household Income from Vegetables	2%	30%	49%
Average Length of Thin Months	4.63 Months	4.7 Months	3.88 Months
Range of Thin Months	6 Months	6.2 Months	9 Months
Coping Strategies Index Score San Ramon	16.83	10.88	17.53

diversity has greatly increased since 2011, with 63% of 24 households producing more than 12 species of vegetables in their gardens and patios, up from 5% in 2011. The number of households producing a higher variety of fruits and fruit trees, roots and tuber, and different categories of herbs has also similarly risen, reflecting a drastically higher availability of, and access to, these foods and plants at the household level.

12.5.3.2 Seed Production and Saving

Access to vegetable seed has been improved through an intense focus on household-level seed saving capacity building as well as the establishment of one new cooperative level seed bank. There was a marked change between 2012 and 2013, with 30% of households saving six or more species of vegetable seed. When asked how they learned seed production and saving skills, 41% of households responded that they had learned through individual exchange of knowledge; 71% from older members of the household; and 29% said they had learned from individual experimentation in their gardens or patios, indicating that tools and skills to improve FSS do exist in the communities and that the project should intensify efforts to facilitate both group and individual knowledge exchange activities to further the dissemination of these practices and raise the proportion of households saving seed of a target number of species.

12.5.3.3 Household Dietary Diversity

Dietary diversity is an indicator of whether higher availability of more diverse foods is actually translating to the consumption of a higher diversity of foods. The household dietary diversity score represents the average number of food groups consumed per day. It is collected by using a 7-day dietary recall, in which respondents list the number of days per week they have eaten each food group. The total number is then divided by seven for estimate of the average number of food groups per day. In San Ramon, the dietary diversity score has improved from 6.61 in 2012 to 7.84 in 2013, indicating a positive change in household diets. A great improvement has also been seen in dietary diversity as demonstrated by the percentage of households consuming more than six food groups per day during the thin months: 100% are consuming more than six food groups per day in 2013, up from 12% in 2011, which is significant. We attribute this to the greatly increased availability of vegetables and fruits in 2013, as compared to 2011 when almost none of the households had vegetable gardens.

12.5.3.4 Soil Fertility Practices

A continued emphasis on building soil fertility for gardens through the production of various kinds of fertilizers (worm compost, traditional compost, and foliar biofertilizers) through capacity building and infrastructure improvement has resulted in 67% of households implementing two or more of these practices on their own farms, up from 50% in 2011. This of course does not include additional practices, such as applying coffee pulp compost, ash, or other already common local practices to improve soil fertility in coffee fields and milpa areas.

12.5.3.5 Water Access

Access to water for irrigation affects availability of food, as it permits more agricultural production and permits production during the dry season. The installation of 34 water storage and irrigation systems has increased the percentage of beneficiary families with year-round access to irrigation water from 15% to 75%, permitting year-round garden production for those households that previously could produce vegetables only during the rainy season. In 2013, 85% of households were producing vegetable gardens year-round, up from 10% in 2011. Clean drinking water is critical to food security as well, and the installation of home water filters has resulted in 47% of households having access to it.

12.5.3.6 Income Diversification

A firm footing was achieved in household income diversification focused on women beneficiaries over the last 3 years through their participation in monthly farmers markets and local sale in their own communities. Fully 49% of households have diversified their income into vegetables, creating a source of income during the times of year when there is no coffee income, and arguably creating more stability in household diets as a result.

12.5.3.7 Thin Months

The thin months are determined by asking a household which months in the last year they experienced scarcity of food. In San Ramon, the thin months decreased from 4.63 months in 2011 to 3.88 months in 2013. The range of thin months jumped from 6 months in 2011 to 9 months in 2013. This means that some families are experiencing scarcity in months when it was not typically experienced (outside the period of the thin months) and can possibly be explained by the onset of the coffee leaf rust in Nicaragua in late 2012. The blight has affected coffee yields among farmers in San Ramon by as much as 60–80%, impacting their household income levels, and leading to increased vulnerability and scarcity as a result.

12.5.3.8 Coping Strategies Index

The Coping Strategies Index (Maxwell et al. 2008) score measures the variety of behaviors that people implement to cope with scarcity. A lower score signifies either the usage of less severe strategies or less frequency of strategies. In San Ramon, the score improved between 2011 and 2012, but then rose significantly in 2013. One explanation for the recent rise is the deleterious impacts on food access from the coffee leaf rust, yields crashed, and income from coffee was drastically diminished. The result was more households substituting their diets with cheaper foods, skipping meals, or using other coping strategies. In interviews in late 2013 during the coffee harvest, women beneficiaries stated that increased food available in their gardens and even small amounts of income generated from selling produce did help to mitigate the effects of the coffee leaf rust on their family economies and diets, indicating that the situation would have been worse without the gardens and fruit production promoted over the last 3 years.

12.6 LESSONS LEARNED AND IMPLICATIONS

Overall, the 24 households surveyed are experiencing shorter periods of food scarcity since this partnership and project started. We can say that they have improved diets and healthier soil, and that households are likely becoming more resilient to future crises as the data shows that they have increased production diversity (and potential incomes sources), increased capacity to produce and save various kinds of vegetable seeds, and improved access to water for irrigation and consumption. In addition to the quantifiable changes in food security indicators, a critical change resulting from the project was the shift in roles of cooperative youth leaders and women from project beneficiaries to leaders of experimentation and innovation within the project context, effectively shifting more control over project direction into their hands. We attribute these positive changes to three important lessons learned during the 3 years of the project, and most crucially, to the project's capacity to integrate these lessons into its ongoing implementation. Through a combination of sharing across individual and collective practices, experimentation and reflection, we argue that all stakeholders engaged in a mutual learning process that adapted the application of an agroecological approach to the local context. In this case, it expanded the original focus on homegardens to incorporate multiple and interrelated components of the food system. The success of the project relied on this collaborative and adaptive approach. The following lessons learned, while derived from this particular case study, can illuminate efforts to build food sovereignty at the community level.

The first lesson learned is that to effect positive change in household FSS using agroecology as a strategy, it is critical to place the emphasis on making more food available through homegarden diversification within the larger production context of staple crops, cash crops, soil fertility, and access to water and land, as well as within the larger food systems context of markets and distribution, and food usage and consumption. No single element of the food system is more important than the other, but rather the synergy of diversified food production with water and soil management and links to markets and social organizations, is what influences increased resiliency. A related and second gap revealed by this case study is that the traditional homegardens approach promoted through rural development projects also tends to ignore in its design broader issues of food production sustainability, including local access to sustainable sources of seed, water, and even land. Agroecology must be promoted not only in "garden" food production, but also simultaneously in coffee and staple crops production. This will ensure a focus on transformation of the food system (Gliessman 2010) that procreates food insecurity itself, rather than just the single element of production.

The implication of these first two lessons is that international development projects with the stated goal of increased food security must go beyond a simple strategy of production diversification and include strategies to build capacity within a social organization to not only support production but also strengthen the entire livelihood of the household. The focus effectively shifts from one of household food security to community food security in which the agroecological knowledge base of the collective must be built or rebuilt from traditional and introduced agroecological technologies.

However, a limitation of diversified production as a livelihood strategy for increasing food security and resiliency might be lack of access to the means of production—land, water, and seed are especially critical. International development projects need to lean away from the "square" garden model toward utilizing available space like the patio and coffee fields themselves as spaces where intensified agroecological production can be enhanced and accompanied by strategies to conserve and improve soil and water quality.

Third, a combination of individual and group horizontal knowledge exchange methodologies that especially embrace and highlight existing local and ancestral knowledge that positively impact FSS at the household and community levels, while also introducing new technologies and fostering the spirit of experimentation and innovation, is essential in that it fosters the development of leadership that may ensure sustainability of initiatives long past the life of the project cycle itself. International development agencies must learn to integrate methodologies to revalue and disseminate ancestral knowledge among project beneficiary families. One major challenge to implementing these methodologies may occur in agencies staffed by people who do not share the same culture or worldview as project beneficiaries and/or do not value local knowledge. One solution is to utilize the model of community promoters; this case study has shown that they serve a critical role in interpreting between beneficiaries and project staff.

The case study showed the importance of closing a final gap in traditional homegarden strategies, which is the linking of food production to food consumption; projects often do not include a component on the preparation and household consumption of the foods being produced in the new homegardens, some of which are often introduced species or varieties, and women often do not know how to prepare them or families do not prefer them. A focus on women as food producers and preparers, as well as marketers—and instilling a cultural pride in the importance of these roles within the household and the community—is critical to translating increased food production into higher levels of household nutrition, echoing current debates around the roles of women in food sovereignty (Agarwal 2014).

In conclusion, this case study presents a model for putting into practice the tenets of food sovereignty to strengthen food security at the local and household levels, within the context of an international development project that works through a local community organization. It shows a possible structure that can be adapted to other contexts, that values local knowledge, encourages innovation as per the FAO State of Food and Agriculture Report 2014, and highlights processes that can address the acknowledged tensions of farmer choice and the role of trade in food sovereignty (Agarwal 2014).

REFERENCES

Agarwal, B. "Food sovereignty, food security and democratic choice: Critical contradictions, difficult conciliations." *The Journal of Peasant Studies* 41 no. 6 (2014): 1247–1268.

Altieri, M.A. "Agroecology: The science of natural resource management for poor farmers in marginal environments." *Agriculture, Ecosystems & Environment* 93 no. 1–3 (2002): 1–24.

Altieri, M.A. "Agroecology, small farms and food sovereignty." *Monthly Review* July–August (2009): 101–113.

Altieri, MA., and P. Koohafkan. "Enduring farms: climate change, smallholders and traditional farming communities." Penang, Malaysia: Third World Network, 2008. http://agroeco.org/wp-content/uploads/2010/11/Enduring-farms.pdf (Accessed April 22, 2012).

Altieri, M.A. and Victor M. Toledo. "The agroecological revolution in Latin America: Rescuing nature, ensuring food sovereignty and empowering peasants." *Journal of Peasant Studies* 38 no. 3 (2011): 587–612.

Baca, M., T. Liebig, M. Caswell, et al. 2013. *Revisiting the Thin Months in Coffee Communities of Guatemala, Mexico and Nicaragua.* Final Research Report. Calí & Vermont: CIAT & University of Vermont ARLG, 2013.

Bacon, C.M. 2008. "Confronting the coffee crisis: Can fair trade, organic, and specialty coffees reduce the vulnerability of small-scale farmers in Northern Nicaragua?" In *Confronting the Coffee Crisis: Fair Trade, Sustainable Livelihoods and Ecosystems in México and Central America.* Christopher M. Bacon, V. Ernesto Méndez, Stephen R. Gliessman, David Goodman, and Jonathan A. Fox. (Eds.). Cambridge: MIT Press, 2008, 155–178.

Bello, W. and M. Baviera. "Food wars." *Monthly Review* July–August (2009): 17–31.

Berkes, F. *Sacred Ecology*, 2nd ed. New York: Routledge, 2008.

Bernal Larrazabal, L. *Speech at GMCR Forum on Food Security in Coffee Regions.* Huehuetenango, Guatemala, October 25, 2012.

Burnett, K. and S. Murphy. "What place for international trade in food sovereignty?" *Journal of Peasant Studies* 41 no. 6 (2014): 1065–1084.

Caswell, M., V. Ernesto Méndez, and Christopher M. Bacon. *Food Security and Smallholder Coffee Production: Current Issues and Future Directions.* ARLG Brief No. 1. Burlington, VT: Agroecology and Rural Livelihoods Group (ARLG), University of Vermont, 2012. www.uvm.edu/~agroecol/?Page=Publications.html.

Cleveland, A.D. and D. Soleri. "Household gardens as a development strategy." *Human Organization* 46 no. 3 (1987): 259–270.

Declaration of Nyéléni. *Declaration of the Forum for Food Sovereignty.* Nyéléni Village, Sélingué Village: Mail, 2007. http://www.nyeleni.org/spip.php?article290 (Accessed January 15, 2014).

De Shutter, O. "The transformative potential of agroecology." In: *Food Movements Unite!*, E. Holt-Giménez (Ed.). Oakland, CA: Food First Books, 2011, 223–242.

Douwe ven der Ploeg, J. "Peasant-driven agricultural growth and food sovereignty." *Journal of Peasant Studies* 41 no. 6 (2014), 999–1030.

Eakin, H., A. Winkels, and J. Senzimir. "Nested vulnerability: Exploring cross-scale linkages and vulnerability teleconnections in Mexican and Vietnamese coffee systems." *Environmental Science & Policy* 12 no. 4 (2009), 398–412.

FAO. *Trade Reforms and Food Security: Conceptualizing the Linkages. Rome: Food and Agriculture Organization of the United Nations.* Rome: Food and Agriculture Organization of the United Nations, 2003. ftp://ftp.fao.org/docrep/fao/005/y4671e/y4671e00.pdf (Accessed April 14, 2012).

FAO. 2008. *An Introduction to the Basic Concepts of Food Security.* Rome: Food and Agriculture Organization of the United Nations, 2008. http://www.fao.org/docrep/013/al936e/al936e00.pdf (Accessed April 11, 2012).

FAO. *The State of Food Insecurity in the World: Addressing Food Insecurity in a Protracted Crisis.* Rome: Food and Agriculture Organization of the United Nations, 2010a. http://www.fao.org/publications/sofi/en/ (Accessed February 7, 2013).

FAO. *FAO Policy on Indigenous and Tribal Peoples.* Rome: Food and Agriculture Organization of the United Nations, 2010b. http://www.fao.org/docrep/013/i1857e/i1857e00.pdf (Accessed April 21, 2012).

FAO. *The State of Food Insecurity in the World 2012.* Rome: Food and Agriculture Organization of the United Nations, 2012. www.fao.org/news/ (Accessed October 10, 2013).

FAO. *The State of Food and Agriculture 2014.* Rome: Food and Agriculture Organization of the United Nations, 2014. http://www.fao.org/publications/sofa/en/?utm_source=faohomepage&utm_medium=web&utm_campaign=featurebar (Accessed December 24, 2014).

FIVIMS. *Food Insecurity and Vulnerability Information and Mapping System.* 2012. www.fivims.org.

Fox, J. and L. Haight. "Mexican agricultural policy: Multiple goals and conflicting interests." In *Subsidizing Inequality: Mexican Corn Policy Since NAFTA.* J. Fox and L. Haight (Eds.). Santa Cruz, CA: Woodrow Wilson International Center for Scholars, 2010, 9–50.

Friedmann, H. 2005. From Colonialism to Green Capitalism: Social Movements and the Emergence of Food Regimes. In Buttel, FH and McMichael, P., eds. *New Directions in the Sociology of Global Development: Research in Rural Sociology and Development,* Volume 1. Amsterdam, Netherlands: Elsevier.

Galhena, D.H., R. Freed, and K.M. Maredia. "Homegardens: A promising approach to enhance food security and well-being." *Agriculture & Food Security* 2 no. 8 (2013), 1–13.

Ghosh, J. "The unnatural coupling: food and global finance." *Journal of Agrarian Change* 10 no. 1 (2010): 72–86.

Gliessman, S.R. "The framework for conversion." In *The Conversion to Sustainable Agriculture: Principles, Processes, and Practices.* S.R. Gliessman and M. Rosemeyer (Eds.). Boca Raton, FL: CRC Press, 2010, 3–14.

Gomiero, T., M.G. Paoletti, and D. Pimentel. "Energy and environmental issues in organic and conventional agriculture." *Critical Reviews in Plant Sciences* 27 (2008): 239–54.

Holt-Giménez, E. "Measuring farmers' agroecological resistance after hurricane Mitch in Nicaragua: A case study in participatory, sustainable land management impact monitoring." *Agriculture, Ecosystems and Environment* 93 (2006a): 87–105.

Holt-Giménez, E. *Campesino a Campesino: Voices from Latin America's Farmer to Farmer Movement for Sustainable Agriculture.* Oakland, CA: Food First Books, 2006b.

International Coffee Organization. 2012. www.ico.org (Accessed November 14, 2014).

Jaffee, D. *Brewing Justice: Fair Trade Coffee, Sustainability, and Survival.* Berkeley, CA: University of California Press, 2007.

Loría Arana, A. *Speech at the Public Forum on Food and Nutritional Security and Sovereignty, August 12, 2012.* Matagalpa, Nicaragua: National Autonomous University of Nicaragua, 2012.

Marsh, Robin. "Building on traditional gardening to improve household food security." *Food, Nutrition and Agriculture* 22 (1998), 4–14.

Martínez-Torres, M.E. and P.M. Rosset. "Diálogo de Saberes in La Vía Campesina: Food Sovereignty and Agroecology." *Journal of Peasant Studies* 41 no. 6 (2014): 979–997.

Maxwell, D., R. Caldwell, and M. Langworthy. "Measuring food security: Can an indicator based on localized food coping behaviors be used to compare across contexts?" *Food Policy* 33 no. 6 (2008): 533–540.

McAfee, K. "Corn culture and dangerous DNA: Real and imagined consequences of maize transgene flow in Oaxaca." *Journal of Latin American Geography* 2 no. 1 (2003): 18–42.

McMichael, P. "Peasants make their own history, but not just as they please..." *Journal of Agrarian Change* 8 no. 2/3 (2008): 205–228.

Méndez, V.E., C.M. Bacon, and R. Cohen. "Agroecology as a transdisciplinary, participatory, and action-oriented approach." *Agroecology and Sustainable Food Systems.* 37 no. 1 (2013): 3–18.

Méndez, V.E., C.M. Bacon, M. Olson, K.S. Morris, and A. Shattuck. 2010. "Agrobiodiversity and shade coffee smallholder livelihoods: A review and synthesis of ten years of research in Central America." *Professional Geographer* 62 no. 3(2010): 357–376.

Pimbert, M. *Towards Food Sovereignty: Reclaiming Autonomous Food Systems*. London: International Institute for Environment and Development, 2009. http://pubs.iied.org/G02268.html (Accessed April 14, 2012).

Pretty, J. "Agroecology in developing countries: The promise of a sustainable harvest." *Environment* 45 no. 9 (2003): 8–20.

Pretty, J., A. Noble, D. Bossio, et al. "Resource-conserving agriculture increases yields in developing countries." *Environmental Science and Technology* 40 no. 4 (2006): 1114–1119.

Ruelle, M.L., Stephen J. Morreale, and Karim-Aly S. Kassam. "Practicing food sovereignty: Spatial analysis of an emergent food system for the standing rock nation." *Journal of Agriculture, Food Systems, and Community Development* 2 no. 1 (2011): 163–179.

United Nations Conference on Trade and Development. *Wake up Before It Is Too Late: Make Agriculture Truly Sustainable Now for Food Security in a Changing Climate* (UNCTAD Trade and Environment Report, 2013), 2013. http://unctad.org/en/pages/PublicationWebflyer.aspx?publicationid=666 (Accessed October 13).

Vanhaute, E. "Peasants, peasantries and (de-)peasantization in the capitalist world-system." In *Routledge Handbook of World-Systems Analysis*. S. Babones and C. Chase-Dunn (Eds.). London: Routledge, 2012.

Wilson, C. and C. Tisdell. "Why farmers continue to use pesticides despite environmental, health and sustainability costs." *Ecological Economics* 39 no. 3 (2001): 449–462.

Windfuhr, M. and J. Jonsén. *Food Sovereignty: Towards Democracy in Localized Food Systems*. (Warwickshire, United Kingdom: ITDG Publishing, 2005. http://www.ukabc.org/foodsovpaper.htm (Accessed April 21, 2012).

CHAPTER 13

The Mesoamerican Agroenvironmental Program
Critical Lessons Learned from an Integrated Approach to Achieve Sustainable Land Management

Isabel A. Gutiérrez-Montes and Felicia Ramirez Aguero

CONTENTS

13.1 MESOAMERICAN AGROENVIRONMENTAL PROGRAM WITHIN THE MESOAMERICAN CONTEXT

During the last decade, global environmental policy documents proposed a comprehensive agenda addressing social, economic, and environmental scenarios in an integrated fashion. The Millennium Development Goals (MDG), the different environmental conventions such as the Convention on Biological Diversity (CBD), the United Nations Convention to Combat Desertification and Land Degradation (UNCCD), and the United Nations Framework Convention on Climate Change (UNFCCC), as well as the recommendations of the Millennium Ecosystem Assessment (MEA), all recognize that the consideration of human needs and opportunities is essential to achieve conservation, restoration, healthy, and sustainable use of natural resources. As suggested by the MEA, economic, social, institutional, political, and environmental issues need to be addressed in a systemic manner to meet the MDG (MEA 2005).

Approaching the deadline to the MDG (year 2015), food insecurity at the global level is still a critical issue that has become a much greater threat to many poor people because of the latest world economic crises. The perceived slowdown of the economy, along with poor management of natural resources and persistent inequalities (social, economic, and cultural), have seriously affected achievement of the MDG, and more specifically the goal of reducing poverty in half by 2015. For example, between 2008 and 2009, and over very short periods, unprecedented increases in the prices of key products (staple foods) for human consumption (e.g., for wheat: 130%, rice: 74%, and corn: 53%) resulted in more than 100 million people who lost access to adequate food and nutrition, and Food and Agriculture Organization (FAO) estimates that 1.02 billion people around the world were undernourished, with 53 million located in Latin America and the Caribbean (FAO 2009).

In recent years, food insecurity in Mesoamerica has particularly affected poor indigenous and *campesino* or peasant farming communities, making it imperative to develop action plans to address food insecurity. These have been done through agroecological approaches to sustainable production (including agroforestry systems), ecosystem services, and rural business development with the goal of achieving poverty reduction in the territories of indigenous people and *campesinos*. The greatest percentage of population in poverty in Central America is located in Honduras (68.9%), Nicaragua (61.9%), Guatemala (54.8%), and El Salvador (47.9%) (PRESANCA II et al. 2011); reversing this trend has been complicated because the areas where vulnerability, poverty, and food insecurity are concentrated are quite diverse. Comparison of the physical, social, economic, and cultural characteristics of these areas with national averages reveals the existence of significant inequalities. However, strengths and weaknesses vary between them, and generic policies and strategies defined with a global and sectorial focus do not respond effectively to the particular needs and demands of these specific disadvantaged areas (IICA 2009). The absence of systematic processes that could help increase the resilience of these areas and thus lessen their social and ecological vulnerability is still a major issue in the region.

13.2 THE MESOAMERICAN AGROENVIRONMENTAL PROGRAM APPROACH

Simultaneously with the development and early implementation of regional strategies to join forces in an intersectorial initiative to face growing social and environmental challenges (e.g., Regional Agro-Environmental and Health Strategy—ERAS from its Spanish title and Central American Strategy for Territorial Rural Development—ECADERT from its Spanish title) (ERAS 2009; CAC SICA 2010), the Tropical Agricultural Research and Higher Education Center (CATIE), along with its main donors (governments of Sweden—SIDA and Norway—NORAD and MFA Norway) developed and implemented the proposal to establish a Mesoamerican Agroenvironmental Program (MAP): an ambitious international intersectorial initiative to develop, test, and communicate, in a diversity of rural territories, methodologies, technologies, and policies designed to integrate production and conservation (CATIE 2008; Gutierrez-Montes 2009). The ultimate goal of the MAP was to improve human well-being by promoting competitive strategies and practices for sustainable land management (SLM). This included a focus on farms and families (household), territory, national and regional levels, including CATIE as a learning institution, which enhanced equity and good governance at the landscape scale. The MAP was aimed to strengthen the institutional role of CATIE within the Mesoamerican region, including CATIE's focus on developing, validating, and communicating concepts, approaches, methods, and technologies in order to assist other organizations to take advantage of new opportunities as well as to resolve challenges in the rural sector.

The MAP integrated the sustainable livelihoods approach and the community capitals framework (Gutierrez-Montes et al. 2009) with a landscape (or territorial) approach, in order to achieve SLM, which aimed to improve the well-being of rural people. The program focused on the production, competitiveness, and environmental issues of the most important agricultural (coffee, cocoa, livestock, and vegetables) and natural resource (forests and water) sectors of the region. MAP identified the contexts (sources of shock and stress) and requirements (enabling environment) that determine how to integrate livelihoods and territorial approaches for the greatest benefits for the rural poor as well as for environmental conservation and management. Both supply- and demand-led mechanisms were used based on the value chain approach, acknowledging that there is no "one size fits all" for policies, technologies, methods and tools, and that appropriate, innovative combinations are needed to provide relevant responses for different contexts and situations.

The basic premises of the program includes (1) the development and use of SLM strategies and technologies can only be achieved with interdisciplinary or transdisciplinary (Francis et al. 2013; Mendez et al. 2013) interventions at all levels from the field to the Minister's office, and back and forth; (2) achieving a positive impact with the international environmental conventions (e.g., CBD, UNCCD, and UNFCCC) depends on applying them in agricultural and forest production as well as in conservation areas (i.e., in managed as well as protected areas); and (3) it is feasible to develop a positive feedback cycle or an upward spiral (Emery and Flora 2006), whereby implementing environmentally friendly and equitable agricultural and natural resource strategies can contribute to reducing poverty, which in turn contributes to reducing pressure on natural resources. This replaces the actual perverse downward cycle where inequality and environmental degradation contribute to greater poverty, leading to more pressure on natural resources and hence increased environmental degradation. Thus, the MAP was designed to contribute to improve environmental management while addressing major production and other concerns of farmers and families, local, national, and even regional organizations.

The MAP (its development and promotion) was an opportunity for an agroecological program to have an impact on mainstream markets, and not just in niche markets such as those provided by organic certification: through (1) more profitable as well as more sustainable land use systems, with an emphasis on quality, local transformation, and certification (and any other option to "add value" with the value chain approach); (2) quantification and valuation of ecosystem goods and services; (3) sustainable rural businesses, involving farmers' cooperatives and associations as well as private companies; (4) innovation, diversification, and low input agricultural and natural resource management technologies; (5) technologies and strategies to maintain and enhance capacities of local and regional populations to adapt to the expected effects of climate change; and (6) collaborative landscape management acknowledging that landscapes are created by mankind and acknowledging that they are a social construction.

The MAP was planned to contribute to the social and political processes that can improve Mesoamerican landscapes, creating a better future for their inhabitants, taking into account that many development experts have stressed the need to strengthen social capital in order that technological solutions can be adapted and implemented even after a project is completed (Flora and Flora 2013). The program aimed to contribute developing an enabling environment and the capacity (including advocacy coalitions in the form of active comanagement platforms) in the region to introduce more productive sustainable land and resource use via the immediate beneficiaries of the MAP who are local, national, and regional organizations. At the field and landscape levels, the projects that make up the MAP developed and tested conceptual and operational frameworks that articulate productive processes and value chains (including processing and marketing) with effective mechanisms for governance allowing the environmental sustainability of these productive schemes.

The MAP had a medium- to long-term perspective where the role of CATIE and its partners changed over time within the different sectorial (or value chain) and environmental (or territorial management) initiatives supported. Program focus included an innovations research focus (e.g., participatory quantification and valuation of ecosystem services resulting from the implementation of different certification schemes); an educational focus (e.g., provide opportunities for postgraduate students to be incorporated in interdisciplinary research for development teams); a training focus (e.g., use and promote participatory action research and training methods such as farmer field schools [FFS]); an entrepreneurial focus to assist in marketing (e.g., identify bottlenecks in value chains and facilitate links to the private sector); a communications focus (e.g., improve two way information flow in the policy–research interface and an effective knowledge management); and a coordination focus (e.g., management of different funding mechanisms as well as institutional collaboration to efficiently and effectively channel financial and human resources to partners).

The MAP also reinforced the capacities of CATIE to work at the research–policy interface contributing to the critical and participatory analysis, assessment and formulation of relevant local, national, and regional policies, as well as communicating results in formats and language appropriate for policy makers specifically in the form of policy briefs (see Gutierrez-Montes et al. 2012b; Laderach et al. 2010; Ramirez et al. 2012d; Rivas et al. 2012a, 2012b; Scheelje et al. 2011; Soares et al. 2012).

The MAP proposal and implementation also reflected the interest and accumulated experience of CATIE and its principal donors (Norway and Sweden) in managing regional projects within which greater integration of resources (or community capitals, including human, social, financial, and political), activities, and knowledge, to guarantee efficient and effective use of the funds provided by International Cooperation, was one of the goals: this interest is directly related to the international agreements to harmonize and align development aid (Paris Declaration, etc.) as well as the recognition of the value of more integrated and collaborative initiatives to address the complex persistent problems faced by the region (Table 13.1).

Research and development work were carried out by CATIE and its partners, with local groups, in two key territories and several pilot zones in the Mesoamerican countries; however, communication of the results has been serving a much wider clientele being carried out by an extended set of local, national, and regional organizations. This new intersectorial program covered a broad range of activities and partners seeking to establish a comprehensive agroenvironmental approach as the basis for rural development in Mesoamerica. The program was prepared specifically for Mesoamerica but after four years of experience and lessons learned, its concepts could easily be extended to other regions, not only in Latin America, but elsewhere, and considering clear opportunities to facilitate south–south exchanges.

Table 13.1   MAP Main Objectives and Results

Development Objective

Mesoamerican societies use SLM strategies that provide ecosystem goods and services that reduce rural poverty.

Program Objective

Local, national, and regional organizations implement SLM technological innovations, policies, and programs.

Main Results				
Result 1. Rural families and farmers organizations in Mesoamerican priority zones adopt sustainable production and natural resource management practices and are integrated into value chains.	Result 2. Local governments implement effective environmental and governance mechanisms.	Result 3. National organizations and decision makers use the production technologies and natural resource management experiences generated by the MAP.	Result 4. Mesoamerican organizations and decision makers use the knowledge, tools, and recommendations from the MAP.	Result 5. CATIE enhances its capacities to collaborate with and support local, national, and regional partners in designing and implementing effective strategies and policies.

Source: CATIE (Centro Agronómico Tropical de Investigación y Enseñanza). "Implementation proposal for the "Mesoamerican Agroenvironmental Program (the MAP)." (Turrialba, Costa Rica: CATIE, 2008).

SLM, sustainable land management; CATIE, Tropical Agricultural Research and Higher Education Center; MAP, Mesoamerican Agroenvironmental Program.

13.3 GEOGRAPHICAL FOCUS WITHIN MESOAMERICA: SECTORIAL AND/OR TERRITORIAL APPROACHES?

MAP was planned to support participatory action research for development work throughout the Mesoamerican region (Belize to Panama), although greater emphasis was given to Nicaragua, Honduras, and Guatemala, because of the greater need (concentration of poverty) and hence donor priority with respect to these three countries. Regional actions were the backbone of the MAP (key territories with a territorial approach), but the foundation stones were exemplary landscapes and farmers' organizations chosen in each country, together with national and regional partners (pilot areas with sectorial approach) (Figure 13.1; Table 13.2).

Pilot areas refer to those zones were MAP projects (coffee, cocoa, vegetables, livestock, and watersheds) conducted work to fulfill specific objectives. Promotion of integration, collaboration, and synergies were strongly recommended but not a requisite; however, there were five requisites for the recognition of a pilot area: (1) importance and potential at national level for the sectorial approach (coffee, cocoa, livestock, vegetables, and water); (2) institutional support and internal (CATIE)/external (local level) critical mass to interact; (3) presence and expressed interest of key partners (local governments, producers cooperatives and associations, NGOs, universities, and committees); (4) a minimum of infrastructure (access roads); and (5) limited identifiable area (municipality or subwatershed).

MAP key territories were defined and conceptualized as ample geographical areas where different projects, initiatives, and CATIE units worked together in a coordinated and integrated manner, leading to a participatory development of systemic solutions (including rural families' livelihoods, territorial approach to environmental issues, and value chain approaches) to the complex and changing challenges in the rural areas.

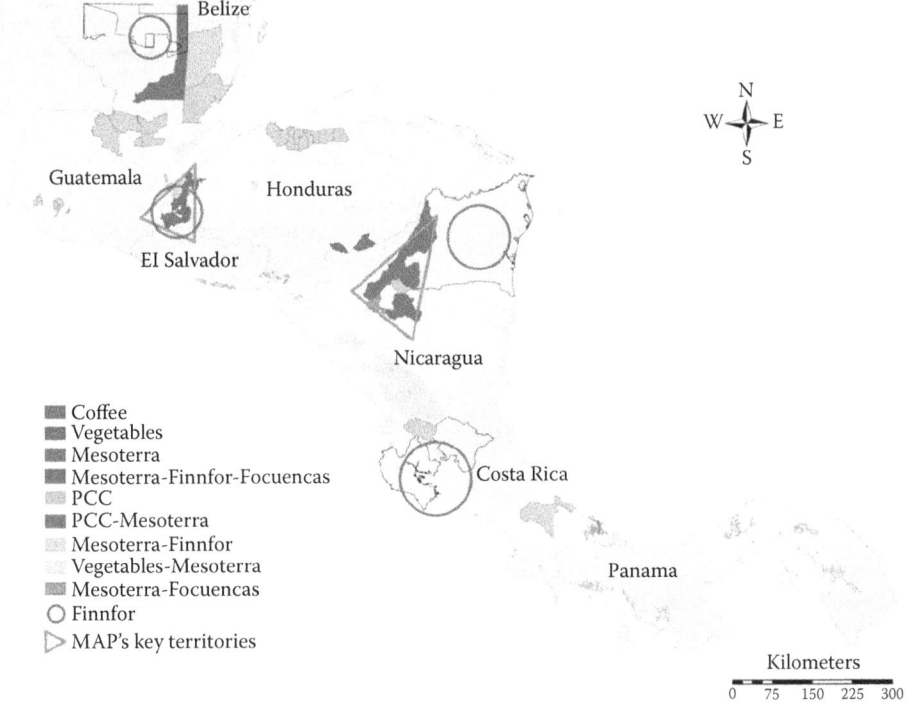

Figure 13.1 MAP key territories and pilot zones.

Table 13.2   Contrasting Characteristics of Sectorial (Coffee, Cocoa, Vegetables, Livestock) and Territorial Approaches to Address Poverty and Environmental Problems in Rural Areas

Sectorial Approach	Territorial Management Approach
Key zones for the sector (e.g., coffee, cacao, livestock, and vegetables) are identified using national statistics on production, beneficiaries, potential, and land areas.	Key zones are chosen using environmental and social statistics such as natural resources of the territory, levels of degradation, importance for water yields, and/or for biodiversity conservation, and social capital.
Focus principally on production chain and brings in local, national, and international actors mostly from the private sector.	Focus principally on environmental governance and brings in local and national actors from the public as well as private sector (e.g., municipalities, central governments, society at large, indigenous communities, academia).
Focus principally on factors limiting production and quality of specific species, for example, how to add value for all actors in the chain (market focus).	Focus principally on factors limiting ecosystem services, for example, how to protect/improve the integrity of the services of the ecosystem and add value for land users through the sale of such services.
Emphasis on strengthening farmer's cooperatives and associations.	Emphasis on strengthening multistakeholder platforms (governance) such as water councils, model forests, and biological corridor councils.
Responds to and is consistent with national sectorial programs (e.g., PRORURAL, Nicaragua) and the formation of clusters (e.g., Cacao, Nicaragua).	Responds to and is consistent with international and regional environmental conventions (UNCCD, biodiversity, climate change, MBC) and national plans in compliance with these conventions.
Strong regional networks (e.g., PROMECAFE) have significant influence in each country and are key components of some CATIE projects (e.g., Proyecto Cacao Centroamerica-PCC).	Regional linkages exist at the political level (CCAD, ERAS, ECADERT) but have a growing influence in each country. Networks are a key component of CATIE's initiatives (RIBM, FOCUENCAS, Mesoamerican Biological Corridor)
Facilitates communication of technologies to specific groups of farmers with similar needs; i.e., horizontal scaling out (e.g., Fedegan, Colombia).	Accelerates local innovation in the management of Natural Resources through the participation of local/national authorities in multistakeholder platforms.
Critical mass of the local, national, or even regional unions (*gremios*) facilitates public private partnerships.	Public funding, including significant international programs (GEF, MDL, bilateral Nordic countries, etc.) together with local commitment, facilitates medium- to long-term interventions.
Some sectors have considerable political influence at the national level (e.g., coffee, livestock).	Growing political influence at local (municipal) and national levels (e.g., importance of safeguarding water supplies) but also at international levels (biodiversity, climate change).
Facilitates the distribution and use of CATIE germplasm (e.g., coffee and cocoa).	Facilitates the distribution and use of CATIE's methodologies and decision-support models to achieve sustainable land management.
Consistent with CAFTA (already signed by Central American countries).	Consistent with the principal environmental conventions (already signed by Central American countries).
Contributes to regional economic integration, for example, Central American countries collaborating to sell the image of quality, certified, sustainably produced coffee or cocoa.	Contributes to regional environmental integration, for example, shared watersheds and biodiversity are strong motivators to develop collaborative projects for themes where competition/national interest groups do not limit implementation.
Focus on specific farmers' unions (*gremios*) [chamber of commerce].	Focus on a broad group of landscape stakeholders.
No territorial limits predefined (apart from ecological limitations for each crop/pasture).	Loosely or strictly predefined territorial limits; key stakeholders define the area of interest.
Emphasis of regulations on quality, innocuity, and certification of specific products.	Emphasis on local agreements/institutions as well as regulations safeguarding ecosystem services

(Continued)

Table 13.2 (*Continued*) Contrasting Characteristics of Sectorial (Coffee, Cocoa, Vegetables, Livestock) and Territorial Approaches to Address Poverty and Environmental Problems in Rural Areas

Sectorial Approach	Territorial Management Approach
Segregated approach.	Holistic approach.
Specific interest groups/limited diversity of stakeholders (sectorial).	Broad range of interest groups/diverse stakeholders (intersectorial).
Usually organized to enhance private goods, financial capital.	Usually organized to protect public goods and cultural/social/political capital

The purpose of this listing is not to attempt to show which approach is correct since both are needed. This listing seeks to identify potential strengths and weaknesses of sectorial and territorial approaches as a basis for improving impact when they are used together.

CAFTA, Central American Free Trade Agreement; CATIE, Tropical Agricultural; Research and Higher Education Center; ERAS, Regional Strategy on Agroenvironment and Health; ECADERT: Central American; MBC, Mesoamerican Biological Corridor; CCAD, Environment and Development Central American Commission; RIBM, Ibero American Model Forests Network; GEF, Global Environmental Fund; MDL, Clean Development Mechanisms; UNCCD, United Nations Convention to Combat Desertification and Land Degradation.

Two key territories, where the MAP supported the development of new governance mechanisms, technologies, and so on, were the pilot areas where CATIE projects, formerly financed by Norway and Sweden, have already developed valuable experiences over a period of time; in other words, the MAP did not "start from scratch." These territories included Trifinio (trans-frontier region including parts of Honduras, Guatemala, and El Salvador) and the Central region of Nicaragua, which are representative of large areas of Central America (e.g., of the Northeast of Guatemala and the Southeast of Nicaragua). These exemplary territories hinged on the number of partners and resources available for the MAP, taking into consideration potential synergies with the existing or planned projects of the possible partners as well as the priorities of national and regional organizations. Key territories selected fulfill at least the following requirements: (1) located within priority areas identified as such in regional (e.g., Mesoamerican Biological Corridor, ERAS, ECADERT) or national strategies; (2) stakeholders have expressed their interest to become one of the key territories and have proven leadership and social capital; (3) potential to support the development of MAP main components (value chains, ecosystem services, and adaptation to climate change); and (4) correspond to CATIE's expertise and located in areas where there is potential for synergies within CATIE and/or with other partners.

Value added from the coordinated and integrated work in the two key territories included the more efficient use of resources (not only financial but human, social, and political capital), consolidation of a single and integrated image for a regional organization as CATIE (working in education, research, and outreach) and moreover an improvement in the deliverables and products.

13.4 CRITICAL LESSONS LEARNED

After four years of implementation of an innovative and urgently needed integrated program as MAP, several lessons are left behind and are effectively used to look forward. The first one is the importance of the agroecological approach (central in the conception and implementation of MAP FFS) as a practical example of a "*transdisciplinary, participatory, and action oriented approach*" (Mendez et al. 2013). The second one is related to the importance of research "in" and research "for" development (Coe et al. 2014) for scaling process (up and out) as the main reasons to promote integration and coordination toward sustainable land management. Last but not least, including gender considerations (encompassing inclusion and equity) for working toward sustainability of an agroecosystem, such that "gender made a difference in terms of access and control over key resources—financial, human, natural, and social capital—is critical for project success" (Flora 2001).

13.5 FARMER FIELD SCHOOLS: STARTING POINT
TOWARD TRANSDISCIPLINARY PAR

According to Flora and Flora (2013), in the past, key stakeholders, including people from the communities, local and external organizations and governmental agencies and institutions, have been focusing their activities on financial and physical/built capital. This often resulted in the decapitalizing of social and natural capital, whereas the authors argue that a systemic focus for development that includes local recognition and commitment might balance these aspects. The vision of a healthy community in the rural context of Latin America is much more than just economic (financial capital) and/or infrastructural investments (built capital). A healthy community reinforces connections and relationships (social capital), respect for and inclusion of different values, points of view and knowledge (cultural capital), access to different levels of power (political capital), sustainable use and care of ecosystems goods and services (natural capital), and development of local skills and knowledge (human capital), all in a synergy that can enhance the overall well-being of individuals and households within the communities. This will, in the end, allow the community to ensure actions toward a healthy managed landscape (Emery and Flora 2006; Gutierrez-Montes et al. 2009).

The MAP, managed in a constructivist and participative manner, was designed to strengthen the livelihoods (and all the community capitals) of rural communities by increasing knowledge and, as a result, help them to develop their territories in a sustainable way. One of the main innovations of the MAP relies on the efforts to establish FFS as the central tool to tackle knowledge management and create capacities at all levels (Figure 13.2).

The MAP recognizes FFS as a new extension method with a participatory action research approach, initially conceived as an approach directed toward the generation of a learning environment in which participants can learn, share, and apply more and better knowledge and skills (human capital) to improve their farms. Knowledge sharing, wisdom dialogue, and reinforcement of cultural capital within the FFS also allows participants to build and strengthen their social capital and conserve the natural capital through the use of innovative sustainable land use management practices. Additionally, the process seeks to empower participants to increasingly take part in local, national, and local decision-making structures (political capital) (Figure 13.3).

Figure 13.2 Cocoa farmer field school-Guatemala.

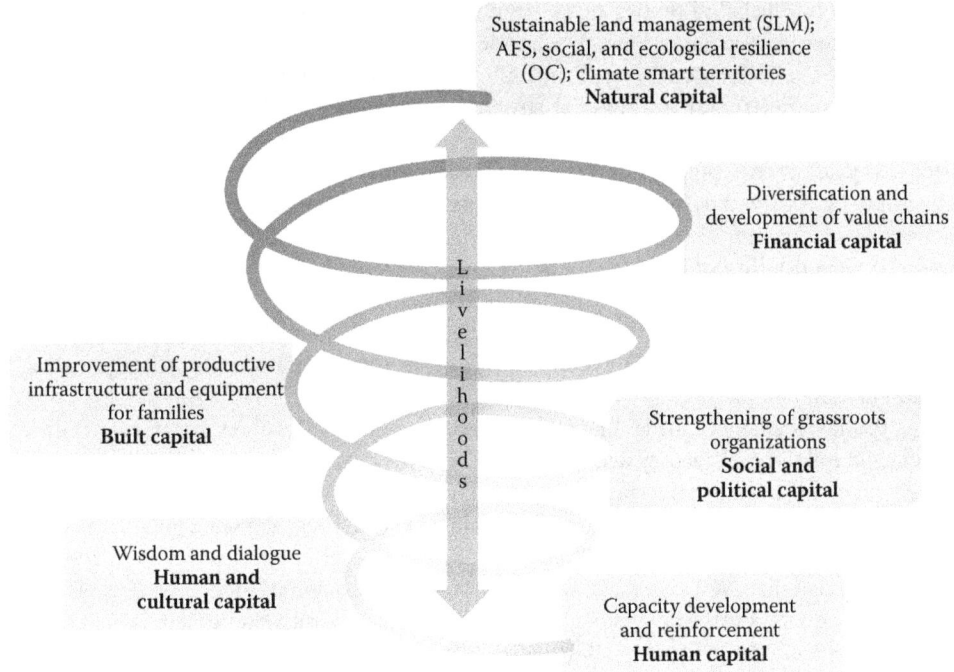

Figure 13.3 Spiraling of capital assets.

Within the program, FFS was the principal tool proposed for the development of capacities at the level of technicians and agricultural promoters (producer leaders/facilitators) as well as rural families (Gutierrez-Montes and Siles 2011; Gutierrez-Montes et al. 2011; Gutierrez-Montes et al. 2012a), so that they could promote and develop agroecological management innovations toward sustainable land management for their particular region. FFS were considered, within MAP, as a key development entry point to promote upward spirals or virtuous cycles of development toward the reduction of rural poverty, while improving the well-being of rural populations that are currently facing increased financial, ecological, and social challenges (Gutierrez-Montes and Siles 2011; Gutierrez-Montes et al. 2012a; Ramirez, Gutierrez-Montes, and Bartol 2012) (Figure 13.3).

13.6 INTEGRATION AND COORDINATION: A MUST TO ACHIEVE AND INSTITUTIONALIZE SUSTAINABLE LAND MANAGEMENT

A major characteristic of the MAP was the focus on integrated projects, resources, organizations, and knowledge, seeking efficiency and impact at different levels through targeted interventions that assisted local, national, and regional partners, to take advantage of new opportunities as well as resolve agroenvironmental problems. The MAP aimed to achieve this integration by planning and implementing, together with a wide range of partners, specific activities in three thematic areas: (1) adaptation to climatic change; (2) ecosystem services (including a strong focus on mitigation, CDM, REDD, etc.); and (3) markets and value chains. The combination of the work in these three areas allowed the MAP to make significant contributions to the all-encompassing theme of SLM and to our overriding goal of improving livelihoods of rural communities.

Despite the efforts to promote exchanges between projects and coordinate activities of different donors and agencies in the region, inefficiencies and missed opportunities can easily be identified.

Initiatives to resolve such limitations of existing research and development programs have been undertaken at international (e.g., Paris Declaration on Harmonization and Alignment), regional (e.g., ERAS, ECADERT), national (e.g., Prorural in Nicaragua), and institutional levels (e.g., establishment of a more horizontal institutional structure, based on interdisciplinary collaboration of CATIE's scientific programs).

After four years of experience, it is recognized that widespread impact of a program such as the MAP can only be achieved by scaling-up and scaling-out of successful experiences as well as lessons learned (including not very successful ones) through organizations mandated to this role. The selection and responsibility of local, national, and regional organizations, for specific components or activities of the MAP, as well as being part of the team guiding this program, was a starting point considering institutional anchoring. CATIE constructed such a foundation for institutional anchoring during the development of the FOCUENCAS II project and more recently when planning the PCC (regional cacao project).

During the development of the MAP proposal, the selection of pilot areas and key territories and the implementation of participatory action research activities, a large number of organizations and individuals were consulted considering their potential contribution to this need. At a national level, these organizations included farmers' organizations (e.g., Chorti Fresca in Guatemala), the National Agricultural Research Institutions (e.g., National Center For Agricultural and Forest Technology in El Salvador), national NGOs (e.g., Nitlapan in Nicaragua), national networks (e.g., Honduran Network for Broadleaf Forest Management in Honduras), national universities (e.g., Universidad de San Carlos en Guatemala), and national agencies (e.g., National Forestry Institute in Guatemala). CATIE's National Technical Offices had a central role in facilitating the links between CATIE programs and projects with national partners. Scaling-up at a regional level was directed through collaboration with regional organizations such as International Union for the Conservation of Nature, FAO, and the multilateral banks (e.g., Central American Bank for Economic Integration, Interamerican Development Bank, and World Bank), but moreover through the active participation and central role within ERAS and ECADERT. Scaling-up and scaling-out at a regional level also included the Central American network of cacao-producing organizations (Central American Cacao Project-PCC) and the Ibero American Network of Model Forests (RIABM) which have developed out of the Regional Network of Model Forests for Latin America and the Caribbean (LAC-NET).

Despite the advances during the implementation, an important lesson learned highlighted that it is not possible to construct a program like MAP through the integration ("*on the move*") of existing projects and initiatives. History, acquired commitments "with partners" at different levels, monitoring and evaluation systems (including reporting and budgeting), thematic and geographical priorities, ways of knowing and doing from the technical personnel, and many other uneven and important facts still limit the possibilities of a truly sustainable integration and oppose a real collaboration and shared vision between initiatives constructed with a territorial vision. Notwithstanding improvements in terms of coordination and integration (internally, within MAP projects and units and CATIE scientific programs, and externally, with partners at all levels: local, national, and regional) this is a topic identified as urgent to attend and resolve to advance and consolidate efforts toward sustainable land management scaling up and out.

13.7 EQUITY AND INCLUSION FOCUS: GENDER, ETHNICITY, AND AGE

Educational and capacity-building processes (within FFS) aimed at both male and female producers, have shown to be key for the economic and social development of rural families. MAP actions were focused on indigenous and *campesino* communities, and in particular on the vulnerability of women, especially considering exposure to climatic shocks and stresses, such as drought or flooding, that have repeatedly struck the Central American region. The impact of climate change on the members of a community is not consistent—not even on members of the same family because individuals possess different abilities and resources (information, rights, power, among others) and/or face cultural norms or rules that limit their capacity to handle a crisis situation (resilience) (Segnestam 2009). The Intergovernmental Panel on Climate Change (IPCC 2007a,b) concluded that the vulnerability of a person to climate change depends on gender and relationships, and that women in developing countries are among the most vulnerable groups. Studies about the vulnerability of specific sectors or populations have been carried out in several countries, but very few of them recognize the different roles that men, women, young people, and older adults within these sectors have in the economies and in the management of natural resources at the familial, territorial, and national levels. This makes it important to identify these roles to determine levels of vulnerability associated with different gender and livelihood strategies (Agrawal and Gibson 1999) and justify the necessity to consider and analyze context-specific conditions, such as social differences, gender, ethnicity, or origins (migration), altogether important factors in understanding vulnerability to environmental and socioeconomic stresses.

The MAP helped CATIE and partners achieve a significant advance in the incorporation of gender considerations, by making gender issues one of the central foci of research, development, training, and communication supported by the agroenvironmental program.

Since its inception phase, MAP promoted several initiatives including the design of a strategy to enable the consideration and addition of gender and equity aspects within projects and MAP units (Siles, Gutierrez-Montes, and Ramirez 2012), comprising: (1) increased consciousness of stakeholders, including CATIE and MAP personnel; (2) emphasis in promoting the progressive inclusion of women in the access of benefits from the project; (3) recording and management of gender-differentiated information for monitoring and evaluation; and (4) systematic documentation of successful experiences and ample distribution of documents (publications) toward and effective management of collectively produced knowledge.

Diagnoses of new pilot zones or sectors comprised the documentation and analysis of the knowledge, opinions, roles, and situations of men and women separately, recognizing their different interests, aspirations, experiences, and actual levels of participation in family decisions, productive activities, access to resources, and community organizations (formal and nonformal) (Figure 13.4).

Dynamic design and planning of the MAP projects encompassed training and continuous adjustments to guarantee the participation of all family members (with a focus on women and young

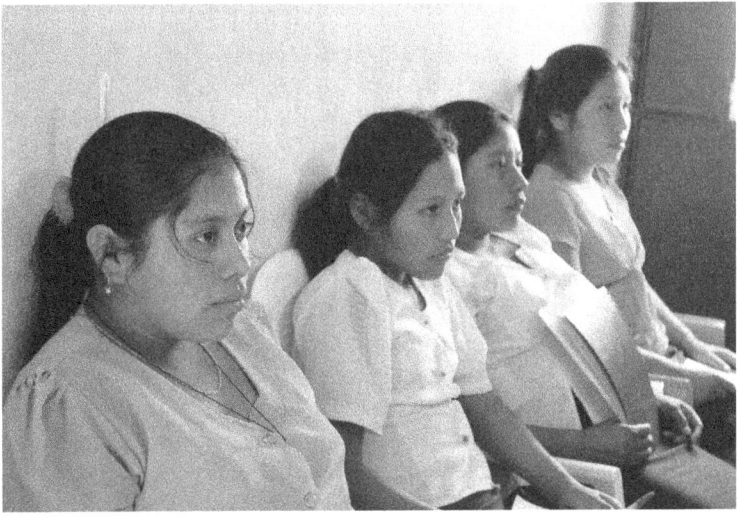

Figure 13.4 Women from a community organization.

people), seeking to establish a balance between numbers of men and women project staff, partners of MAP projects (e.g., balanced numbers of men and women beneficiaries within communities), and attention to different necessities and solutions (for men and/or women).

Training of the MAP and partner staff on gender issues, based on CATIE's gender policy (CATIE 1996) and MAP gender strategy (Siles, Gutierrez-Montes, and Ramirez 2012), sought equal inclusion of men and women in all the MAP activities, starting with training courses on communication with inclusive language and the production and distribution of training materials (Calvo 2013; Mora 2013). In terms of formal education and research, MAP was central in the development of thesis projects that included a focus on gender issues as part of the sustainable management of agriculture and natural resources (Martinez 2012; Posada 2012; Toruño 2012).

Inclusion of gender aspects in MAP research and development projects with the ultimate aim of providing equal opportunities for all family members (e.g., with respect to MAP education and training activities); equal opportunities to generate and manage income (e.g., facilitating the management of resources by women's groups); equal access to productive activities and resources (e.g., value added activities); and in general equal participation in family and community activities, had an impact related to (1) opening of spaces toward rural women empowerment through FFS, and reinforcing their human, social, and political capital (Gutierrez-Montes et al. 2012; Ramirez et al. 2012c); and (2) an increased participation of women in projects activities under two considerations: (i) women, men and young people have equal access, opportunities, benefits, and economic, social, and cultural responsibilities within projects; and (ii) use of the family concept as an integrating category that promotes equal participation of persons in productive, reproductive, and community activities (Ramirez et al. 2012b,c; 2013).

There are also important advances regarding awareness of different stakeholders, including MAP personnel and partner organizations at different levels, starting with the publication and ample distribution of technical documents (Gutierrez-Montes and Ramirez 2013; Padilla et al. 2013; Ramirez et al. 2012b,c), policy briefs (Ramírez et al. 2012d; Rivas-Platero et al. 2012a,b), and research related to gender and development, as well as the implementation of strategies and the use of methods and tools to advance in its practical application (Gutierrez-Montes and Ramirez 2013).

MAP also considered the development of impact, effect, and process gender indicators for the monitoring and evaluation scheme of the MAP, for example, numbers of female-headed vs. male-headed households with increased income (due at least partially to the MAP's activities); percentage of women in leadership positions; and number of community initiatives/projects that incorporate gender aspects, respectively.

Out-migration of labor, especially youth, is a contextual condition that historically has been affecting the success and impact of an ambitious and innovative program as MAP, but which is largely beyond the program's control. Nevertheless, out-migration (permanent and circular) is closely related with livelihood strategies in rural areas of Mesoamerica. According to the two evaluations of the program (midterm and final), MAP achieved a positive impact on livelihoods in rural areas (a basic premise of the program), increasing incentives for the rural population to remain in their communities. MAP projects, such as cacao (PCC), vegetables, coffee, and livestock (Mesoterra) had a clearly defined and specific emphasis on education and training at different levels (but especially within FFS), thus offering rural youth new opportunities, perspectives, and rewards from maintaining rural life styles (Figure 13.5).

MAP had a clear goal of reducing poverty in rural areas, one of the key factors (along with security issues and climate conditions) involved in the decision to migrate. One advantage of the decision to use the sustainable livelihoods approach and the community capitals framework within the implementation of the MAP pointed to ensuring a clear focus on incentives (tangible and intangible) for the rural population to remain in their communities. The MAP experience with FFS showed a key role of these efforts of capacity building of women, men, and young people (human capital) toward the generation of a learning and enabling environment (social capital) that resulted in improved learning, sharing, and application of more and better knowledge and skills to improve their natural

Figure 13.5   Farmer field school-Honduras.

capital. Knowledge sharing, including the recognition and celebration of local and traditional knowledge, or wisdom dialogues (cultural capital) within FFS, allowed participants to build and strengthen relationships (social capital) and conserve the ecosystem services (natural capital) through the use of innovative sustainable land management practices, such as agroforestry systems and other agroecological approaches. Additionally, the process intended to empower participants (especially historically marginalized and alienated groups, such as women and young people) to increasingly take part, not only in family but community organizations (cooperatives and associations) and decision-making structures (political capital). Finally, there is also a perceived enhancement of the productive infrastructure (physical capital) and an increase of income and/or savings (financial capital) coming from the improvement of the productivity, not only benefiting men, but women and youth (Figure 13.3).

13.8 CONCLUDING REMARKS

Innovative, integrated, and transdisciplinary efforts like the MAP have the potential to promote environmental and socioeconomical sustainable changes in rural areas. With a capacity building focus, which was the reason behind the FFS, and based on gender inclusion and equity considerations, reversing the vicious circle of poverty and natural resources degradation is possible; and an upward spiral toward sustainable land management can be started. Despite the recognized advances in terms of integration and coordination, basic premises within processes of scaling-up and out, there are still gaps to bridge to anchor and institutionalize the results and lessons learned from this research *for* and *in* a development program.

REFERENCES

Agrawal, A., and C.C. Gibson. "Enchantment and disenchantment: The role of community in natural resource conservation." *World Development* 27 no. 4 (1999): 629–649.
CAC (Consejo Agropecuario Centroamericano), SICA (Sistema de Integración Centroamericana). "Estrategia Centroamericana de Desarrollo Rural Territorial 2010–2030." (San Jose, Costa Rica: ECADERT, 2010).
Calvo, Y. 2013. "Redacción con Lenguaje Inclusivo." (Turrialba, Costa Rica: CATIE, 2013).

CATIE (Centro Agronómico Tropical de Investigación y Enseñanza). "CATIE's gender policy." (Turrialba, Costa Rica: CATIE, 1996).

CATIE (Centro Agronómico Tropical de Investigación y Enseñanza). "Implementation proposal for the "Mesoamerican Agroenvironmental Programme (the MAP)." (Turrialba, Costa Rica: CATIE, 2008).

CBD (Convention on Biological Diversity). *The Convention on Biological Diversity.* Montreal, Canada. 1992. http://www.cbd.int/convention/ (Accessed February 20, 2014).

Coe, R., F. Sinclair, and E. Barrios. "Scaling up agroforestry requires research in rather than for development." *Current Opinion in Environmental Sustainability* 6 (2014): 73–77.

Emery, M., and C. Flora. "Spiraling-up: Mapping community transformation with community capitals framework." *Community Development* 37 no. 1 (2006): 19–35.

ERAS (Estrategia Regional Agroambiental y de Salud). "Estrategia Regional Agroambiental y de Salud: Un Instrumento Estratégico de la Integración Regional-Centroamérica 2009–2024." (San Jose, Costa Rica: CAC, COMISCA; CCAD, SICA, 2009).

FAO (Food and Agriculture Organization). "The State of Food Insecurity in the World: Economic Crises— Repercussions and Lessons Learned." (Rome: FAO, 2009.)

Flora, C. B. "Access and control of resources: Lessons from the SANREM CRSP." *Agriculture and Human Values* 18 (2001): 41–48.

Flora, C. B., and J. Flora. *"Rural Communities: Legacy and Change."* (Boulder, CO: Westview Press, 2013).

Francis, C., T.A. Breland, E. Østergaard, G. Lieblein, and S. Morse. "Phenomenon-based learning in agroecology: A prerequisite for transdisciplinary and responsible action." *Agroecology and Sustainable Food Systems* 37 no. 1 (2013): 60–75.

Gutierrez-Montes, I. "Programa Agroambiental Mesoamericano: Una Estrategiapara la Coordinación y Acción Conjunta Hacia el Desarrollo Sostenible." *Recursos Naturales y Ambiente* 56–57 (2009): 4–7.

Gutierrez-Montes, I., M. Emery, and E. Fernandez-Baca. "The sustainable livelihoods approach and the community capitals framework: The importance of system-level approaches to community change efforts." *Community Development* 40 no. 2 (2009): 106–113.

Gutierrez-Montes, I.A., and J. Siles. *"Achieving Sustainable Land Use in Rural Mesoamerica: Mesoamerican Agroenvironmental Program (MAP)."* Joint Newsletter: IACD and CDS, VII: 5 (2011).

Gutiérrez-Montes, I.A., D. Padilla, and G. G Rivas. "Experiencia de Investigación Agrícolapara el Desarrollo de las Escuelas de Campo (ECAS): Una Apuesta Innovadora Hacia la Investigación, Acción Participativa del Programa Agroambiental Mesoamericano (MAP) en Trifinio." (Turrialba, Costa Rica: CAITE, 2011).

Gutierrez-Montes, I.A., P. Bartol de Imbach, F. Ramirez, J. Lopez Payes, E. Say, et al. "Las Escuelas de Campo del MAP-CATIE: Práctica y Lecciones Aprendidas en la Gestión del Conocimiento y la Creación de Capacidades Locales para el Desarrollo Rural Sostenible." Serie técnica, Boletín Técnico, No. 52. 1st ed. (Turrialba, Costa Rica: CAITE, 2012a).

Gutierrez-Montes, I.A., D. Soares, M. Thiebault, G.G. Rivas-Platero, G. Pinto, F. Ramirez Agüero, R. Romero, and R. Lopez. "Vulnerabilidad Social ante el Cambio Climático: Retos y Propuestas de Políticas Desde un Enfoque de Equidad Social." Síntesis para decisores. PB 15. (Turrialba, Costa Rica: CAITE, 2012b).

Gutierrez-Montes, I.A., and F. Ramirez. "Herramientas para el Análisis de Género en el Ciclo de los Proyectos: Listas de Verificación e Instrumentos de Análisis." Serietécnica, Boletín Técnico, No. 63. 1st ed. (Turrialba, Costa Rica: CATIE, 2013).

IICA (Instituto Interamericano de Cooperaciónpara la Agricultura). "Seguridad Alimentaria en las Américas: Se Necesita un Nuevo Modelo de Desarrollo." (San Jose, Costa Rica: IICA, 2009).

IPCC (Intergovernmental Panel on Climate Change). *"Fourth Assessment Report (Synthesis Report),"* 2007a. http://www.ipcc.ch/publications_and_data/publications_ipcc_fourth_assessment_report_synthesis _report.htm (Accessed February 20, 2014).

IPCC (Intergovernmental Panel on Climate Change). *Climate Change 2007: The Physical Science Basis. Contribution of Working Group I to the Fourth Assessment Report of the Intergovernmental Panel on Climate Change.* (Cambridge: Cambridge University Press, 2007b).

Laderach, P., J. Haggar, C. Lau, A. Etzinger, O. Ovalle, M. Baca, A. Jarvis, and M. Lundy. *"Mesoamerican Coffee: Building a Climate Change Adaptation Strategy."* Policy brief no. 2 (Managua, Nicaragua: CIAT, 2010).

Martinez, V. "Mujer, Manejo de la Agrobiodiversidad y su Relación con los Medios de Vida en dos Localidades del Municipio de San Juan Cancuc, Chiapas, México." MSc Thesis. (Turrialba, Costa Rica: CATIE, 2012).

Mendez, V.E., C. M. Bacon, and R. Cohen. "Agroecology as a transdisciplinary, participatory, and action oriented approach." *Agroecology and Sustainable Food Systems* 37 no. 1 (2013): 3–18.

Millennium Ecosystem Assessment. *"Living Beyond Our Means: Natural Assets and Human Well Being,"* 2005. http://www.millenniummassessment.org/documents/document.429.aspx.pdf (Accessed February 20, 2014).

Mora, E. "El LenguajeInclusivo en Textos Técnicos y Científicos: Guíapara Personal Técnico del CATIE, y Organizaciones Socias, Promotores y Promotoras," 1ª. Ed., 2012, Serie técnica. Boletín técnico no. 68. (Turrialba, Costa Rica: CATIE, 2013).

Padilla, D., L. Hernández. F. Ramirez, I.A. Gutiérrez-Montes. "Buenas Prácticas de Género a Nivel Organizacional." 1ª. Ed. Serietécnica. Manual técnico no. 117. (Turrialba, Costa Rica: CATIE, 2013).

PESA, IALCSH, PRESANCA II, PRESISAN, FAO, AECID, SICA. "Centroamérica en Cifras: datos de seguridadalimentaria, nutricional y agricultura familiar," 2011. http://www.fao.org/fileadmin/user_upload /AGRO_Noticias/docs/CentroAm%C3%A9ricaEnCifras.pdf (Accessed February 28, 2015).

Posada, K. "Impacto del Sistema Agroforestal Kuxur Rum en la Sostenibilidad de los Medios de Vida de las Familias Rurales en Camotán y Jocotán, Guatemala." MSc Thesis. (Turrialba, Costa Rica: CATIE, 2012).

Ramírez Agüero, F., I. A. Gutierrez-Montes, and P. Bartol de Imbach. "Las Escuelas de Campo del MAP: Dialogo de Saberes Hacia el Empoderamiento de las Familias Rurales." Serie divulgativa no. 12. 1 ed. (Turrialba, Costa Rica: CATIE, 2012).

Ramirez, F., L. Hernández, I.A. Gutiérrez-Montes, G. Rivas-Platero, and D. Padilla. "La Perspectiva de Género en los Procesos de Desarrollo Comunitario y Sostenible." 1ª. Ed. Serie técnica. Manual técnico no. 108. (Turrialba, Costa Rica: CATIE, 2012a).

Ramirez, F., L. Hernández, D. Padilla, I.A. Gutiérrez-Montes, and G. Rivas-Platero. "El Género en las Escuelas de Campo: Capsulaspara el Aprendizaje y la Inclusión." 1ª. Ed. Serie divulgativa no. 16. 10 hojas sueltas. (Turrialba, Costa Rica: CATIE, 2012b).

Ramírez Agüero, F., I.A. Gutiérrez-Montes, L. Hernández, A. Escobedo, and D. Padilla. "El Empoderamiento de las Mujeres en las Cadenas de Valor: Un Retoparalas Políticas de Desarrollo Rural." Síntesis para decisores, PB 16. (Turrialba, Costa Rica: CATIE, 2012c).

Ramirez, F., L. Hernández, D. Padilla, I.A. Gutiérrez-Montes, and G. Rivas-Platero. "El Género en las Escuelas de Campo: Capsulaspara el Aprendizaje y la Inclusión." 1ª. Ed. Serie divulgativa no. 18. 11 hojas sueltas. (Turrialba, Costa Rica: CATIE, 2013).

Rivas-Platero, G.G., I.A. Gutierrez-Montes, F. Ramirez, D. Padilla, J.G. Suchini, L. Hernández, L. Rodriguez, et al. "Hacia el Fortalecimiento de una Agricultura Familiar: Pilar de los Territorios Rurales." Síntesis para decisores. PB 12. (Turrialba, Costa Rica: CATIE, 2012).

Rivas-Platero, G.G., I.A. Gutierrez-Montes, F. Ramirez, D. Padilla, J.G. Suchini, L. Hernandez, L. Rodriguez, et al. "Las Escuelas de campo del Programa Agroambiental Mesoamericano en la Región de Trifinio: Una Plataforma para la Gestión de Conocimiento y la Creación de Capacidades Locales Hacia el Desarrollo Territorial Sostenible." Síntesis para decisores. PB 13. (Turrialba, Costa Rica: CATIE, ECADERT, Plan Trifinio, ICRA, EIRAD, 2012).

Scheelje, M., G. Detlefsen, and M. Ibrahim. "Costa Rica: Oportunidad esparauna Legislación Forestalque Facilite el Aprovechamiento del Potencial Maderable en Fincas Agropecuarias." Síntesis para decisores. PB 11. (Turrialba, Costa Rica: CATIE, 2011).

Segnestam, L. "Division of capitals—What role does it play for gender—differentiated vulnerability to drought in Nicaragua?" *Community Development* 40 no. 2 (2009): 154–176.

Siles, J., I. Gutierrez-Montes, and F. Ramirez. "Estrategia de Equidad e Igualdad de Género. 1ª." Ed. Serie técnica. Boletín técnico no. 57. (Turrialba, Costa Rica: CATIE, 2012).

Soares, D., I.A. Gutierrez-Montes, R. Romero, R. Lopez, G.G. Rivas-Platero, and G. Pinto. "Vulnerabilidad Social ante el Cambio Climático: Retos y Propuestas de Políticas desde un Enfoque de Género." Síntesis para decisores. Policy Brief. (Turrialba, Costa Rica: CATIE, 2011).

Toruño, I. "Análisis Financiero-Económico de Fincas con Varias Actividades Productivas y el Rol de la Familia en la Producción y Toma de Decisiones en el Centro Norte de Nicaragua." MSc Thesis. (Turrialba, Costa Rica: CATIE, 2012).

CHAPTER 14

Analysis of Tropical Homegardens through an Agroecology and Anthropological Ecology Perspective

Alba González-Jácome

CONTENTS

14.1 INTRODUCTION

This chapter discusses the history and dynamics of homegarden agroecosystems through the integration of perspectives from the fields of agroecology and anthropological ecology, with an emphasis on cultural ecology.[*] To do this, I use two case studies of Mexican agroforestry homegardens in two different geographical spaces, both rural landscapes and ecosystems: one in a tropical region in the southeastern Mexico next to the Gulf of Mexico and the other in a temperate region in the Central Mexican Plateau. By linking disciplines, a transdisciplinary and

---

[*] Cultural ecology is the study of how human adaptation to the environment takes place by way of cultural mechanisms, particularly social organization, economy, and technology (Moran 2000: 343).
Cultural ecology can function as a model that can be applied to the study of the dynamic interaction of humans, environment, technology, social structure, the organization of work, and culture. The methodology can include synchronic and/or diachronic studies according to the main objectives of the specific research. This model emphasizes subsistence and production in contrast to a political economy approach.

interdisciplinary participatory research emerges that seeks to understand how social aspects integrated with cultural processes act in the interrelationship between local agrifood systems and larger regional and national economies. Second, through this anthropological/agroecological perspective, we seek to analyze and engage the organization of local societies and their related social and cultural processes at different scales, in two different regions, with a focus on landscapes where homegardens are still functioning (Aké et al. 1999; Alayón-Gamboa and Gurri-García 2008; Blanco-Rosas 2006; Álvarez-Buylla et al. 1989; Greenberg 2003; Herrera 1994; Porter-Bolland et al. 2008).

Finally, this integrated anthropological–ecological perspective seeks to better understanding of processes linked to sustainability and relationships with biodiversity, cultural resilience, and the transmission of ecological and cultural knowledge from one generation to the next, using as an example the relation between homegarden plants and food. This element involves the interrelationship between natural, social, and cultural aspects of two case studies and their ways of adjusting to specific environments through time (Moran 2000; Wilk 1991). Data from the two cases are compared with information taken from several studies of homegardens in Mexico (Aké et al. 1999; Alayón-Gamboa and Gurri-García 2008; Blanco-Rosas 2006; Álvarez-Buylla et al. 1989; González-Jácome 2007; Greenberg 2003; Herrera 1994; Mariaca-Méndez et al. 2007; Porter-Bolland et al. 2008).

A cultural ecology approach applies interdisciplinary research, which according to the specific study can include agroecology and cultural geography through the concept of landscape, but also uses information about physiographic characteristics of regions. It can include basic elements and concepts of demography, local ethnobotany, local soils, interrelationships with nearby regional markets, or peasant economy (McC Netting 1993; Roseberry 1989, 1995; Wilk 1991). Cultural ecology uses basic cultural anthropology concepts and methodologies (ethnography) applied to examining information about human societies. It also integrates other academic disciplines such as environmental history and different forms and social components of the transmission of cultural knowledge, as well as attention to cosmovisions of the world from different rural societies where homegardens are still present (Haenn and Wilk 2005; McC Netting 1993).

A cultural ecology approach and methodology can be used to study, describe, and understand important parts of daily life, family economy, and the daily foods of rural families. This chapter includes some aspects and related concepts that are basic in the relationship between agroecology and cultural ecology: sustainability of the agroecosystem, some aspects of resilience,* and the relation between plants of the homegarden and diet as one element of traditional ecological knowledge (Alayón-Gamboa and Gurri-García 2008: 395–407; Cahuich-Campos 2012).

In Mexico, the articulation between agroecology and cultural ecology has a long history (Palerm 1952, 1954). However, not all anthropological schools of thought, their topics of study, or their basic concepts, are compatible with the goals of agroecological concepts or methodologies. From my point of view, ecological anthropology is the closest approach to achieving objectives and targets related to agroecology and its scope. The anthropological approaches of human ecology, cultural ecology, and political ecology are in closer proximity with the research interests and perspectives for both disciplines. Thus, with the case study of homegardens, and using the cultural ecology

---

* Resilience in ecology is considered to be "the capacity of an ecosystem to respond to a perturbation or disturbance by resisting change and recovering quickly. Such perturbations and disturbances can include stochastic events such as fires, flooding, windstorms, insect population explosions, and human activities such as deforestation and the introduction of exotic plant or animal species. Disturbances of sufficient magnitude or duration can profoundly affect an ecosystem and may force it [...] to reach a threshold beyond which a different regime of processes and structures predominates" (Folke et al. 2004). However, to understand this ecological concept, it is necessary to include the human roleacting together with the environment: "Interdisciplinary discourse on resilience now includes consideration of the interactions of humans and ecosystems through socioecological systems, and the need for shift from the maximum sustainable yield paradigm to environmental resource management, which aims to build ecological resilience through [...] resilience analysis, adaptive resource management, and adaptive governance" (Peterson et al. 1998; Walker et al. 2004).

theoretical and methodological approach (McC Netting 1993; Moran 2000; Palerm 1954; Roseberry 1989; Steward 1955), we will begin this discussion about some of the issues that also are the core of contemporary studies on agroecosystems.

14.2 SUSTAINABILITY AND ANTIQUITY OF HOMEGARDENS IN MEXICO

Homegardens all over the world have recently attracted more attention of scientists because these systems have the ability to mitigate environmental problems such as loss of biodiversity or elevated levels of CO_2 in the earth atmosphere (Kumar and Nair 2006). Much less focus has been placed on the historical, social, and anthropogenic features that make them sustainable (Mariaca-Méndez et al. 2007: 119). In Mexico, recent archeological studies and environmental historical research (García-Martínez 1999; García-Cool 2014; González-Jácome 2011: 96–103; MacNeish 1997; Vanderwarker 2006) suggest that homegarden and cornfield (milpa) agroecosystems emerged in Prehistoric times (9000 BC–7000 BC).

Regardless of the characteristics that are attributed to homegardens, this agroforestry system is one of the oldest in the world due to its resemblance to natural ecosystems and their incidental inception (Rindos 1984). Although the homegardens of the Mexican Central Plateau and the *solar** of the Maya region have prehistoric origins, the oldest indications are based on evidence of the first eaten fruits whose remains were throughout the place the people was living at that time (incidentally domestication). This fact would correspond to the formation of the first homegardens in the Tuxtlas region of southern Veracruz and the Tehuacan Valley, Puebla, in Prehistoric times (MacNeish 1967, 1997; Vandermaker 2006). With a less ancient date archaeological remains of what appear to be homegardens in the Coba Mayan area of Quintana Roo have also been found and dated to the Classical Maya period by archaeologist Barba-Pingarrón (1987: 71–86).

We envision that the first homegardens were not clearly distinguishable from the surrounding forest, except for the large amount of fruit and edible plants occurring in them. An abundance of paleobotanical and archaeological remains of fruits and edible plants as well as the remains of animals have been found (MacNeish 1967, 1997; Vanderwarker 2006). There were also areas that were annually used in the first permanent settlements. The early inhabitants created an incidental homegarden near the housing settlement as a by-product of their taste for eating certain fruits and throwing their seeds on the ground. This "dump heap" or "midden" process was incidentally convenient because as useful plants appeared and were protected (from pests and predatory fauna); they also served as decoys to attract wild animals, which were in turn used for food, skins, and fur (Emslie 1981; Linares 1976; Neusius 1996; Vanderwarker 2006).

---

* The *solar* is a homegarden which is sourrounded by a wall made of stones called an *albarrada* or *tecorral*. Inside the terrain there are the family house, the outdoor kitchen, an outhouse, a place for domestic animals, at least one table for pots and plants which are being grown or experimented with, and places where trees for shade, wood, and edible fruit are grown.

The Mayan homegardens without the wall also became known as a *solar* between 1550 and 1560, when much of the Yucatan indigenous population abandoned their ancestral settlements and settled on land which Hispanics called "ordered peoples." The space was organized in a grid plan, around a square where the church was the main building together with the council and other public buildings. Small communities were absorbed by larger ones, losing their identity in the process. For water supply, the new settlements and the "ordered peoples" or "nations formed" used the old sinkholes or *cenotes*, which were created in pre-Hispanic times as water sources. Forced resettlement on new sites was not always successful (Roys 1957; Scholes and Adams 1959: 199). This state policy was called "the congregation of peoples" (Moreno Toscano 1987: 351). The *solar* as a form of land tenure, legally born with these congregations (Indian pueblos), were units of distribution and organization of the land: a plot measuring about 50 m × 50 m, with four lots in a block of 100 m × 100 m, except for the houses of the nobles, which were 100 m × 100 m. The first settlement laws and regulations and also the establishing settlements were based on the idea that each neighbor would have spaces for agriculture and homegarden organization (González-Jácome 2011: 207–208), and as spaces for commercial agriculture and consumption (8th Law That Montes Sean Common Fruit/Signed by Doña Juana, in Monsoon, Spain, the June 15, 1510).

Several studies demonstrate homegarden usefulness as a hunting area (creating access for food sources also known as *garden game*). A good example is presented by Vanderwarker (2006, 23: 148–181) in her study of the Tuxtlas. Archeological data show the Tehuacan Valley is also an important place in the evolution of the homegarden (Smith 1967: 220–255). The plant research of Smith (1967: 231) in the Coxcatlan Cave of the Tehuacan Valley found that in about 7000 BC, food plant remains included mesquite (*Prosipis juliflora*), fragments of the palm *Brahea dulcis*, other edible plant fragments such as seeds of Setaria cf. macrostachya, *Amaranthus* spp., prickly pear seeds (*Opuntia* spp.), one avocado (*Persea americana*) cotyledon, and a seed of chupandilla (*Cyrtocarpa procera*). These incidental gardens in both the Tehuacan Valley and the Tuxtlas region had a mix of wild plants and incidentally domesticated plants. Gradually, these became irrigated homegardens as part of the agriculture based on irrigation that was practiced in Tehuacan between 850 BC and AD 150 (MacNeish 1967: 290–309, 1997).

As humans altered the natural vegetation by introducing new plants around settlement areas, they cleared and planted new habitaos* created for the establishment of a wide range of pioneering weeds. These plants grew in abundance in open places, attracting insects and their predators. These habitaos included both wild and cultivated plants, encouraging greater diversity and thus density of small animals, since larger animals (such as deer) preferred undisturbed environments. The process had an anthropogenic origin, generating a pool of animals: a quick and readily available source of protein for the human population. The new predatory cycle was established in disturbed habitaos and was qualitatively and quantitatively different from the primary and undisturbed environments. This shortening the time needed to hunt and fish, which in turn provided extra time that was dedicated to reorganizing the major livelihood systems (Vanderwarker 2006: 148).

The proposed Vanderwarker (2006: 149) "garden hunting model" is based on studies of Emslie (1981: 306), Linares (1976: 331), and Neusius (1996: 276), who propose that many of the animals were pests of crops, so hunting in the gardens had the dual purpose of providing people with animal protein in addition to protecting their crops. Vanderwarker (2006: 151, 181) supports the feasibility of this model with examples of hunters in the Amazon, archeological studies in the Southwestern United States, and her analysis of zooarchaeological materials obtained in La Joya and Bezuapan sites in the region of Los Tuxtlas, Veracruz, corresponding to a period between the Maya Early Formative and Early Classic (2000 BC and AD 250).

In the Tuxtlas region, this type of homegarden system has persisted over time until today. Studies about current homegardens, such as the one by Blanco-Rosas (2006, 2011), show that until recent years, the Popoluca people of Soteapan in southern Veracruz obtained wild edible birds like partridges, pigeons, and small mammals in homegardens. This gave garden hunting importance to daily diet as a source of animal protein throughout the year. Homegardens have undergone changes along their long history. However, they are still functioning as agroecosystems that provide a historical record of permanence through time; evidently this is partial proof of their sustainability.

In this long historical process, vegetation plays a key role because it is the factor permitting the existence of this agroecological system; some trees or some plants in the system can be changed, but the structure remains based on the rest of the arboreal content of the homegarden. Furthermore, the homegarden increases in plants diversity when people live far away from food stores or commercial centers and the homegarden is the main provider of food for the families. Blanco-Rosas' (2006: 55) research based on historical records on the Tuxtlas region found that in 1580 the homegardens had five different types of sapotes (*Manilkara zapota*), avocados (*P. americana*), bananas (*Musa* spp.), pineapples (*Ananas comosus*), pitahayas (*Hilocereus undatus*), and toloches (no data about this tree).

The same author (Blanco-Rosas 2006: 58), using data from the *Geographical Relations of Oaxaca*, written almost 200 years later (1777–1778), found that fruit trees in homegardens were

---

* From Latin: *habitat* singular and plural *habitaos*.

avocado (*P. americana*), five varietes of sapote (*M. zapota*), oranges (Citrus sinensis), limes (*Citrus aurantifolia*), lemon trees (*Citrus limon*), two types of annona (*Annona squamosa* and *Annona muricata*), coconut palms (*Cocos nucifera*), guava (*Psidium guajava*), and banana trees (*Musa paradisiaca*), all of which were combined with other fruits such as pineapples (*Ananas cosmusus*), papaw (*Carica papaya*), watermelon (*Citrullus lanatus* var. lanatus), melons (*Cucumis melon*), sugarcane (*Saccharum officinarum*), sweet potato (*Ipomoea batatas*), manioc (*Manihot esculenta*), squash (*Cucurbita americana*), nances (*Byrsonima crassifolia*), and pitahayas (*H. undatus*). These fruit trees and plants were intermixed with tropical cedar (mahogany), palms, tree types of hicacos (*Chrysobalanus icaco*), *Bursera simaruba*, guayacán (*Tabubeuia guayacan* Seem. Hemsl.), and other trees.

14.3 TIME AND SUSTAINABILITY: MEXICAN MODERN AGRICULTURE AND STATE POLICIES

Although this chapter is not dedicated to the linear history of Mexican homegardens, the above sections discussed studies that show that this type of agroecosystem is very old. These homegardens have been sustained through long periods of time and in different types of landscapes and geographical and cultural spaces. During the last century, however, they have been undergoing drastic changes. In order to take a historical jump to current cases, it is necessary to review some data related to the past century.

Mexican anthropology has touched agricultural issues in its studies related to mobilization of human groups as a result of natural disasters and the building of hydroelectric and irrigation dams (Palerm 1954, 1970). Agriculture has been very important to these studies because dams were constructed in rural areas all over the country and they affected agriculture and agroecosystems drastically (González Jácome 1988b: 55–189; González-Jácome and Velasco Orozco 2008, 215–238; Moctezuma-Pérez et al. 2010; Palerm 1970; Robles-Linares 2014).

Mexican homegardens have undergone major transformations over time up to the present. Many of these changes are a consequence of land tenure changes, increasing population and migration (González-Jácome 2009: 71–111). Many of these changes have made homegardens disappear as agroecosystems because people have transformed them into housing plots or into gardens used for family leisure time. However, although some of their characteristics as a permanent food source and monetary earnings for families are still functioning—many times on a minor scale—homegarden ecological resilience is greatly weakened without sociocultural resilience (Alayón-Gamboa and Gurri-García 2008: 395–407; González-Jácome 2010, 2012a; Moctezuma-Pérez 2013a, 2013b).

Homegardens can disappear when land is insufficient to maintain their agricultural purposes, or when family labor is not sufficient. Out-migration of young people is changing agriculture in communities such as the ones located in the states of Tabasco, Oaxaca, Yucatan, and Quintana Roo, because people prefer to work in the service sector instead of working their family lands (González-Jácome 1988b: 253–261, 2010: 141–158). Regardless, there are many ecological, social, and cultural factors suggesting that homegardens are resilient and sustainable, including the traditional ecological knowledge of the people and the ways through which this knowledge is transmitted from one generation to the next (Alayón-Gamboa and Gurri-García 2008: 395–407; González-Jácome 2010, 2012a; Moctezuma-Pérez 2013a, 2013b).

During the last three decades in San Francisco Tepeyanco, Tlaxcala, and also in Yaxcabab, Yucatán, some homegardens shifted to a predominance of ornamental plants. These spaces were transformed into residential areas for married sons. These two cases are important because the two towns are settled in very different environs, and they also have very diverse altitudinal, ecological, and climatological characteristics. Furthermore, both towns are suffering from the lack of human labor due to out-migration and young people's daily migration to neighboring cities for employment in the service sector (González-Jácome 2010; Mariaca-Méndez et al. 2007: 119–138; Moctezuma-Pérez 2013a, 2013b; Kumar and Nair 2006).

In Hopelchén, Campeche, homegardens still have very important and basic functions for families to obtain edible plants for their daily diet, for medicines, and for social and cultural functions of local society by helping people to construct social networks that join families and communities (Cahuich-Campos 2012). The Hopelchen case is important because this Maya region was uninhabited for several centuries, although it has repopulated in the second half of the twentieth century. Homegardens were a good way for the recently arrived population to obtain food for daily consumption (Cahuich-Campos 2012; Molina-Rosales 2010). In this case, the rebuilding of homegardens was based on traditional ecological knowledge both from Maya people from the north of the State of Campeche, as well as outsiders from different regions of the country (Molina-Rosales 2010) (Figure 14.1).

Many recent inhabitants of southern Mexico migrated from different regions of the country where environs were also diverse, such as Central Mexico and the north of the country. People carried with them their traditional agroecological knowledge, which permitted the reconstruction of the homegarden system on new lands (González-Jácome 2010: 141–158; Mariaca-Méndez 2002; Molina-Rosales 2010).

Homegardens are widely distributed in Mexico; they are found both in cold climates as well as semidesert, temperate, and warm ecosystems. In the latter, we find, to a greater extent, cases where homegardens are still very important for rural societies. By converting them into commercial systems where the crop combination is trees, shrubs, and medicinal plants characterized by a multilayered arrangement, high biodiversity is maintained, as well as the provision of food, products, and economic resources for family consumption that is distributed throughout the year (Robles-Cervantes 2008: 142–150).

Figure 14.1   A homegarden in the community of Chan Kom, Paso de Ovejas, Veracruz, Mexico. The garden forms a complex agroecosystem around the home.

The combination of traditional and commercial elements in homegardens meets various economical, social, cultural, and ideological needs in the daily lives of its practitioners (Robles-Cervantes 2008: 142–150). Homegardens in the Municipality of Álamos, Sonora, in the north of Mexico, are located on small plots from 25 to 100 m², but they still have fruit trees, herbs, and flowers, as occur in other parts of the country. Complementary market production in this northern case is achieved through cattle ranching in the savanna grasslands, which means the use of several interrelated ecological and economical systems (Moctezuma-Pérez et al. 2010; González-Jácome 2012a, 2012b, 2013).

The agricultural policies of the Mexican State in the twentieth century arose by the end of the Mexican Revolution (1910–1921). Commercial agriculture to produce exports became the main goal, and Agrarian Reform programs began to separate small-scale peasants from medium- and large-scale agriculture. Government programs added other goals, such as the introduction of modern agricultural machinery and fertilizers, and massive irrigation and hydrological development projects to boost regional economic growth and industrialization (Anonymous 1985: 33; Palerm 1970; Rutsch and González-Jácome 2011; González-Jácome and Velasco Orozco 2008: 215–238).

Ministries within the government promoted these developments, from the Ministry of Water Resources (SRH) to the Ministry of Agriculture (SAG), all focused on channeling natural resources and using modern inputs and technologies to raise yields of economically valuable crops and commodities. Impacts on small-scale farmers and traditional peasant agriculture and their related human societies were not considered. With commercial agriculture as the policy goal, modernization began to displace centuries of local, traditional, and indigenous agriculture (Ornelas 1993; Palerm 1954, 1970; Rutsch and González-Jácome 2011).

From the 1940s to the 1970s, the stated intention of government action on the economy was to promote "industrialization at all costs" (Ornelas 1993). The promotion of industrialization was supported by import substitution and the search for "a growth rate of high and sustained long-term GDP by the concentration of capital resources to stimulate private savings and domestic investment, using this public investment to boost private capital formation" (Ornelas 1993: 39). Starting in 1948, the National Commission for Colonization (CNC) was given the task of handling the surplus rural population living in densely populated regions where people had no more land for agriculture (Anthology 1985: 33). This process began a massive relocation of small-scale farmers and peasants from the central and northern regions of the country to the tropics of southeastern and southern Mexico (Anonymous 1985: 33). Each relocation project had its own specific characteristics and objectives, and people's desires were not taken into account. This relocated population was sent to live in tropical regions of south and southern Mexico, where the new settlers constructed homegardens near their houses and used them for growing food plants, medical herbs, and fuel to cover necessities which were tied to the lack of nearby stores and markets (Blanco-Rosas 2006, 2011).

14.4 HOMEGARDENS, SUSTAINABILITY, AND CLIMATOLOGICAL RISKS

Homegardens have been described as sustainable systems and studied since the early 1970s for their ability to remain as agroecosystems over time (Kumar and Nair 2006). This sustainable characteristic is due to the dynamics of homegarden vegetation and also to its capacity for resilience, which is based upon a combination of ecological characteristics, family labor (social organization), and traditional knowledge. However, despite the importance of homegardens, several fundamental issues have not been analyzed, especially in relation to social and cultural components. These are linked to the local economy and the ability to generate food, cash, and other important products, such as medicinal plants, shade for the house, ornamental plants, a provision of wood and fuel, and also native animals that could be collected and hunted (Mariaca-Méndez et al. 2007: 119–138; Kumar and Nair 2006).

Figure 14.2   Stacked firewood harvested from trees in a homegarden system in X-Mejia, Yucatan, Mexico. Homegarden agroecosystems provide multiple products, from food to firewood.

Homegarden agroecosystems are proposed as being resilient because they have the capacity to adjust to severe changes in the environment caused by external forces such as climatological disasters, or processes caused by humans such as deforestation of the region, burning of the forest, and so on. Family organization tied with local cultural knowledge and needs act together to rebuild the society and the environment. Furthermore, homegardens, the same as cornfields (milpas), are subject to different types of risks, including environmental, climatological, political, economical, and demographical. These could include emigration and the lack of a family labor force and so on. It has been considered of resilience in homegardens has to do with agrobiodiversity, biodiversity, and the role of traditional ecological knowledge (Alayón-Gamboa and Gurri-García 2008; Blanco-Rosas 2006; Kumar and Nair 2006) (Figure 14.2).

The agrobiodiversity in Mexican homegardens has been studied as a space for the production of food and holding of festive rituals that follow annual cycles of climatological or environmental events. Both of these activities require management by peasant families (Christie 2004; Estrada Lugo et al. 1998; Greenberg 2003; Herrera 1994; Juan-Pérez, Uribe and Madrigal 2005; Márquez Mireles 2000; Terán and Rasmussen 1994; Terán et al. 1998; Stuart 1993). In tropical regions of southern Mexico, both homegardens and cornfields have suffered from variable climatological conditions and events with destructive impacts (Ingold 1992, 2000; Márquez Mireles 2000; Oliver-Smith and Hoffman 1999). Reconstruction of towns, agricultural parcels, homegardens, and daily life of rural people are helped by the capacity of the agroecosystems to be rebuilt on the basis of previous design and organization (resilience). Historical studies about forest destruction for agriculture, abandonment, followed by successional development over time, are important for showing the role of resilience in ecosystems (García-Martínez 1999: 55–74).

14.5 DIVERSITY IN TWO CASE STUDIES OF MEXICAN HOMEGARDENS

In this section, I introduce two case studies of Mexican homegardens to illustrate some of the characteristics outlined up to now in this chapter, and then included in further descriptions of this agroecosystem below. The cases were recently studied, one of them in the tropical region of the State of Campeche in southern Mexico (Calakmul, Mexican Peten), and the other one in San Francisco Tepeyanco, Tlaxcala, in the Central Mexican Plateau (Cahuich-Campos 2012; Moctezuma Pérez 2013a, 2013b).

14.5.1 Homegardens in X-Mejía in Hopelchén, Campeche

Diana Cahuich-Campos' study (2012) of homegardens (called *solares* in the Maya región) in the village of X-Mejía in Hopelchén, Campeche, in southern Mexico, is recent and important because she introduced social and cultural elements in her system analysis. In her research, homegardens are defined as "those areas adjacent to the family home formed by trees, crops, fruit, vegetables, medicinal plants, ornamental plants, and all-purpose crops with different harvest times, which ensures production during the year and a place for raising domestic animals and wildlife, with different use values." The researcher found a high agrobiodiversity in small areas (half an acre). The structure of these homegardens varied according to their age, environmental, and climatic conditions, but also varied in terms of cultural, social, political, and economic organization of the families managing each garden system (Buchmann 2009; Cahuich-Campos 2012; González-Jácome 2012a; Mariaca-Méndez et al. 2007) (Figure 14.3).

These homegardens are located in a tropical region west of the city of Campeche and north of Calakmul Biosphere Reserve. The town is small of recent creation and is under the Hopelchén Municipality. In 2010, X-Mejía had a population of 417 inhabitants (213 men and 204 women).[*] A percentage of 87.77 of adult people speak Maya. The population of the community lives in 73 houses, each of them having a homegarden. Accepted local history is that people have lived in the area since the Prehispanic Yucatec Maya times, and there are still remains of the old wells that gave the name to the region (Los Chenes), which continue today to play an important part in the survival of the inhabitants. However, during the second half of the seventeenth century, the region was abandoned and it remained unoccupied until the twentieth century (Gerhard 1986, 1991). Some of the inhabitants come from old towns north of the community, and also north of Campeche, and most are of Mayan origin.

Floristic composition in 10 X-Mejía homegardens include 159 plants, of which 50 were trees (31.4%), 34 shrubs (21.4%), 61 herbs (38.4%), three aquatic plants (1.9%), five tubers (3.1%), one palm (.63%), three agaves (1.9%), one prickly pear Opuntia (.63%), one vine (.63%), plus an area dedicated to corn cultivation. The uses[†] of plants (209) in this homegardens were 56 for medicinal

Figure 14.3 A view of a typical homegarden agroecosystem in the town of Yaxcabab, Yucatán, Mexico. The home is surrounded by a high species as well as structural diversity.

---

[*] The percentage of illiteracy among adults is 12.47% (7.98% in men and 17.16% women) with female/male ratio of 0.958. The fertility rate of the female population is 2.76 children per woman, and the education level is grade 5.32 (5.80 in men and 4.88 in women). The town is 150 m above sea level.

[†] Some of the plants have multipurpose uses.

purposes (26.8%), 51 for ornamental goals (24.4%), 44 for food (21%), 29 plants used for ceremonies and rituals (13.9%), 9 for condiments (4.8%), 7 for construction purposes (3.3%), 6 for washing clothes, cleaning houses, and soap for washing hair (2.9%), 3 for providing shade for the family home and garden (1.4%), 2 for fuel (.96%), and 1 for obtaining sweet (.5%).

14.5.2 Homegardens in San Francisco Tepeyanco, Tlaxcala

San Francisco Tepeyanco is a town with Prehispanic origins, which is located 9 km south of the city of Tlaxcala, the capital of Tlaxcala state. This village is part of the Southwest Tlaxcala region and is located on the border of three natural regions: (1) the western slopes of the Matlalcueye (La Malinche) volcano, (2) the hills and hillocks that are based in the southern region of the rugged natural orographic Tlaxcala Block, and (3) the northeastern edge of the Zacatelco plain. This town lies in a natural depression surrounded by hills that protect the zone from climatic elements like wind, frost, and hail (González-Jácome 1985, 2004).

Tepeyanco's altitude is 2,246 m above sea level. Its climate is temperate and humid, has deciduous forest vegetation, and an annual rainfall between 800 and 900 mm. This moisture is adequate for seasonal crops, but there is an important zone with irrigation from a river, and also a canal system derived from the old hacienda Molino de Flores de Santa Ana. There is a marked seasonality of precipitation, of which over 90% is concentrated from May to October. The average annual temperature ranges from 15°C to 17°C (González-Jácome 1985, 2004).

The research on homegardens in Tepeyanco dates at least from 1969. This town has a long agricultural tradition that goes back many years before the arrival of the Spanish conquerors to the New World. After the Spanish arrival, homegardens underwent great changes, both in cultivated plants and techniques to manage them, due largely to the teachings of the Franciscan friars who settled in the region in the sixteenth century. They brought to the New World technologies such as grafting; cultivating plants such as wheat, barley, and oats; and they used metal tools to complement the ones made of wood or obsidian that already existed (González-Jácome 1995, 2004, 2011).

From 1969 to 1972, the first studies with a cultural ecology approach of homegardens in Tepeyanco were carried out by González-Jácome (1976), with an emphasis on corn cultivation. Before that time, several studies on raised field farming systems in Atlamajac (then a barrio of Tepeyanco, and now an independent town) were organized and published by Wilken (1969). Later Roldán-Botello (1979) studied Tepeyancan homegardens from 1977 to 1979. Between 1981 and 1983, Stephen Gliessman (1990a:160-168) and his students from the University of California, Santa Cruz, continued with the ecological study of homegardens in Tepeyanco (Allison 1983). I also undertook studies with graduate students Gerardo Mariscal and Jesús Álvarez between 1985 and 1995 (Figure 14.4).

In 1977 and 1978, the settlement of the town was organized in reticular blocks with houses and their homegardens surrounded by water canals, windbreaks, and shade trees (Roldán-Botello 1979). Within these areas, vegetables and flowers were planted. The farmers had other agricultural plots planted with maize, tomato, pigweed, potato, pumpkin, and alfalfa. The vegetation of the town included *Alnus* spp., casuarinas, oaks, junipers, Buddleia spp., conifers, *Salix* spp., and scrub. Crops and other vegetation depended on water resources located in the village, and the inhabitants developed an intensive set of farming systems. Shelterbelts of trees were used to protect smaller plants from winds and frost. According to Roldán-Botello (1979), out-migration was taking place, especially by people leaving to study or conduct business. The migrant population, through the bonds of kinship and community membership, sent remittances to fund religious festivals and improve local agriculture (González-Jácome 1988b: 253–261).

During 1970, there were five Tepeyanco agroecosystems on 1,410 ha of land, of which only 52 (4%) were dedicated to homegardens. The homegarden system was the smallest of them, but it was also the most productive agricultural system of the entire region (González-Jácome 1985, 2004). Its inhabitants had good income and the production levels allowed them to reach the distant

Figure 14.4 Homegardens along a road in the community of San Francisco Tepeyanco, Tlaxcala, Mexico. People's homes are covered by the homegarden agroforestry system.

markets of Puebla, Atlixco, and Mexico City. These homegardens were irrigated agroecosystems, which allowed for more diversification of marketable products. Their location in the close vicinity of houses permitted homegardens to receive the care and attention of the families who owned them. These homegardens did not require the use of machinery, but a complex network of agricultural techniques, such as the use of natural fertilizer, planting and transplanting tree seedlings, and mixing soil with mud, among others, was practiced. The low capitalization of homegardens required a high degree of knowledge for their maintenance. In Tepeyanco, the homegarden system provided economic surpluses that farmers invested in improving their homes, their land, for agricultural technology, or savings to legally migrate to the United States (González-Jácome 1985: 529).

14.6 DOMESTIC FAUNA IN HOMEGARDENS

Research on homegarden agroecosystems has focused on vegetation composition and structure, but the faunal diversity has been rarely studied (Aké et al. 1999; Cano 2003; Cuanalo-de la Cerda and Guerra 2008; López 2005; Medina 2005). Among the reasons for this omission is that plant species predominate in frequency and diversity and provide a larger amount of food, medicinal plants, ornaments, fuel, shade, and aesthetic satisfactions. Also social and cultural aspects must be included, such as family organization, food, and rituals, among others (Buchmann 2009; Del Ángel-Pérez and Mendoza 2004; Mariaca-Méndez et al. 2007; Palacios Sánchez 2009). While in some current homegardens, domestic animals are not always present due to limitations on space or cultural issues, in the peninsular Maya area, where X-Mejía is located, it has been reported that 95% of family homegardens or solares have animals that are raised to meet various family needs (Cahuich-Campos 2012; Estrada Lugo et al. 1998; Mariaca-Méndez et al. 2007) (Figure 14.5).

Homegardens are considered integrated systems where the relationship between humans and nature is very important. It is also very important to conceptualize homegardens as traditional agroecosystems. They require few external inputs, since labor is provided by family members and is not paid. Production may become more or less seasonal, and it is mainly directed to subsistence (Fernandes and Nair 1986, Gaytán et al. 2001; Gliessman 1999; Mariaca-Méndez et al. 2007). However, when there is a surplus, it can be sold in local or regional markets, providing income to supplement the household economy (Ake et al. 1999; Berkes et al. 2003; Bellon et al. 2004; Shagarodsky et al. 2003).

Current homegardens vary from the exuberant with high biodiversity, to very small plots with one or two trees, and a few medicinal and ornamental plants, some of them cultivated in pots. A good example of this type is found in the Tlaxcalan towns of San Tadeo Huexoyucan, San Mateo Huiloapan, San Francisco Temetzontla, and all the communities along the Totolapan River and ravine in central Tlaxcala (Bilbao-Ercoreca 1979; González-Jácome 2011) (Figure 14.6).

Figure 14.5 Interior view of a homegarden in the community of Chan Kom, Yucatan, Mexico. Mulitple activities take place in the garden, from raising small livestock to hanging clothes to dry.

Figure 14.6 A view of the structural diversity of a homegarden in Chan Kom, Yucatan, Mexico. As in a forest, gardens show mulitple layers in their vegetative structure.

14.7 THE MULTIPLE USES OF NATURAL RESOURCES AND SOCIAL NETWORKS

In X-Mejía, homegarden products result from various internal production systems that are interconnected within the peasant family and are the key for a strategy of multiple uses of natural resources. This diversification strategy reduces vulnerability and ensures food security throughout the year (Cahuich-Campos 2012; Mariaca Méndez et al. 2007). Biodiversity in these homegardens is the result of the intergenerational transmission of traditional ecological knowledge about natural, social, and cultural resources to which the rural families have access. Such a process has led researchers to consider homegardens as laboratories of biotic interactions, as biodiversity banks *in situ*, and as educational fields where peasant experience in managing the selected varieties of plants and animal species and fungal germplasm is key (Leff 2003; Mariaca-Méndez et al. 2007; Reinhardt 2007). The knowledge obtained from daily practices and experimentation, which is transmitted from the homegardens managers to their children, together with an important group of beliefs, customs, and values, helps shape the cultural identity of the peasant family (Barrera-Vázquez 1999; Gonzalez-Jácome 2007, 2012a; Mariaca-Méndez et al. 2007).

Homegardens in X-Mejía are considered by rural families to be a biocultural heritage to be passed on to their descendants as well as be shared with neighbors and relatives (Cahuich-Campos 2012; Cahuich-Campos and Mariaca-Méndez 2010). This feature is observed equally in migrant families who re-create in their homegardens what they learned from their parents about managing flora and fauna, and adapted this knowledge to the new natural environments in which they now live. A social network is created based on the enrichment of knowledge and takes place through the exchange of new species and varieties of plants and animals with the new regional communities with which they have come into contact (Cahuich-Campos 2012; Flores 2003; Guerrero 2007; Vogl and Vogl-Lukasser 2003). In several towns, such as Yaxcabab and Chan Kom in Yucatán, small communities in Marqués de Comillas, Chiapas, and San Francisco Tepeyanco, in Tlaxcala, people still remember the history of each one of the trees growing in their homegardens (González-Jácome 2007; Mariaca-Méndez 2002).

In social and cultural aspects, homegardens contribute to connecting people to the land, making the residence permanent, encouraging and strengthening the links between families, promoting the exchange of goods and knowledge, and cooperation and mutual assistance (Buchmann 2009; Cuanalo-de la Cerda and Guerra 2008; Ellis and Porter-Bolland 2007:213-242; González-Jácome 2012: 522–534; Mariaca-Méndez et al. 2007). With regard to ideological aspects, homegardens provide quiet, aesthetic, and recreational value, as well as being the place where children are reared and older people are taken care of (Gliessman 1999; González-Jácome 2012a; Mariaca-Méndez et al. 2007). In tropical regions, homegardens ameliorate the atmosphere by mitigating sudden changes in temperature and humidity, reducing the temperature of the house, protecting the home from strong winds and hurricanes, and generating less extreme microclimates (Cahuich-Campos 2012; González-Jácome 2004; Lok 1998).

Due to all these features, homegardens have persisted through time, meeting multiple needs, and reducing the vulnerability of peasant families and farmers from different forms of environmental, social, and economic risk. They have enabled self-sufficiency and supplemented family needs for food, good health, and income, among other material and nonmaterial social and cultural values, which are mainly dependent of social organization and cultural traits of each local community (Alayón-Gamboa and Gurri-García 2008: 12; Christie 2004; Ellis and Porter-Bolland 2007:213-242; González-Jácome 2012; Herrera 1994; Juan-Pérez and Madrigal-Uribe 2005; Mariaca-Méndez et al. 2007; Maroyi 2009; Rojas 2008).

However, when towns increase in total population and density, when available agricultural land diminishes, when daily migration and emigration take away the young members of the family, when monetary earnings are not enough to cover the family's necessities and young people abandon agricultural activities, a process of homegarden degradation occurs. Homegardens are gradually disappearing to be converted into land for new houses for sons and daughters, or new additions to the houses, such as garages or rooms. They often look like abandoned plots with domestic garbage all around and domestic animals such as chickens, turkeys, and pigs running loose, as is happening in Chan Kom, Yucatán (González-Jácome 2007, 2012a).

14.8 HOMEGARDENS, FOOD, AND DIET

The recent literature on Maya ethnobotany (Anderson et al. 2003; Atran and Ucan 2004; Balick et al. 2000; Fedick 2010: 953–954) indicates that presently Mayans make use of more than 500 species of food plants. These plants are native to the Yucatan peninsula lowlands, and before the sixteenth century they were supplemented by other species introduced from northern Mesoamerica and from South America. These food plants included 92 different taxonomic families. Ford (2008: 179–199), using data collected from 18 gardens cultivated by Maya in west-central Belize lowlands, finds that of the 20 dominant species from the forest—which were also cultivated in the gardens—only ramon (*Brosimum alicastrum*) was wind pollinated, while the others species were pollinated by bats, beetles, bees, moths, and various insects. Furthermore, of the 37 species of cultivated plants in the gardens, seven of them were pollinated by wind (Ford 2008: 179-199).

From the case study sites, as well as other sites presented in this chapter, we find additional data on homegardens. For the preparation of food, plants in homegardens of X-Mejía (solares) are an important strategy for rural women to maintain agrobiodiversity adapted to local environmental conditions and fulfilling the social and cultural preferences of the family and the local society. Use of this agrobiodiversity is dynamic and strongly marked by the seasons, including times of plenty and lean times, in terms of products for consumption and monetary earnings (Anderson 1993; Caballero 1992; Cahuich-Campos 2012; Christie 2004; Gonzalez-Jácome 2007, 2012a; Greenberg 2003; Juan-Pérez 2003). In the case of tropical Maya homegarden, the agroecosystem not only provides food, but is also the site where other production system's activities converge, such as the preparing of seed for planting in cornfields (milpa), beekeeping, and traditional hunting, gathering, and extraction (Mariaca-Méndez et al. 2007), helping to maintain the livelihoods of Mayan farm families despite social, economic, and cultural pressures (Ellis and Porter-Bolland 2007:213-241; Porter-Bolland et al. 2008; Toledo et al. 2008: 345–352).

One of the means for maintaining cultural identity is passed from mothers to daughters through traditional knowledge related to food preparation that is transmitted, fostering relationships of kinship and friendship (Juan-Pérez 2003; Juan-Pérez and Madrigal-Uribe 2005). If the homegarden belongs to a migrant family, the preparation of food is learned in a woman's place of origin and then is adapted to new plants and animals of the new homegarden location (Flores 2003; Greenberg et al. 2003). Greenberg et al. (2003) mention that women who are responsible for family care in migrant households spend most of their time in their gardens, and that the homegardens are the primary sites for the biological conservation of traditional plant species used in the Yucatan cuisine (Greenberg et al. 2003).

Cahuich-Campos (2012) discussed the importance of the culinary tradition as part of the adjustment processes in migrant families of Calakmul, Campeche. She mentions that corn provides 62% of everyday ingredients to make 25 dishes, with the additional 38% of ingredients coming from the homegardens. This agroecosystem also provides edible species such as fresh fruits that are consumed directly. Estrada Lugo et al. (1998) mentioned that Quintana Roo homegardens have an average of 62.71% of edible species that are consumed by families, including fresh fruit such as *Annona* spp., *P. guajava*, and *Citrus* spp.

Another important dietary element is the satisfaction of caloric energy and protein needs from the domestic animals present in the *solar*. These animals represent a continuous production of protein at low cost, which does not require much capital for households with limited resources (Hernández-X 1995; Wieman and Leal 1998; Estrada Lugo et al. 1998; Porter-Bolland et al. 2008). Animals are considered a back up food in times of need (Herrera 1994), some of which can be consumed directly or certain products derived from them, such as eggs (Cuanalo-de la Cerda and Guerra 2008; Mariaca-Mendez et al. 2007) or lard (Wieman and Leal 1998). In homegardens of southeastern Mexico, chickens, pigs, and turkeys are found (Estrada Lugo et al. 1998). An important feature of this is the small size of animals—they are easy to kill and prepare. They supply meat that does not require refrigerated storage and is consumed by the family on the same day the animals are killed. This aspect is very important in tropical and remote areas, where small animals have a short life cycle, reach maturity quickly, and reproduce easily (Wieman and Leal 1998). Besides adapting to local conditions, animals such as creole pigs in the Yucatan are appreciated for their unique gastronomic qualities and attributes (Scarpa et al. 2003).

Kitchens, where food is prepared, have a special place in the homegarden, as cooks like to socialize at the time of food preparation and tasting (Barrera-Vázquez 1999; Abu-Shams 2008). Traditionally, the kitchen is the woman's place, where they decide which dishes will be developed based on the resources to which they have access, often keeping the forms of preparation true to traditional rules and secrets that they have inherited and which they in turn pass on to their daughters through the oral tradition and everyday practice, forming culinary techniques that are part of their cultural heritage (Abu-Shams 2008; Nájera 2009; Pérez Contreras and Alcaraz 2007). Hence it is said that people eat what they eat depending on the society and culture to which they belong and that the selection of food is governed by following certain cultural patterns (Abu-Shams 2008; Nájera 2009).

Each population and every household has its techniques for preparing food, including roasting, boiling, frying, and grilling, as well as various traditional flavoring sauces and spices (Rebato 2009). These forms of food preparation identify people and distinguish them from other human groups and are considered a marker of ethnic identity (Abu-Shams 2008; Greenberg 2003; Rebato 2009). This cultural characteristic can be lost when products come into contact with other groups of society, and along with it, much local agrobiodiversity (Abu-Shams 2008; De Garine and Vargas 1997; Flores 2003; Greenberg 2003).

Recent studies of Popolucan homegardens show that one of their important functions is to attract local wildlife, which could be easily hunted and contributed to the daily diet of the inhabitants. This also meant the consumption of meat by villagers when it might otherwise be scarce. Varderwarker (2006: 160–161) found that hunting in the garden overlooking the Prehispanic settlements of the late Formative Olmec site at La Joya, meat came from two types of squirrel, two types of rabbits, collared peccaries, white-backed deer, two types of field rat, and other animals. Such wildlife food favored the creation of anthropogenic habitaos through the clearing of the forest, in which hunting could more easily occur. Blanco-Rosas (2006) estimated that in the Popoluca region of Soteapan, in Veracruz, catching partridges, pigeons, and other wild birds in the homegarden provided families enough animal protein to meet their basic nutritional needs.

The number of plants cultivated by the Maya population has apparently always been high, with many of them concentrated in the homegardens. Up until the mid-sixteenth century, native plants cited in historical sources were similar to a contemporary ethnobotany study in 2008 (Mariaca-Méndez et al. 2011), with species grown around houses reaching about 50, not including cultivars or varieties. These species included *Cordia dodecandra* D.C., *Pileus mexicana* (D.C.) Johnston, *C. papaya* L., *Theobroma cacao* L., *Apoplanesia paniculata* Presl, *B. crassifolia* (L.) HBK, *Cedrela mexicana* Roem, *P. guajava* L., *Pimienta dioica* (L.) Merr, *Sabal yapa* (Wrigth.) Stand, *Sapindus saponaria* L., *Chrysophylum cainito* L., *Lucuma hypoglauca* Stand, and *Nicotiana tabacum* L.

14.9 HOMEGARDENS BIODIVERSITY AND TRADITIONAL MEDICINE

The presence of flora with medicinal uses in homegardens has been reported by Alburquerque et al. (2005); Buchmann (2009); Caballero (1992); Del Ángel-Pérez and Méndez (2004); Mariaca-Méndez et al. (2007); Herrera (1994); Ortega et al. (1993); Rao and Rao (2006); Robles-Cervantes 2008; Vogl and Vogl-Lukasser (2003); and Wezel and Bender (2003). Some of the medicinal plants in homegardens are deliberately cultivated from material collected from the wild, so these agroeco-systems are important areas both for production and for conservation *in situ* (Rao and Rao 2006). The presence of medicinal plants allows rural families to have resources for curing sickness and disease and is most often seen in homegardens of households that do not have easy access to health services (Buchmann 2009). In homegardens of the town of Purification Tepetitla, Texcoco, Mexico, the cultivation of medicinal plants in homegardens has become a good business for families, and plants are sold in stands in the Sonora market of Mexico City (Robles-Cervantes 2008).

Many medicinal plants are multipurpose. In addition to their medicinal characteristics, they are also used for food, as decorative ornaments, as fiber, or as condiments (Moctezuma-Pérez 2013a; Rao and Rao 2006). In this sense, the work of De Clerck and Negreros-Castillo (2000) reported that in homegardens in Quintana Roo, of the 35 medicinal plants they found only nine were exclusively used medicinally, whereas 26 were multipurpose. Meanwhile, Rico-Gray et al. (1991) mention 210 species obtained from forests and homegardens that could be used for medicinal purposes, but only 16 of them were uniquely for medicinal use. Medicinal plants are generally used for curing, prevention, placebos, palliative care, and nutritional supplements (Rao and Rao 2006). Not only are they used to treat human diseases and conditions, but also to address livestock diseases, and for fishing, as they can be used as bait or to stun fish for harvest. Some are also used as bio-pesticides (Rao and Rao 2006). The parts of medicinal plants used for therapeutic purposes cover all plant parts, including young shoots, flowers, leaves, stems, seeds, bark, pods, rhizomes, bulbs, fruits, roots, and inflorescences, depending on the species (Rao and Rao 2006).

In La Purificación Tepetitla, Texcoco, Maribel Robles-Cervantes' (2008: 145–150) study of 10 homegardens found 27 trees (seven for shade, one with edible flowers, 19 fruit trees) and 26 trees used for medicinal purposes, 30 medicinal shrubs or bushes (seven of them were also for food and four were used for condiments), 20 cultivated medicinal herbs, and five additional plants that were collected for human food (one of them was medicinal). Furthermore, 21 plants and the leaves of two trees with medicinal purposes were collected in the neighboring area outside of the homegardens and used for selling in the region, or taken to the market in Mexico City, which specializes in medicinal plants.

The peninsular Mayan groups have long had important knowledge of traditional medicine (Barrera-Vázquez 1999; Mariaca-Méndez et al. 2008; Navarijo-Ornelas 2004), based on historical manuscripts considered as sources for the study of the flora, fauna, and minerals used in traditional Mayan medicine (e.g., Chilam Balam of Ixil, Tekax Manuscripts and Nah, Chan Cah Manuscripts, Jewish Books, the Ritual of Bacabes; Song of Dzitbalché). This indicates that the Maya had a medical system that was part of their culture and their daily lives (Redfield and Villa Rojas 1990; Roys 1957; Standley 1990; Steggerda 1941). On the diseases and conditions for which peninsular Mayan families used medicinal plants, Ankli et al. (1999) have documented the following categories: gastrointestinal disorders in young children (147 species); respiratory disorders such as cough, bronchitis, and asthma (87 species); for uses as diuretics (77 species); for bites and stings of venomous animals (44 species); all eye infections (27 species); an assortment of species used for gynecological conditions, andrology, and diabetes; and cultural diseases like "bad air," evil eye, witchcraft, or illnesses caused by nocturnal birds and bats.

Although in many traditional systems plants are used more frequently and in greater numbers than animals, faunal resources, parts, and products (urine, fat, and so on) are also important elements of traditional medicine (Alves and Alves 2011; Alves and Rosa 2005; Ankli et al. 1999).

The faunal resources used in the treatment of disease or infirmity have been less studied compared with medicinal plants used by indigenous groups and the rural population (Alves and Alves 2011; Enríquez et al. 2006; Navarijo-Ornelas 2004). In the tropical regions of the Yucatan Peninsula, mostly women are involved in the cultivation of medicinal plants, keep the traditional knowledge about their use, and even do the selling in community markets (Howard 2006). In the case of Texcoco and the Sierra de Puebla in central Mexico, where planting homegardens and plots is purely commercial, the men are the ones who cultivate, harvest, and prepare the plants to be combined with other herbs or taken to the market (Robles-Cervantes 2008).

14.10 THE SOCIOECONOMICS OF HOMEGARDENS

The homegarden systems studied in X-Mejía show how income is used by families in two ways: (1) it generates savings by producing food that the family does not have to buy and (2) surplus production can be sold for additional income (Aké et al. 1999; Cahuich-Campos and Mariaca Méndez 2010:123-140; Lok 1998; Marsh and Hernández 1998). The role of the gardens in the economy of the family is dynamic, by complementing the household economy in times of relative security and prosperity, but also playing a predominant role in times of crisis when the gardens can become the main source of food for peasant families (Marsh and Hernández 1998). This situation occurs generally in the months preceding the harvest when a family is very poor, when they are affected by climatological events like hurricanes, droughts, disease, or if there is a death in the family (Estrada Lugo et al. 1998; Mariaca-Méndez et al. 2007; Marshall 1992).

Items with economic value produced by homegardens include fruits and vegetables sold within communities or at local markets, such as nance (*B. crassifolia*), banana (*Musa* sp.), orange (*Citrus* sp.), soursop (*Annona muricata*), chile (*Capsicum* sp.) (Aké et al. 1999; Ortega et al. 1993; Ponette-González 2007; Shagarosky et al. 2003); flowers to decorate altars in homes, or for religious celebrations (Ortega et al. 1993); and ornamental palms like *Chamedora elegans* (Ponette-González 2007). These sales are part of a diversification strategy characteristic of some indigenous agroecosystems for reducing risk (Ponette-González 2007). Juan-Pérez and Madrigal-Uribe (2005) and Alejandra Palacios-Sánchez (2009) reported the presence of cultivated flowers in the Totonac homegardens that families used in rituals such as the Day of the Dead altars, and religious ceremonies in the use of foods like pumpkin, bananas, and oranges, which are prepared with brown sugar, or are hung from the altar on the Day of the Dead.

There are also studies that have quantified the monetary contribution of homegarden products. Pulido et al. (2008) report significant income for households, ranging from 10% to 100% of total income from homegardens in Nicaragua (Mendez et al. 2001). In Honduras, this contribution varies between 10% and 26% (Lok 1998). In Belize, 62% of the studied homegardens provided most of the family income, whereas in the community of San José the major income comes from the sale of agroforestry products (Levasseur and Olivier 2000). But there is also an indirect contribution that cannot be evaluated as part of the economic productivity. This is the labor force, individual and collective of the family members (González-Jácome et al. 2007; Lok 1998), or the decline in economic involvement by changes in market prices and the bartering and exchange within the communities (Lok 1998; Ban and Coomes 2004; Vera 2005:176-189). The open areas of the homegarden can be used as zones for men working as carpenters, making furniture, working in apiculture, or making other items that are sold within the community or at external markets, as seen in some peninsular Mayan communities like Chan Kom (Mariaca-Méndez et al. 2007; Porter-Bolland et al. 2008).

Homegardens also provide a place for socialization and coexistence of the family household where kin and friends can talk. Men make crafts and items for everyday activities. Women prepare food and hang the clothes to dry in the sun. Children play. The garden is a space for receiving visitors and offering food and drink and exchanging the daily life stories of the residents. In these spaces, the family maintains and reinforces kinship and friendship. The local cosmovision and culture of the family are reflected in the structure of the homegarden. as well as in selected species and their associations (Cahuich-Campos 2012; Lazos and Alvarez-Buylla 1988; Pulido et al. 2008). Social interactions allow for visits from other families, creating a social network, which is very important to ensure solidarity (Buchmann 2009; Valera 2001).

Plants and animals from the homegardens are used in making ritual dishes that mark annual cycles in life, maintain the identity of ethnics groups, and promote social learning and cultural reproduction, as seen in the case of semiurban families with pre-Hispanic origins of Xochimilco, Ocotepec, and Tetecala, in the State of Mexico (Christie 2004) or peninsular Mayan communities (Estrada Lugo et al. 1998). The homegarden is as much a physical space as it is a place for family ceremonies. The rituals are sociocultural aspects that are part of the identity and survival of human groups. They also are one of the ways to transmit traditional ecological knowledge from one group to another.

14.11 CONCLUSIONS

The literature on homegardens focuses on a broad variety of concepts where each author empha-sizes one or more of the system's components (Allison 1983; Caballero 1992:135-155; Cuanalo-de la Cerda and Guerra 2008; Toledo et al. 2008). From an agroecological perspective, homegardens are a small physical area of cultivated land located around a home, constructed by humans, and having both plants and animals, all of this forming a type of agroecosystem. Homegarden size can be as small as 25 m² to many hectares, depending on several factors such as being simple mono-cultures with very few species or very complex polycultural gardens; the type of trees and crops found there; if the garden is exclusively worked with family labor or with a combination of family, nonpaid, and paid seasonal labor; if it is irrigated or not; if production is directed to subsistence or to the market; and the types of markets from local to distant.

The fact that homegardens have persisted through time is related to their ability to provide an important part of domestic food for the family group, but also to provide products such as fruits, medicinal plants, firewood, timber, and animal protein, most of which can also be marketed. Homegarden trees provide shade and function as windbreaks to protect the home and modify the local environment. The gardens are biodiverse systems containing plants from the Old and New World, whose management generally uses simple technologies. They do not require external inputs such as synthetic chemical fertilizers or pesticides.

Studies emphasizing biodiversity in homegardens describe them as *in situ* gene banks. Homegardens are also conceptualized as possessing adaptive traits that are specialized and adapted to local environmental factors and variation. Biodiversity is an important element in the system and it is found in homegardens of different sizes and also in different stages of development. Other researchers include cultural and economic factors and describe the family in the homegarden as a reflection of the cultural identity of a human group in relation to nature, as well as an economic unit of consumption. The homegarden is considered a low-risk agroecosystem, cushioning the impact of periods of shortage due to the continuous production of goods for consumption or for trade.

Ethnobotanists often define homegardens as complete agroforestry systems, dominated by a tree overstory, where energy flow and nutrient cycles create a "closed" system, while providing diverse plant and animal resources for human consumption. The homegarden is also characterized by the use of local knowledge for the management of biodiversity and use of successional processes for both garden development and recovery from disturbance (Gliessman 1990, 2015).

Anthropology and agroecology join together to study homegardens as complex human–nature relationships. Archaeology and historical studies seek to understand the historical processes of sub-sistence in agricultural societies, and the evolutionary changes in human societies related to popu-lation increases and the appearance of agriculture. In particular, ecological anthropology defines homegardens to include the human factor, their biological components, and the physical space they occupy. In this sense, ecological anthropology defines the homegarden as an agroecosystem with historical and traditional roots, where human beings simultaneously live and produce necessities. Homegardens consist of trees, crops, and animals that occupy spaces located in the vicinity of the family dwelling. Homegardens are also integrated production systems in which people create social networks where traditional ecological knowledge is passed from one generation to the other. This traditional ecological knowledge is closely tied to sustainability, with aspects of resilience that characterize homegarden agroecosystem's permanence through time.

ACKNOWLEDGMENT

The author is grateful to Stephen R. Gliessman, Ernesto Méndez, and William L. Crothers for their generous language review and comments to this chapter.

REFERENCES

Abu-Shams, L. "La Alimentación Como Signo de Identidad Cultural Entre los Inmigrantes Marroquíes." *Zainak* 30 (2008):177–193.

Aké, G., M. Ávila, and J.J. Jiménez-Osornio. "*El Valor de los Productos Directos que se Obtienen en el Agroecosistema Solar; El Caso de Hocabá, Yucatán.*" Universidad Autónoma de Yucatán, Facultad de Economía, 1999. www.itanconkal.edu.mx/agroecologia/material/articulos/6Impacto/Doc/new/valor_pro_agroe.pdf.

Alayón-Gamboa, J.A., and F.D. Gurri-García. "Homegarden production and energetic sustainability in Calakmul, Campeche, México." *Human Ecology* 36 no. 3 (2008): 395–407.

Alburquerque, U.P., L.H.C. Andrade, and J. Caballero. "Structure and floristics of homegardens in Northeastern Brazil." *Journal of Arid Environments* 62 (2005): 491–506.

Álvarez-Buylla R., M.E. Lazos-Chavero, and J.R. García-Barrios. "Homegardens of a humid tropical region in Southeast México: An example of an agroforestry cropping system in a recently established community." *Agroforestry Systems* 8 (1989): 133–156.

Alves, R.R., and H.N. Alves. "The faunal drugstore: Animal-based remedies used in traditional medicines in Latin America." *Journal of Ethnobiology and Enthnomedicine* 7 (2011): 1–43.

Alves, R.R., and I.L. Rosa. "Why study the use of animal products in traditional medicines?" *Journal of Ethnobiology and Enthnomedicine* 1 (2005): 5.

Allison, J.L. *An Ecological Analysis of Homegarden (Huertos Familiares) in Two Mexican Villages.* Masters Thesis in Biology. California: University of California Santa Cruz, 1983.

Anderson, E.N. "Gardens in Tropical America and Tropical Asia." *Biótica, Nueva Época* 1 (1993): 81–102.

Anderson E.N., et al. *Those Who Bring the Flowers: Maya Ethnobotany in Quintana Roo, Mexico.* Chetumal, Mexico: El Colegio de la Frontera Sur, 2003.

Ankli, A., O. Sticher, and M. Heinrich. "Medical ethnobotany of the Yucatec Maya: Healers' consensus as a quantitative criterion." *Economic Botany* 53 no. 2 (1999):140–160.

Anonymous. *Antología de la Planeación en México.* México, Gobierno de México, 1985.

Atran S.L.X., and E.E. Ucan. *Plants of the Peteén Itza' Maya.* Ann Arbor: University of Michigan Museum, 2004.

Balick M.J., M.H. Nee, and D.E. Atha. *Checklist of the Vascular Plants of Belize.* New York: New York Botanical Garden Press, 2000.

Ban, N., and O.T. Coomes. "Homegardens in Amazonian Peru: Diversity and exchange of planting material." *Geographical Review* 94 (2004): 348–367.

Barba-Pingarrón, L. "Estudio de áreas de actividad." In *Coba, Quintana Roo. Un Análisis de dos Unidades Habitacionales Mayas del Horizonte Clásico.* N.L. Manzanilla (Ed.). México: Universidad Nacional Autónoma de México (UNAM): I.I.A., *Serie Antropológica* 82 (1987): 69–115.

Barrera-Vázquez, A. "Las Fuentes para el Estudio de la Medicina Nativa de Yucatán." *Biomed* 10 (1999): 253–261.

Bellon, M.R., G.J.A. Aguirre, M. Smale, J. Berthaud, R.I. Manuel, J. Mendoza, et al. "Intervenciones participativas para la conservación del maíz en fincas de los Valles Centrales de Oaxaca, México." In *Manejo de la Diversidad de los Cultivos en los Agroecosistemas Tradicionales*. J.L. Chávez Servia, J.Y.Tuxill, and D.I. Jarvis (Eds.). Cali, Colombia: Instituto Internacional de Recursos Fitogenéticos, 2004.

Berkes, F., J. Colding, and C. Folke. *Navigating Social-Ecological Systems: Building Resilience for Complexity and Change*. Cambridge: Cambridge University Press, 2003.

Bilbao-Ercoreca, J.A. *Fieldwork Report. Tlaxcala: Huexoyucan, Mexico*. Mexico: Universidad Iberoamericana (UIA), 1979.

Blanco-Rosas, J.L. *La Erosión de la Agrodiversidad en la Milpa de los Zoque-Popoluca de Soteapan: Xutuchincon y Aktevet*. PhD Thesis in Social Anthropoogy. Mexico, Universidad Iberoamericana (UIA), 2006.

Blanco-Rosas, J.L. "Los Afectados Chinantecos de la Presa Cerro de Oro: Resistencia de los Conflictos y Riesgos Socioambientales." In *Culturas y Políticas del Agua en México y un Caso del Mediterráneo*. M. Rutsch and A. González-Jácome (Eds.). México: Instituto Nacional de Antropología e Historia (INAH) y UIA, (2011): 177–202.

Buchmann, C. "Cuban homegardens and their role in social-ecological resilience." *Human Ecology* 37 (2009): 705–721.

Caballero, J. "Maya homegardens: Past, present and future." *Etnoecológica* 1 (1992):135–155.

Cahuich-Campos, D.R. *La Calidad de Vida y el Huerto Familiar, Desde la Percepción Ambiental de las Familias de X-Mejía, Hopelchén, Campeche*. PhD Thesis in Ecology and Sustainable Development. San Francisco de Campeche, Campeche: El Colegio de la Frontera Sur, 2012.

Cahuich-Campos, D.R., and R. Mariaca-Méndez. "El Huerto Familiar Maya en Campeche como Patrimonio Cultural y Biológico de las Familias Campesinas." In *Patrimonio Biocultural de Campeche. Experiencias, Saberes y Prácticas desde la Antropología y la Historia*. G.L. Huicochea and M.B. Cahuich-Campos (Eds.). San Francisco de Campeche, Campeche: El Colegio de la Frontera Sur, (2010), 123–140.

Cano, R.M. *Los Huertos Familiares de Tepango, Guerrero*. BA Thesis. México: Universidad Nacional Autónoma de México, 2003.

Christie, M.E. "Kitchenspace, fiestas and cultural reproduction in Mexican house-lot gardens." *Geographical Review* 94 no. 3 (2004): 368–390.

Cuanalo-de la Cerda, H., and R. Guerra. "Homegarden production and productivity in a Mayan community of Yucatan." *Human Ecology* 36 (2008): 423–433.

De Clerck, F.A.J., and P. Negreros-Castillo. "Plants species of traditional Mayan homegardens of México as analogs for multistrata agroforests." *Agroforestry Systems* 48 (2000): 303–317.

De Garine, I., and L.A. Vargas. "Introducción a las Investigaciones Antropológicas sobre Alimentación y Nutrición." *Cuadernos de Nutrición* 20 no. 3 (1997): 21–28.

Del Ángel Pérez, A.L., and M.A. Mendoza. "Totonac homegardens and natural resources in Veracruz, Mexico." *Agriculture and Human Values* 21 (2004):329–346.

Ellis, E.A., and L. Porter-Bolland. "Agroforestería en la Selva Maya: Antiguas Tradiciones y Nuevos Retos." *Los Nuevos Caminos de la Agricultura. Procesos de Conversión y Perspectivas*. México: UIA y Ed. Plaza y Valdés (2007): 213–242.

Emslie, S.D. "Birds and prehistoric agriculture: The new Mexican pueblos." *Human Ecology* 9 no. 3 (1981): 305–329.

Enríquez, V.P., R. Mariaca-Méndez, G.O.G. Retana, and P.E.J. Naranjo. "Uso Medicinal de la Fauna Silvestre en los Altos de Chiapas, México." *Interciencia* 31 no. 7 (2006): 491–499.

Estrada Lugo, E., E. Bello, and P.L. Serralta. "Dimensiones de la Etnobotánica: El Solar Maya como Espacio Eocial." In *Lecturas en Etnobotánica. Publicaciones del Programa Nacional de Etnobotánica*. S. Cuevas (Ed.). Texcoco, México: Universidad Autónoma Chapingo (1998): 457–474.

Fedick, L.S. "The Maya Forest: Destroyed or cultivated by the ancient Maya?" *PNAS* 107 no. 3 (2010): 953–954.

Fernandes, E.C.M., and P.K.R. Nair. "An evaluation of the structure and function of tropical homegardens." *Agricultural Systems* 21 no. 4 (1986): 279–310.

Flores, M.V.I. *Importancia de la tradición culinaria como parte de los procesos de adaptación, en Calakmul, Campeche*. BA Thesis in Physical Anthropology. México: Escuela Nacional de Antropología e Historia (ENAH), 2003.

Folke, C., S. Carpenter, B. Walker, M. Scheffer, T. Elmqvist, L. Gunderson, et al. "Regime shifts, resilience, and biodiversity in ecosystem management." *Annual Review of Ecology, Evolution, and Systematics* 35 (2004): 557–581.

Ford, A. "Dominant plants of the Maya forest and gardens of El Pilar: Implications for paleoenvironmental reconstructions." *Journal of Ethnobiology* 28 (2008): 179–199.

García Cook, Á. *Arqueología e Historia de Tlaxcala*. México: Gobierno del estado de Tlaxcala, Instituto Tlaxcalteca de la Cultura (ITC), 2014.

García-Martínez B. *El Monte de Mixtlán: Una Reflexión sobre el Contrapunto Entre Poblamiento y Naturaleza en el México Colonial*. Estudios sobre Historia y Ambiente en América. Argentina, Bolivia, México, Paraguay. Mexico: El Colegio de México (COLMEX) e Instituto Panamericano de Geografía e Historia (IPGH), (1999): 55–74.

Gaytán, A.C., H. Vibrans, H.G. Navarro, and V.M. Jiménez. "Manejo de Huertos Familiares Periurbanos de San Miguel Tlaixpan, Texcoco, Estado de México." *Boletín de la Sociedad Botánica de México* 69 (2001): 39–62.

Gerhard, P. *Geografía Histórica de la Nueva España 1519–1821*. México: UNAM, 1986–1972.

Gerhard, P. *La Frontera Sureste de la Nueva España*. México: UNAM, 1991.

Gliessman, S.R. (Ed.). *Agroecology: Researching the Ecological Basis for Sustainable Agriculture*. New York: Springer-Verlag, 1990.

Gliessman, S.R. "Un enfoque agroecológico en el estudio de la agricultura tradicional." In *Agricultura y Sociedad en México: Diversidad, Enfoques, Estudios de Caso*. A. González-Jácome, S. del Amo-Rodríguez, and F.D. Gurri-García (Eds.). México: UIA y Plaza y Valdés Consejo Nacional para la Enseñanza de la Biología, 1999.

Gliessman, S.R. *Agroecology: The Ecology of Sustainable Food Systems*. 3rd ed. Boca Raton, FL: CRC Press/Taylor & Francis Group, 2015.

González-Jácome, A. "Homegardens in Central Mexico." In *Prehistoric Intensive Agriculture in the Tropics*. I.S. Farrington (Ed.). (England, Manchester: BAR International Series 232, 1985), 521–537.

González-Jácome, A. "La Agricultura Mesoamericana." In *La Antropología en México. Panorama Histórico, Vol. IV, Las cuestiones medulares*. México: INAH, (1988a), 55–189.

González-Jácome, A. "Migration as a socioeconomic factor of change in agricultural systems." In *Global Perspectives on Agroecology and Sustainable Agriculture*. Santa Cruz, CA: University of California Santa Cruz, 1988b, 253–261.

González-Jácome, A. *Cultura y Agricultura: Transformaciones en el Agro Mexicano*. México, Universidad Iberoamericana (UIA), 2004.

González-Jácome, A. "El Huerto Familiar En México. Un Agroecosistema Antiguo que Puede Ser Sustentable." In *Avances en Agroecología y Ambiente* 1. Jesús Francisco López Olguín et al. (Eds.). (México: Universidad Autónoma de Chapingo [UACH] y Benemérita Universidad Autónoma de Puebla [BUAP], 2007), 119–138.

González-Jácome, A. "Mexico's agriculture under NAFTA: The passing of traditional agroecosystems." In *North America at the Crossroad: NAFTA after 15 Years*. I. Hussain (Ed.). (Mexico: Universidad Iberoamericana [UIA], 2009), 71–111.

González-Jácome, A. "Natural resources and out-migration in local communities of Southern Mexico: Non-NAFTA issues impacting NAFTA." In *The Impacts of NAFTA on North America. Challenges outside the Box*. I. Hussain (Ed.). (Basingstoke, England: Palgrave MacMillan, 2010), 141–158.

González-Jácome, A. *Historias Varias. Un Viaje en el Tiempo con los Agricultores Mexicanos*. México, Universidad Iberoamericana (UIA), 2011.

González-Jácome, A. "Del Huerto a los Jardines y Vecindades: Procesos de Cambio en un Agroecosistema de Origen Antiguo." In *El Huerto Familiar del Sureste de México*. R. Mariaca-Méndez (Ed.). (México: Secretaría de Recursos Naturales y Protección Ambiental del Estado de Tabasco y ECOSUR, 2012a).

González-Jácome, A. *Fieldwork Report on the Río, Mayo River Basin, Sonora, Mexico*. Mexico: Universidad Iberoamericana (UIA), 2012b.

González-Jácome, A. *Fieldwork Report on Hermosillo-Puerto Peñasco Routh, Sonora, Mexico*. Mexico, Universidad Iberoamericana (UIA), 2013.

González-Jácome, A., and J.J. Velasco-Orozco. "Agua y Agricultura en sociedades rurales: El Desarrollo por Cuencas Hidrológicas en México." In *Nuevas Rutas para el Desarrollo en América Latina. Experiencias Globales y Locales*. J. Maestre Alfonso, A. Casas-Gragea y, and A. González-Jácome (Eds.). (México: Universidad Iberoamericana [UIA], 2008), 215–238.

Greenberg, L.S.Z. "Women in the garden and kitchen: The role of cuisine in the conservation of traditional house lot crops among Yucatec Mayan immigrants." In *Women and Plants: Gender Relations in Biodiversity Management and Conservation*. Howard, P.L. (Ed.). (Canada: Zed Books, 2003), 51–65.

Guerrero, P.A.G. *El Impacto de la Migración en el Manejo de los Solares Campesinos, Caso de Estudio La Purísima Concepción, Mayorazgo, San Felipe del Progreso, Estado de México*. México: Boletín del Instituto de Geografía, UNAMno. 63 (2007): 105–124.

Haenn, N., and R.R. Wilk. *The Environment in Anthropology. A reader in Ecology, Culture, and Sustainable Living*. New York: New York University Press, 2005.

Hernández-Xolocotzi, E. "La Producción Agrícola en Yucatán [...] La Tierra de Menos Tierra." In *La Milpa en Yucatán. Un Sistema de Producción Agrícola Tradicional*. X.E. Hernández, E. Bello, and T.S. Levy (Eds.). (México: El Colegio de Postgraduados [COLPOS], 1995), 1–5.

Herrera, C.N.D. *Etnoflora Yucatanense: Los Huertos Familiares Mayas en el Oriente de Yucatán*. Mérida, México: Universidad Autónoma de Yucatán (UADY), Fascículo 9, 1994.

Howard, P.L. "Gender and social dynamics in swidden and homegardens in Latin America." In B.M. Kumar and P.K.R. Nair (Eds.). *Tropical Homegardens a Time-Tested Example of Sustainable Agroforestry*. New York: Springer, 2006, 159–182.

Ingold, T. "Culture and perception of the environment." In *Bush Base: Forest Farm*. E. Croll and D. Parkin (Eds.). (London: Routledge, 1992), 39–55.

Ingold, T. *The Perception of the Environment. Essays on Livelihood, Dwelling and Skill*. London and New York: Routledge, 2000.

Juan-Pérez, J.I. *Tiempo con Dinero y Tiempo sin Dinero: De la Agricultura Tradicional a la Comercial en Progreso Hidalgo, Estado de México*. PhD Thesis in Social Anthropology. Mexico, Universidad Iberoamericana (UIA), 2003.

Juán-Pérez, J.I., and D. Madrigal-Uribe. "Huertos, Diversidad y Alimentación en una Zona de Transición Ecológica del Estado de México." *Ciencia Ergo Sum* 12 (2005): 54–63.

Kumar, B.M., and P.K.R. Nair (Eds.). *Tropical Homegardens. A Time-Tested Example of Sustainable Agroforestry*. New York: Springer, 2006.

Lazos, C.E., and R.M.E. Álvarez-Buylla. "Ethnobotany in a tropical-humid region: The homegardens of Balzapote, Veracruz, Mexico." *Journal of Ethnobiology* 8 no. 1 (1988): 45–79.

Leff, E. *Ecología y Capital. Racionalidad Ambiental, Democracia Participativa y Desarrollo Sustentable*. México: Siglo XXI editores, 2003.

Levasseur, V., and A. Olivier. "The farming system and traditional agroforestry systems in the Maya community of San Jose, Belize." *Agroforestry Systems* 49 (2000): 275–288.

Linares, O.F. "Garden hunting in the American tropics." *Human Ecology* 4 no. 4 (1976): 331–349.

Lok, R. *Huertos Caseros Tradicionales de América Central: Características, Beneficios e Importancia, Desde un Enfoque Multidisciplinario*. Turrialba, Costa Rica: Centro Agronómico Tropical de Investigación y Enseñanza (CATIE), 1998.

López, A.M.H. *Diversidad y Manejo de los Solares Familiares: Su Contribución al Diseño de una Estrategia de Desarrollo Comunitario en la Zona Central de Veracruz, México*. MA Thesis.Veracruz, México: COLPOS, 2005.

MacNeish, R.S. "A summary of the subsistence." In *The Prehistory of the Tehuacan Valley. Environment and Subsistence*. C.D. Byers (Ed.). Vol. 1. (Austin, TX: University of Texas Press, 1967), 390–309.

MacNeish, R.S. "El Origen de la Civilización Mesoamericana Visto desde Tehuacán." In *Simposium Internacional Tehuacán y su Entorno: Balance y Perspectivas*. E. Lama (Ed.). (México: INAH, Colección Científica no. 313, 1997), 80–93.

Mariaca-Méndez, R. *Marqués de Comillas, Chiapas, Procesos de Inmigración y Adaptabilidad en el Trópico Húmedo de México*. PhD Thesis in Social Anthropology. Mexico: Universidad Iberoamericana (UIA), 2002.

Mariaca-Méndez, R., A. González-Jácome, and T. Lerner-Martínez. "El Huerto Familiar en México. Un Agroecosistema Antiguo que Puede ser Sustentable." In *Avances en Agroecología y Ambiente*. J.F. López Olguín et al. (Eds.). Vol. 1. (Texcoco: México, UACH and BUAP, 2007), 119–138.

Maroyi, A. "Traditional homegardens and rural livelihoods in Nhema, Zimbabwe: A sustainable agroforestry system." *International Journal of Sustainable Development and World Ecology* 16 (2009): 1–8.

Márquez Mireles, L.E. *Mayas Yucatecos en Quintana Roo. Agricultura de Roza en El Naranjal.* MA Thesis in Social Anthropology. México: Universidad Iberoamericana (UIA), 2000.

Marsh, R., and I. Hernández. "El aporte económico del huerto a la alimentación y la generación de ingresos familiars." In *Huertos Caseros Tradicionales de América Central: Características, Beneficios e Importancia, Desde un Enfoque Multidisciplinario.* R. Lok (Ed.). (Turrialba, Costa Rica: CATIE, 1998), 151–183.

Marshall, N.M. "The contemporary role of women in lowland Maya livestock production." In *Maya Subsistence. Studies in Memory of D.E. Puleston.* K.V. Flannery (Ed.). (New York: Academic Press, 1992), 313–325.

McC Netting R. *Smallholders, Householders: Farm Families and the Ecology of Intensive, Sustainable Agriculture.* Redwood City, CA: Stanford University Press, 1993.

Medina, E. *Conocimiento por Género y Edades en el Uso de Solares Mayas en Halachó, Yucatán, México.* BA Thesis. Mexico: Oaxaca, Instituto Tecnológico Agropecuario de Oaxaca, 2005.

Moctezuma-Pérez, S. *San Francisco Tepeyanco: Ambiente, Cultura y Agricultura.* PhD Thesis in Social Anthropology. Mexico: Universidad Iberoamericana (UIA), 2013a.

Moctezuma-Pérez, S. "Cambios y Continuidades en el Manejo de Huertos Familiares del Suroeste de Tlaxcala, México." In *Perspectivas Latinoamericanas.* Japan: Nanzan University Centro de Estudios Latinoamericanos no. 10 (2013b): 83–101.

Moctezuma-Pérez, S., J. Sales-Colín, and J.M. Pérez-Sánchez. *Fieldwork Report; Basin of the Mayo River, San Bernardo, Alamos Municipality, Mexico.* Mexico, Universidad Iberoamericana (UIA), 2010.

Molina-Rosales, D.O. *Colonización y Estrategias Adaptativas entre Campesinos del Sur de Calakmul, Campeche.* PhD Thesis in Social Anthropology. Mexico, Universidad Iberoamericana (UIA), 2010.

Moran, F.E. *Human Adaptability. An Introduction to Ecological Anthropology.* Boulder, CO: Westview Press, 2000.

Moreno Toscano, Al. "El Siglo de la Conquista." In *Historia General de México. Tomo 1.* D. Cosío-Villegas (Ed.). (México: Harla y El Colegio de México, 1987), 289–369.

Nájera, C.A.J. *Prácticas Alimentarias en Comunidades del Pueblo Tojolabal.* MA Thesis in Sciences (Natural Resources and Rural Development). México: ECOSUR, 2009.

Navarijo-Ornelas, M.L. "Presencia e Importancia de los Animales en la Medicina Tradicional de los Grupos Otopames." In *Estudios de Cultura Otopame.* (México: UNAM, IIA, 2004), 197–214.

Neusius, S.W. "Game procurement among temperate horticulturalists: The case for garden hunting by the Dolores Anazasi." In *Case Studies in Environmental Archaeology.* E.J. Reitz, A. Newson, and S.J. Séudder (Eds.). (New York: Plenum Press, 1996), 273–287.

Oliver-Smith, A., and S.M. Hoffman. *The Angry Earth. Disaster in Anthropological Perspective.* New York: Routledge, 1999.

Ornelas, D.J. *Ordenamiento Territorial y Política Regional en México.* Tlaxcala, México: Universidad Autónoma de Tlaxcala (UAT), 1993.

Palacios Sánchez, A. *La Muerte: Símbolo de Vida Entre los Totonacas de Papantla, Veracruz.* PhD Thesis in Social Anthropology. Mexico: Universidad Iberoamericana (UIA), 2009.

Palerm, Á. "La Civilización Urbana." In *Historia Mexicana* II no. 2 (México: El Colegio de México, 1952), 184–209.

Palerm, Á. "La Distribución del Regadío en el Area Central de Mesoamérica." In *Ciencias Sociales* 5 no. 25 (Washington, DC: Unión Panamericana, 1954), 2–15.

Palerm, Á. *Informe de los Aspectos Socioculturales de la Población Afectada por el Proyecto de la Angostura: Estudio y Recomendaciones.* México: Comisión Federal de Electricidad, 1970.

Pérez Contreras, T.F., and G.M. Alcaraz. "Transiciones y Nostalgias: El Sistema Alimentario de los Moradores de Acandi, Colombia." *Revista de la Facultad Nacional de Salud Pública* 25 (2007): 65–74.

Peterson, G., C.R. Allen, and C.S. Holling. "Ecological resilience, biodiversity, and scale." *Ecosystems* 1 no. 1 (1998): 6–18.

Ponette-González, A.G. "A household analysis of Huastec Maya agriculture and land use at the height of the coffee crisis." *Human Ecology* 35 (2007): 289–301.

Porter-Bolland, L., M.C. Sánchez-González, and E.A. Ellis. "La Conformación del Paisaje y el Aprovechamiento de los Recursos Naturales por las Comunidades Mayas de La Montaña, Hopelchén, Campeche." *Boletín del Instituto de Geografía* 66 (2008): 65–80.

Pulido, M.T., E.M. Pagaza-Calderón, A. Martínez-Balleste, B. Maldonado-Almanza, A. Saynes, and R.M. Pacheco. 2008. "Homegardens as an alternative for sustainability: Challenges and perspectives in Latin America." In *Current Topics in Ethnobotany*. U.P. De Alburquerque and R.M. Alves (Eds.). (India, Kerala, 2008), 1–25.

Rao, M.R., and R. Rao. "Medicinal plants in tropical homegardens." In *Tropical Homegardens: A Time-Tested Example of Sustainable Agroforestry*. B.M. Kumar and P.K.R. Nair (Eds.). (New York: Springer, 2006), 205–232.

Redfield, R., and A. Villa-Rojas. *Chan Kom: A Maya village*. 2nd ed. Chicago: University of Chicago Press, 1990 [1962].

Reinhardt, S. *Huertos Familiares; Tesoros de Diversidad.Hojas Temáticas People and Biodiversity in Rural Areas*. Echbom, Alemania: 2007. www2.gtz.de/dokumente/bib/04-5108ª4.pdf (Accessed November 2, 2010).

Rico-Gray, V., A. Chemás, and S. Mandujano. "Uses of tropical decidous forest species by the Yucatecan Maya." *Agroforestry Systems* 14 (1991):149–161.

Rindos, D. *The Origins of Agriculture.An Evolutionary Perspective*. Orlando: Academic Press, 1984.

Robles-Cervantes, M. *Purificación Tepetitla: Organización Familiar para la Producción de Plantas Medicinales en los Huertos del Somontano*. MA Thesis on Social Anthropology. Mexico: Universidad Iberoamericana (UIA), 2008.

Robles-Linares, G.M.G. *Agua, Sociedad y Cultura en la Cuenca Media del Río Mayo. Los Guarijíos del Sureste de Sonora*. PhD Thesis on Social Anthropology. Mexico: Universidad Iberoamericana (UIA), 2014.

Rojas, H.M. "Experienced poverty and income poverty in Mexico: A subjective well-being approach." *World Development* 36 (2008):1078–1093.

Roldán-Botello, D.P. "Un Caso de Desarrollo Agrícola en Tlaxcala: San Francisco Tepeyanco." BA Thesis in Social Anthropology. Mexico: Universidad Iberoamericana (UIA), 1979.

Roseberry, W. *Anthropologies and Histories: Essays in Culture, History and Political Economy*. New Brunswick, NJ: Rutgers University Press, 1989.

Roseberry, W. *Coffee, Society and Power in Latin America*. Balitomore, MD: John Hopkins University Press, 1995.

Roys, R. *The Political Geography of the Yucatan Maya*. Washington, DC: Carnegie Institution, 1957.

Rutsch, M., and A. González-Jácome. *Culturas y Políticas del Agua en México y un Caso del Mediterráneo*. México: INAH y Universidad Iberoamericana (UIA), 2011.

Scarpa, R., A.G. Drucker, S. Anderson, N. Ferraes-Ehuan, V. Gómez, C.R. Risopatrón, et al. "Valuing genetic resources in peasant economies: The case of 'Hairless' Creole pigs in Yucatán." *Ecological Economics* 45 (2003): 427–443.

Scholes F.V., and E.B. Adams. "Documents relating to the Mirones expedition to the interior of Yucatan." *Maya Research* 3 1959 [original 1936]: 153–176, 251–276.

Shagarodsky, T., V. Fuentes, O. Barrios, L. Castiñeiras, Z. Fundora, Z., P. Sánchez, et al. "Diversidad de Especies Alimenticias en Tres Mercados Agrícolas de la Habana, Cuba." *Revista de Agronomía Mesoamericana* 14 (2003): 27–39.

Smith, C.E. Jr. "Plant Remains." In *The Prehistory of the Tehuacan Valley. Environment and Subsistence*. D.S. Byers (Ed.). *Vol 1*. (Austin: University of Texas Press, 1967), 220–255.

Standley, R.S. "Demographic Aechaeology in the Maya Lowlands." In *The Classic Maya Collapse*. T.P. Culbert and O. S. Rice (Eds.). (Albuquerque: University of New Mexico Press, 1990), 325–365.

Steggerda M. *Maya Indians of Yucatan*. Washington DC: Carnegie Institution, 1941.

Steward, J.H. *Theory of Culture Change: The Methodology of Multilinear Evolution*. Chicago: University of Illinois Press, 1955.

Stuart, J.W. "Contribution of dooryard gardens to contemporary Yucatan Maya subsistence." *Biótica, Nueva época* 1 (1993): 53–61.

Terán, S., and C. Rasmussen. *La Milpa de los Mayas*. Mérida Yucatán, México: Editorial Danida, 1994.

Terán, S., C. Rasmussen, and O. May-Cauich. *Las Plantas de la Milpa entre los Mayas*. Yucatán, Mérida, México: Fundación Tun Ben Kin AC, 1998.

Toledo, V.M., N. Barrera-Bassols, E. García-Frapolli, and P. Alarcón-Chaires. "Uso Múltiple y Biodiversidad entre los Mayas Yucatecos (México)." *Interciencia, Mayo* 33 no. 5 (2008): 345–352.

Valera, S.J.L. "El Huerto: Bienestar de la Familia Campesina." *LEISA* (2001): 19–20.

Vanderwarker, M.A. *Farming, Hunting, and Fishing in the Olmec World*. Austin: University of Texas Press, 2006.

Vera, N.R. "Elementos Constitutivos para Medir la Pobreza y la Calidad de Vida." In *Espacios Públicos* no. 8. México: Universidad Autónomadel Estado de México (UAEM), 2005, 176–189.

Vogl, C.R., and B. Vogl-Lukasser. "Tradition, dynamics and sustainability of plant species composition and management in homegardens on organic and non-organic small scale farms in Alpine Eastern Tyrol, Austria." *Biological Agriculture and Horticulture* 21 (2003): 349–366.

Walker, B., C.S. Holling, S.R. Carpenter, and A. Kinzig. "Resilience, adaptability and transformability in social–ecological systems." *Ecology and Society* 9 no. 2 (2004): 5.

Wezel, A., and S. Bender. "Plant species diversity of homegardens of Cuba and its significance for household food supply." *Agroforestry Systems* 57 (2003): 39–49.

Wieman, A., and D. Leal. "La Cría de Animales Menores en los Huertos Caseros." In *Huertos Caseros Tradicionales de América Central: Características, Beneficios e Importancia, desde un Enfoque Multidisciplinario.* R. Lok (Ed.). (Turrialba, Costa Rica: CATIE, 1998), 85–115.

Wilk, R. *Household Ecology: Economic Change and Domestic Life among the Kekchi Maya in Belize.* Tucson: University of Arizona Press, 1991.

Wilken, G.C. "Drained-field agriculture: An intensive farming system in Tlaxcala, Mexico." *Geographical Review* 59 (1969): 215–241.

Index